AF510287

VASOPEPTIDES

Chemistry, Pharmacology, and Pathophysiology

ADVANCES IN EXPERIMENTAL MEDICINE AND BIOLOGY

Editorial Board:

Nathan Back — *Chairman, Department of Biochemical Pharmacology, School of Pharmacy, State University of New York, Buffalo, New York*

N. R. Di Luzio — *Chairman, Department of Physiology, Tulane University School of Medicine, New Orleans, Louisiana*

Alfred Gellhorn — *University of Pennsylvania Medical School, Philadelphia, Pennsylvania*

Bernard Halpern — *Collège de France, Director of the Institute of Immuno-Biology, Paris, France*

Ephraim Katchalski — *Department of Biophysics, The Weizmann Institute of Science, Rehovoth, Israel*

David Kritchevsky — *Wistar Institute, Philadelphia, Pennsylvania*

Abel Lajtha — *New York State Research Institute for Neurochemistry and Drug Addiction, Ward's Island, New York*

Rodolfo Paoletti — *Institute of Pharmacology and Pharmacognosy, University of Milan, Milan, Italy*

Volume 1
THE RETICULOENDOTHELIAL SYSTEM AND ATHEROSCLEROSIS
Edited by N. R. Di Luzio and R. Paoletti • 1967

Volume 2
PHARMACOLOGY OF HORMONAL POLYPEPTIDES AND PROTEINS
Edited by N. Back, L. Martini, and R. Paoletti • 1968

Volume 3
GERM-FREE BIOLOGY: Experimental and Clinical Aspects
Edited by E. A. Mirand and N. Back • 1969

Volume 4
DRUGS AFFECTING LIPID METABOLISM
Edited by W. L. Holmes, L. A. Carlson, and R. Paoletti • 1969

Volume 5
LYMPHATIC TISSUE AND GERMINAL CENTERS IN IMMUNE RESPONSE
Edited by L. Fiore-Donati and M. G. Hanna, Jr. • 1969

Volume 6
RED CELL METABOLISM AND FUNCTION
Edited by George J. Brewer • 1970

Volume 7
SURFACE CHEMISTRY OF BIOLOGICAL SYSTEMS
Edited by Martin Blank • 1970

Volume 8
BRADYKININ AND RELATED KININS: Cardiovascular, Biochemical, and Neural Actions
Edited by F. Sicuteri, M. Rocha e Silva, and N. Back • 1970

Volume 9
SHOCK: Biochemical, Pharmacological, and Clinical Aspects
Edited by A. Bertelli and N. Back • 1970

Volume 10
THE HUMAN TESTIS
Edited by E. Rosemberg and C. A. Paulsen • 1970

Volume 11
MUSCLE METABOLISM DURING EXERCISE
Edited by B. Pernow and B. Saltin • 1971

Volume 12
MORPHOLOGICAL AND FUNCTIONAL ASPECTS OF IMMUNITY
Edited by K. Lindahl-Kiessling, G. Alm, and M. G. Hanna, Jr. • 1971

Volume 13
CHEMISTRY AND BRAIN DEVELOPMENT
Edited by R. Paoletti and A. N. Davison • 1971

Volume 14
MEMBRANE-BOUND ENZYMES
Edited by G. Porcellati and F. di Jeso • 1971

Volume 15
THE RETICULOENDOTHELIAL SYSTEM AND IMMUNE PHENOMENA
Edited by N. R. Di Luzio and K. Flemming • 1971

Volume 16A
THE ARTERY AND THE PROCESS OF ARTERIOSCLEROSIS: Pathogenesis
Edited by Stewart Wolf • 1971

Volume 16B
THE ARTERY AND THE PROCESS OF ARTERIOSCLEROSIS: Measurement and Modification
Edited by Stewart Wolf • 1972

Volume 17
CONTROL OF RENIN SECRETION
Edited by Tatiana A. Assaykeen • 1972

Volume 18
THE DYNAMICS OF MERISTEM CELL POPULATIONS
Edited by Morton W. Miller and Charles C. Kuehnert • 1972

Volume 19
SPHINGOLIPIDS, SPHINGOLIPIDOSES AND ALLIED DISORDERS
Edited by Bruno W. Volk and Stanley M. Aronson • 1972

Volume 20
DRUG ABUSE: Nonmedical Use of Dependence-Producing Drugs
Edited by Simon Btesh • 1972

Volume 21
VASOPEPTIDES: Chemistry, Pharmacology, and Pathophysiology
Edited by N. Back and F. Sicuteri • 1972

Volume 22
COMPARATIVE PATHOPHYSIOLOGY OF CIRCULATORY DISTURBANCES
Edited by Colin M. Bloor • 1972

Volume 23
THE FUNDAMENTAL MECHANISMS OF SHOCK
Edited by Lerner B. Hinshaw and Barbara G. Cox • 1972

Volume 24
THE VISUAL SYSTEM: Neurophysiology, Biophysics, and Their Clinical Applications
Edited by G. B. Arden • 1972

Volume 25
GLYCOLIPIDS, GLYCOPROTEINS, AND MUCOPOLYSACCHARIDES OF THE NERVOUS SYSTEM
Edited by Vittorio Zambotti, Guido Tettamanti, and Mariagrazia Arrigoni • 1972

Volume 26
PHARMACOLOGICAL CONTROL OF LIPID METABOLISM
Edited by William L. Holmes, Rodolfo Paoletti, and David Kritchevsky • 1972

Volume 27
DRUGS AND FETAL DEVELOPMENT
Edited by Marcus A. Klingberg, Armand Abramovici, and Juan Chemke • 1972

Volume 28
HEMOGLOBIN AND RED CELL STRUCTURE AND FUNCTION
Edited by George J. Brewer • 1972

VASOPEPTIDES

Chemistry, Pharmacology, and Pathophysiology

Proceedings of the Symposium on Vasopeptides held at the Villa Medici (Auditorium Hoechst), Fiesole (Florence), Italy, July 20–22, 1971

Edited by

Nathan Back

Department of Biochemical Pharmacology
School of Pharmacy
State University of New York
Buffalo, New York

and

F. Sicuteri

Department of Clinical Pharmacology
University of Florence
Florence, Italy

PLENUM PRESS · NEW YORK – LONDON · 1972

Library of Congress Catalog Card Number 78-190395
ISBN 978-1-4684-7441-1 ISBN 978-1-4684-7439-8 (eBook)
DOI 10.1007/978-1-4684-7439-8

© 1972 Plenum Press, New York
Softcover reprint of the hardcover 1st edition 1972

A Division of Plenum Publishing Corporation
227 West 17th Street, New York, N.Y. 10011

United Kingdom edition published by Plenum Press, London
A Division of Plenum Publishing Company, Ltd.
Davis House (4th Floor), 8 Scrubs Lane, Harlesden, London, NW10 6SE, England

All rights reserved

No part of this publication may be reproduced in any form without
written permission from the publisher

PREFACE

The dynamic state of developments within the field of the vasoactive polypeptides is reflected in the continuous series of publications appearing on this topic. Thus, the proposal to convene another symposium only two years after the 1969 Fiesole Symposium (Plenum Press, 1970) was received enthusiastically and with high expectation. Enthusiasm was based, undoubtedly, in part on meeting together once again in the very pleasant surroundings of Florence with most gracious and enchanting hosts. The 25th International Congress on Physiological Sciences held in Munich the following week also provided the impetus for the organization of this satellite symposium on vasopeptides.

However, the accelerated pace of research in the "kinin" field was the major motivating force for the symposium. It was felt necessary to inform each other of the many advances that were taking place in the varied aspects of this specialized but important field. The role of kinins in both physicologic and pathologic processes still remains to be elucidated definitively.

Thus, these proceedings contain new and unpublished information on the chemistry, biochemistry, physicology, pharmacology, as well as pathophysiology of vasopeptides. Recent data on structure-activity relationships of kinin analogues are complemented by a description of radioimmunoassay techniques for kinins, and studies on the site of synthesis of substrates for kinin-forming enzymes. Under intensive study, and securing an important place in the symposium, are the studies of the interaction of the kinin-forming system with components of the fibrinolysin, blood coagulation, and complement systems. The primary role of Hageman factor in the release of kinins receives additional support.

Included in the proceedings are studies on tissue localization of kinin-forming enzymes, the metabolic fate of various vasopeptides and prostaglandins, and identification of a bradykinin-potentiating peptide. The pharmacologic actions of several vasopeptides also are highlighted. The pathophysiology of such conditions as toxemia of pregnancy, inflammation, cold injury, shock, malignancy, hypertension, myocardial infarction, and liver cirrhosis is reconsidered in terms of the role of vasopeptides. The more informal and intimate discussions with colleagues comprised a very delightful and informative part of the meeting.

The success of the symposium can be attributed to many factors that require acknowledgment and our thanks. The members of the Organizing Committee have earned our indebtedness for their generous help and advice: E. Werle, M. Rocha e Silva, F. Sicuteri, N. Back, A. Giotti, L. Corti, P. L. Del Bianco, M. Fanciullacci, and G. Franchi. We acknowledge with many thanks the competent and skillful assistance of the secretaries of the Organizing Committee, S. Bowsher and M. Masi. The Organizing Committee also is grateful to the Mayor of Fiesole, Mr. B. Latini, and the Fiesole Tourism Board for their kind interest and collaboration, and to Hoechst Italia for their kind hospitality and many courtesies extended at their facility, the site of the symposium, the Villa Medici Auditorium.

This international symposium was under the sponsorship of the Italian Research Council and the Italian Clinical Pharmacology Society.

The organizers and participants acknowledge gratefully the following pharmaceutical companies, without whose generous financial support, this Symposium could not have taken place: Abbott, Compoverde (Latina); Bayer Ilalia, Milan; Carlo Erba, Florence; Cilag-Chemie Italiana, Milan; Crinos, Villaguardia (Como); Guidotti, Pisa; Hoechst, Milan; Lepetit, Milan; Lirca, Milan; Malesci, Florence; Manetti & Roberts, Florence; Marxer, Loranze d'Ivrea (Turin); Merck Sharp & Dohme, Rome; Organon, Rome; Pfizer, Rome; Roche, Milan; Schiaparelli, Turin; Sigma Tau, Rome; Wellcome, Pomezia (Rome).

LIST OF CONTRIBUTORS

Burton M. Altura, Departments of Anesthesiology and Physiology, Albert Einstein College of Medicine, Bronx, New York

F. M. Antonini, Department of Clinical Pharmacology and of Gerontology, University of Florence, Florence, Italy

M. P. O. Antonio, Department of Pharmacology, School of Medicine, Ribeirão Prêto, São Paulo, Brazil

R. L. Araujo, Instituto de Ciências Biológicas da U.F.M.G., Belo Horizonte, Brazil

D. Armstrong, Duhring Laboratories, University of Pennsylvania School of Medicine, Philadelphia, Pennsylvania

H. Arold, Institute of Pharmacology and Toxicology, University of Rostock, Rostock, Germany

R. B. Arora, Department of Pharmacology, All-India Institute of Medical Sciences, Ansari Nagar, New Delhi, India

Nathan Back, Department of Biochemical Pharmacology, School of Pharmacy, State University of New York at Buffalo, Buffalo, New York

E. Bassenge, Department of Physiology and Department of Clinical Chemistry, University of Munich, Munich, Germany

N. Basu, Department of Pharmacology, All-India Institute of Medical Sciences, Ansari Nagar, New Delhi, India

G. Benelli, Research Laboratories ZAMBON S.p.A., Bresso-Milan, Italy

W. T. Beraldo, Departments of Physiology and Morphology, ICB. Universidade Federal M. G., Brazil

N. Bhargava, Departments of Pharmacological Research of N. V. Organon, Oss, The Netherlands

K. D. Bhoola, Department of Pharmacology, University of Bristol Medical School, Bristol, England

I. L. Bonta, Department of Pharmacology, Medical Faculty Rotterdam, The Netherlands

Marisa A. Boucault, Servico de Farmacologia de Instituto Butantan, São Paulo, Brazil

John Bradley, Departments of Medicine and Surgery, University of Liverpool, Liverpool, England

A. H. Brady, University of Miami School of Medicine, Miami, Florida

F. T. Bryan, Departments of Medicine and Surgery, University of Miami School of Medicine, Miami, Florida

Clement E. Burrowes, Division of Experimental Pathology and Institute of Immunology, University of Toronto, Toronto, Ontario, Canada

V. M. Campese, Department of Pharmacology, University of Bari, Bari, Italy

L. Caramelli, Department of Angiology, Main Regional Hospital Santa Maria Nuova, Florence, Italy

L. Caramelli, Department of Angiology, Main Regional Hospital Santa Maria Nuova, Florence, Italy

A. Castania, Department of Pharmacology, School of Medicine of Ribeirão Prêto, São Paulo, Brazil

C. Corsi, Department of Angiology, Main Regional Hospital, Santa Maria Nuova, Florence, Italy

H. Croxatto, Laboratorio Fisiologia Universidad Catolica de Chile, Santiago, Chile

C. Curradi, Department of Clinical Pharmacology and of Gerontology, University of Florence, Florence, Italy

Alfred Cuschieri, Department of Surgery, Liverpool University, Liverpool, England

V. L. d'Ancona, Division of Clinical Pharmacology, Faculty of Medicine, University of Florence, Florence, Italy

N. Dao Hai, Organon R & D, Eragny-sur-Epte, France

Glaci R. DaSilva, Department of Pharmacology, Faculty of Medicine, University of Sao Paulo, Ribeirao Preto, Sao Paulo, Brazil

C. Deidda, Department of Angiology, Main Regional Hospital Santa Maria Nuova, Florence, Italy

P. L. Del Bianco, Department of Clinical Pharmacology and of Gerontology, University of Florence, Florence, Italy

D. Della Bella, Research Laboratories ZAMBON S.p.A., Bresso-Milan, Italy

C. Deidda, Department of Angiology, Main Regional Hospital Santa Maria Nuova, Florence, Italy

C. J. deVos, Research Laboratories of N. V. Organon, Oss, The Netherlands

J. J. Dias, Department of Pharmacology, School of Medicine of Ribeirão Prêto, São Paulo, Brazil

W. Dias da Silva, Department of Biochemistry, Instituto de Ciências Biológicas da U.F.M.G., Belo Horizonte, Brazil

G. Dorey, Department of Pharmacology, University of Bristol School of Medicine, Bristol, England

L. Dorigotti, Farmitalia, Istituto Ricerche, Milan, Italy

V. Eisen, Rheumatology Research Department, Middlesex Hospital Medical School, London, England

E. G. Erdös, Department of Pharmacology, University of Oklahoma School of Medicine, Oklahoma City, Oklahoma

M. Fanciullacci, Department of Clinical Pharmacology, University of Florence, Florence, Italy

Mina Fichman, Servico de Farmacologia de Instituto Butantan, Sao Paulo, Brazil

G. Franchi, Department of Clinical Pharmacology, University of Florence, Florence, Italy

Penelope J. Fuller, Division of Experimental Pathology and Institute of Immunology, University of Toronto, Toronto, Ontario, Canada

Á. Gecse, Institute of Pathophysiology and Dermatologic Clinic, University of Szeged School of Medicine, Szeged, Hungary

A. H. Glässer, Istituto Ricerche di Base Farmitalia, University of Milan, Milan, Italy

D. W. Goodwin, Department of Medicine, University of Miami School of Medicine, Miami, Florida

S. K. Gupta, Department of Pharmacology, All-India Institute of Medical Sciences, Ansari Nagar, New Delhi, India

S. G. Hershey, Departments of Anesthesiology and Physiology, Albert Einstein College of Medicine, Bronx, New York

Elisa Hiraichi, Servico de Farmacologia de Instituto Butantan, Sao Paulo, Brazil

Yoshio Hojima, The Laboratory of Physiological Chemistry, Faculty of Pharmaceutical Sciences, Science University of Tokyo, Tokyo, Japan

K. Horiuchi, Institute for Protein Research, Osaka University, Osaka, Japan

R. Igic, Department of Pharmacology, University of Oklahoma School of Medicine, Oklahoma City, Oklahoma

H. Juan, Institute of Pharmacology, University of Graz, Graz, Austria

S. Karády, Institute of Pathophysiology, University of Szeged School of Medicine, Szeged, Hungary

H. Kato, Institute for Protein Research, Osaka University, Osaka, Japan

M. Komiya, Institute for Protein Research, Osaka University, Osaka, Japan

M. Kucharczyk, Department of Physiology and Department of Clinical Chemistry, University of Munich, Munich, Germany

Alba A. C. Lavras, Servico de Farmacologia de Instituto Butantan, Sao Paulo, Brazil

F. Lembeck, Institute of Pharmacology, University of Graz, Graz, Austria

J. Lonovics, Institute of Pathophysiology and Dermatologic Clinic, University of Szeged School of Medicine, Szeged, Hungary

Antii Louhija, The Wihuri Research Institute, Helsinki, Finland

C. Loveday, Rheumatology Research Department, Middlesex Hospital Medical School, London, England

M. Luz San Martin, Laboratorio Fisiologia, Universidad Catolica de Chile, Santiago, Chile

C. R. S. Machado, Departments of Physiology and Morphology, ICB. Universidade Federal M. G., Brazil

M. L. Mashford, Department of Pharmacology, University of Melbourne, Parkville, Australia

D. Mitolo-Chieppa, Department of Pharmacology, University of Bari, Bari, Italy

Hiroshi Moriya, The Laboratory of Physiological Chemistry, Faculty of Pharmaceutical Sciences, Science University of Tokyo, Tokyo, Japan

H. Z. Movat, Division of Experimental Pathology, University of Toronto, Toronto, Canada

S. Nagasawa, Institute for Protein Research, Osaka University, Osaka, Japan

T. Nakajima, Department of Pharmacology, University of Oklahoma School of Medicine, Oklahoma City, Oklahoma

L. C. Neves, Department of Pharmacology, School of Medicine of Ribeirão Prêto, São Paulo, Brazil

R. S. Niemeyer, Departments of Medicine and Surgery, University of Miami School of Medicine, Miami, Florida

A. Nuti, Department of Angiology, Main Regional Hospital Santa Maria Nuova, Florence, Italy

Lyioko Okino, Department of Pharmacology, Faculty of Medicine, Ribeirão Prêto, São Paulo, Brazil

Ayse Özge-Anwar, Division of Experimental Pathology and Institute of Immunology, University of Toronto, Toronto, Ontario, Canada

I. Paegelow, Institute of Pharmacology and Toxicology, University of Rostock, Rostock, Germany

E. Pajola, Research Laboratories ZAMBON S.p.A., Bresso-Milan, Italy

M. Paradiso, Division of Clinical Pharmacology, Faculty of Medicine, University of Florence, Florence, Italy

Pentti Peltola, Kivelä Hospital, Helsinki, Finland

P. Periti, Division of Clinical Pharmacology, Faculty of Medicine, University of Florence, Florence, Italy

F. Porcelli, Istituto di Chimica, Facolta di Medicina, Universita Cattolica, Rome, Italy

G. Porcelli, Istituto di Chimica, Facolta di Medicina, Universita Cattolica, Rome, Italy

F. Pratesi, Department of Angiology, Main Regional Hospital Santa Maria Nuova, Florence, Italy

Helen A. Pyle, Division of Experimental Pathology, University of Toronto, Toronto, Ontario, Canada

K. Rajanna, Department of Pharmacology, All-India Institute of Medical Sciences, Ansari Nagar, New Delhi, India

A. S. Ramaswamy, Department of Pharmacology, All-India Institute of Medical Sciences, New Delhi, India

Marina L. Reis, Department of Pharmacology, Faculty of Medicine, USP, Ribeirão Prêto, São Paulo, Brazil

S. Reissmann, Department of Biochemistry, Friedrich Schiller University Jena, Germany

M. L. Roberts, Department of Medicine, University of Melbourne, Parkville, Australia

J. Roblero, Laboratorio de Fisiologia, Universidad Catolica de Chile, Santiago, Chile

J. A. A. Rodrigues, Departments of Physiology and Morphology, ICB. Universidade Federal M. G., Brazil

M. Rocha e Silva, Department of Pharmacology, Faculty of Medicine, University of São Paulo, Ribeirão Prêto, São Paulo, Brazil

G. Rostock, Institute of Pharmacology and Toxicology, University of Rostock, Rostock, Germany

A. M. Rothschild, Department of Pharmacology, School of Medicine of Ribeirao Preto, Sao Paulo, Brazil

J. W. Ryan, Department of Medicine, University of Miami, Miami, Florida

M. Schlaeger, Department of Biochemical Pharmacology, Max-Planck Institut für experimentelle Medizin, Göttingen, Germany

P. Schmuziger, Servico de Farmacologia de Instituto Butantan, Sao Paulo, Brazil

G. Seidel, Department of Biochemical Pharmacology, Max-Planck Institut für experimentelle Medicin, Göttingen, Germany

S. Severi, Division of Clinical Pharmacology, Faculty of Medicine, University of Florence, Florence, Italy

F. Sicuteri, Division of Clinical Pharmacology and Coronary Unit of the Division of Gerontology, Faculty of Medicine, University of Florence, Florence, Italy

H. H. Siddiqui, Department of Pharmacology, All-India Institute of Medical Sciences, Ansari Nagar, New Delhi, India

R. Sipilä, The Wihuri Research Institute, Helsinki, Finland

G. Siqueira, Departments of Physiology and Morphology, ICB. Universidade Federal M. G. Brazil

U. Smith, Department of Medicine, University of Miami, Miami, Florida

M. J. Soltay, Division Experimental Pathology and Institute of Immunology, University of Toronto, Toronto, Ontario, Canada

K. Sorrells, Department of Pharmacology, University of Oklahoma School of Medicine, Oklahoma City, Oklahoma

R. Steger, Department of Biochemical Pharmacology, School of Pharmacy, State University of New York, Buffalo, New York

J. M. Stewart, University of Colorado School of Medicine, Denver, Colorado

H.-U. Stücker, Department of Biochemical Pharmacology, Max-Planck Institut für experimentelle Medizin, Göttingen, Germany

K. Sundaram, Quantum Biology Group, Department of Biophysics, All-India Institute of Medical Sciences, New Delhi, India

T. Suzuki, Institute for Protein Research, Osaka University, Osaka, Japan

L. Szekeres, Institute of Pathophysiology and Dermatologic Clinic, University of Szeged School of Medicine, Szeged, Hungary

H. Takahashi, Institute for Protein Research, Osaka University, Osaka, Japan

S. V. Talekar, Quantum Biology Group, Department of Biophysics, All-India Institute of Medical Sciences, New Delhi, India

M. Tarantelli, Department of Angiology, Main Regional Hospital Santa Maria Nuova, Florence, Italy

M. Tariq, Department of Pharmacology, All-India Institute of Medical Sciences, Ansari Nagar, New Delhi, India

M. Tesi, Department of Angiology, Main Regional Hospital Santa Maria Nuova, Florence, Italy

T. Tijs, Department of Pharmacology, Medical Faculty, Rotterdam, The Netherlands

T. Tijs, Research Laboratories of N. V. Organon, Oss, The Netherlands

M. P. Treloar, Pharmakologisches Institut, Justus Liebig Universität, Giessen, Germany

G. Vaghi, Istituto Ricerche di Base Farmitalia, University of Milan, Milan, Italy

P. Valli, Research Laboratories ZAMBON S.p.A., Bresso-Milan, Italy

B. B. Vargaftig, Organon S. A., Eragny-sur-Epte, France

M. N. R. Vieira, Department of Biochemistry, Instituto de Ciências Biológicas da U.F.M.G., Belo Horizonte, Brazil

W. Vogt, Department of Biochemical Pharmacology, Max-Planck Institut für Experimentelle Medizin, Göttingen, Germany

W. von Restorff, Department of Physiology and Department of Clinical Chemistry, University of Munich, Munich, Germany

U. Wendel, Department of Biochemical Pharmacology, Max-Planck Institut für experimentelle Medizin, Göttingen, Germany

E. Werle, Department of Physiology and Department of Clinical Chemistry, University of Munich, Munich, Germany

B. Wiegershausen, Institute of Pharmacology and Toxicology, University of Rostock, Rostock, Germany

M. Yano, Institute for Protein Research, Osaka University, Osaka, Japan

H. P. Zach, Institute of Clinical Chemistry and Clinical Biochemistry, University of Munich, Munich, Germany

I. J. Zeitlin, Wolfson Gastrointestinal Laboratories, Gastrointestinal Unit, Western General Hospital, Edinburgh, Scotland

E. Zsilinszky, Institute of Pathophysiology and Dermatologic Clinic, University of Szeged School of Medicine, Szeged, Hungary

CONTENTS

SECTION III - PATHOLOGY

BIOCHEMISTRY

HYDROXYPROLINE ANALOGS OF BRADYKININ: BIOLOGICAL ACTIVITIES AND SOLUTION STRUCTURES.

John Morrow Stewart*, A. H. Brady** and James W. Ryan**

*University of Colorado School of Medicine, Denver, Colorado 80220 and **University of Miami School of Medicine, Miami, Florida 33152

Our previous studies by circular dichroism (CD) indicate that bradykinin, Lys-bradykinin, Met-Lys-bradykinin, and Polistes kinin exist in solution as disordered chains (1,2), and examination of the CD spectra in the 250-300 nm range yields no evidence of intra- or inter-molecular association of the phenylalanine residues. In brief, spectra of all of the biologically active analogs and homologs of bradykinin show maxima at 220 nm and minima at 235 nm which are independent of changes in solvent and temperature. None of the analogs, e.g. D-Pro^{2}-BK or D-Phe^{5,8}-bradykinin, having different spectra possessed biological activity. This suggests that it is unlikely that highly ordered analogs of amino acid composition and sequence similar to bradykinin would possess biological activity. The conclusion that bradykinin exists as a disordered chain has recently received support from an independent molecular orbital computation of the molecular conformation of minimum energy (3).

Chemical, physical and biological reasons prompted the synthesis and study of analogs of bradykinin in which each of the three proline residues was replaced in turn with 4-hydroxyproline. The chemical motivation was the desire for better understanding of solid phase peptide synthesis with hydroxy amino acids. Experience with threonine had shown that branches were formed in the peptide chain by acylation of the threonine hydroxyl when it was unprotected (4). For this reason Boc-O-benzyl threonine was developed and has been widely used in solid phase synthesis. Throughout the development of the solid phase method, O-benzyl serine was available, and has been routinely used. It was of interest to learn whether similar problems of hydroxyl acylation would arise with hydroxyproline.

Since proline-of which there are three in the nonapeptide bradykinin-plays such an important role in determining the conformation of peptide chains, it was of interest to see if addition of hydroxyl groups to these prolines would alter the conformation of the peptide, either by virtue of the size or polar character of these groups.

The biological activity of bradykinin also depends to a remarkable degree on the proline residues. Individual replacements of proline by alanine showed that the ring structure of the residues in position 2 and 7 was important for activity, while that of the 3-proline was not (5). The relative inactivity of [D-Pro^3]-bradykinin, as compared to the Ala^3-analog, probably indicates that available space on the receptor at this locus is severely limited, not allowing the molecule to fit when the proline ring is on the opposite side of the peptide backbone (5). An alternate explanation is that inversion of the configuration of this residue displaces the important N-terminal arginine from its binding site, although it would seem that if this were the case, the activity of [D-Pro^2]-bradykinin should be lower. Addition of a hydroxyl group to the proline rings should yield information about the availability of space on the receptor at these loci, especially if the hydroxyproline analogs were shown not to differ grossly in overall solution conformation.

EXPERIMENTAL SECTION

Synthesis. The peptides were synthesized by the Merrifield solid-phase method as described by Stewart and Young (6), using an automatic instrument similar to one previously described (7). The resin support had 2% crosslinking, and deprotection of Boc-groups was done with 4 M HCl in dioxane containing 1 mg/ml of 2-mercaptoethanol. DCI-Mediated coupling reactions were done in $CHCl_3$, with 2.5 moles of DCI and Boc amino acids per mole of peptide. Peptides were cleaved from the resin by treatment with anhydrous HF in the presence of anisole, and were purified by CCD in the system 1-butanol:1% TFA. Boc-amino acids were purchased from Schwarz Bio-Research, except for Boc-hydroxyproline, which was purchased from Fox Chemical Co.

Circular dichroism measurements were performed using a Cary 60 spectropolarimeter with a circular-dichroism attachment. In general, peptides were dissolved in 0.03 K_2HPO_4 at pH 7.2. The spectropolarimeter was standardized with an aqueous solution of d-10-camphorsulphonic acid, giving a coefficient of dichroic absorption ($\Delta\varepsilon = \varepsilon_L - \varepsilon_R = 2.20$ at 290 nm). Spectra were checked for conformity to Beer's law to rule out spurious effects such as aggregation.

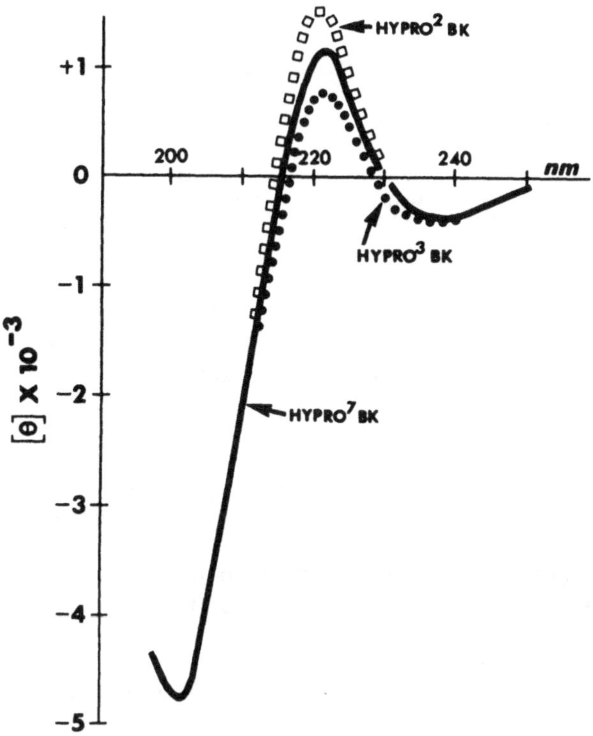

Figure 1. Circular dichroism spectra of Hypro^1-, Hypro^2-, and Hypro^7-bradykinin.

Assays for biological activity were performed on rat uterus, guinea pig ileum, and the intact rat (blood pressure and pulmonary inactivation) as previously described (8).

RESULTS AND DISCUSSION

Synthesis. The solid phase synthesis of these three hydroxyproline peptides proceeded without difficulty, despite the lack of any blocking group on the hydroxyl. A search among the minor byproducts separated by CCD failed to uncover any branched-chain peptides formed by acylation of the hydroxyl groups.

Circular Dichroism. Over the wavelength range of 200-300 nm the analogs, Hypro^2-, Hypro^3- and Hypro^7-bradykinin, produced spectra like that of bradykinin (Fig. 1). Specifically, the phenylalanyl transitions (250-300 nm) were identical. Furthermore, each had minima at 235 nm and maxima at 220 nm. The spectrum of Hypro^3-BK was precisely the same as that of bradykinin and the spectra of the other two did not differ significantly.

TABLE I

Biological Activities

(Quantities required to produce equivalent responses)

Peptide	GP Ileum [a]	Blood Pressure [b]	Pulmonary Inactivation [c]
Bradykinin (BK)	50 ng	80 ng	97-99%
$HyPro^2$-BK	50	80	85-95
$HyPro^3$-BK	80	200	50-80
$HyPro^7$-BK	>500	2000	<30

[a]The concentration (ng/ml) required to produce a half maximal contraction. Results obtained on rat uterus were similar to ileum.

[b]The amount required by intraaortic injection into a 200 g rat to produce a fall in blood pressure of 25 mm.

[c]Calculated from the difference in the amounts needed by intravenous and intraaortic injection to produce the same depressor response in the rat.

Biological Activities. By intra-aortic injection, $Hypro^2$-BK was fully active in producing the major bradykinin responses, namely initial lowering of blood pressure and the subsequent rise owing to catecholamine release. By intravenous injection, $Hypro^2$-BK was marginally more active than bradykinin, possibly because of a slightly reduced rate of pulmonary inactivation. $Hypro^2$-BK was fully active on the guinea pig ileum.

The apparent biological activity of $Hypro^3$-BK varied from 30-50% on blood pressure and guinea pig ileum. $Hypro^7$-BK was virtually inactive (<1%) on blood pressure and guinea pig ileum.

Apparent Rates of Pulmonary Inactivation. The hydroxyproline analogs were compared for their effects on blood pressure when given intravenously and by injection into the ascending aorta. By finding the quantities needed to give equivalent responses, we obtained an index of the rates of inactivation in the lungs. $Hypro^2$-BK was inactivated at a rate marginally less than that of bradykinin. $Hypro^3$-BK was inactivated at a rate similar to that of Met-Lys-bradykinin. $Hypro^7$-BK may be inactivated at a slower rate, but because of its relative inactivity it was necessary to inject μg quantities by either route of administration--and could possibly have exceeded saturation limits of the pulmonary enzymes.

In conclusion, our results show that substitution of hydroxyl groups on the three prolines of BK does not affect the conformation of BK in solution. However, these substitutions can affect the interaction of BK with its receptor. It appears that the ring of Pro^2 does not interact with the receptor, or if it does, the requirements for interaction are not very stringent. We regard it as much more likely that Pro^2 functions to fix the conformation of adjacent residues, possibly Arg^1. Results with Pro^3 are less clear. However, in solution the rings of the three prolines lie in different planes, making it possible that the ring of Pro^3 does not interact with the receptor but is close enough that the 4-OH group adds a weak steric hindrance to the interaction of the rest of the bradykinin molecule. HO-Pro in position 7 clearly interferes, possibly because of steric hindrance of a ring interaction or because of H-bonding to a nearby nucleophilic group, which would tend to distort the peptide on the receptor. We cannot exclude the possibility that Pro^7 has two functions: 1) ring interaction with the receptor, and 2) restriction of possible conformations of adjacent residues.

Finally, we believe that our studies bring out the point that all future studies of the biological effects of residue modification, replacement or deletion, must take into account the effects of the given change on conformation. Our data of our previous study and the present study indicate that those modified bradykinins having different spectra are unlikely to have bradykinin-like biological effect. Thus, as a step in determining whether lack of activity is owing to the loss or substitution of a specific side-chain or ring, it should be established that the change has not affected conformation.

Acknowledgement. These studies were supported in part by grants from the National Institutes of Health (HE 14360, HE 12843, HE 12325, HE 11035) and from the Population Council.

REFERENCES

1. A. H. Brady, J. M. Stewart and J. W. Ryan, in "Bradykinin and Related Kinins", F. Sicuteri, M. Rocha e Silva and N. Back, eds., Plenum Press, New York, 1970, p. 47.

2. A. H. Brady, J. W. Ryan and J. M. Stewart, Biochem. J., 121, 179 (1971).

3. L. B. Kier and J. M. George, in "Molecular Orbital Studies in Chemical Pharmacology", L. B. Kier, ed., Springer-Verlag, New

York, 1970, p. 82.

4. J. M. Stewart, T. Mizoguchi and D. W. Woolley, Abstracts of American Chemical Society 153rd Meeting, Miami, Fla., March 1967, Abstract O-206.

5. J. M. Stewart, in "Structure-Activity Relationships of Protein and Polypeptide Hormones", M. Margoulies and F. C. Greenwood, eds., Excerpta Medica, 1971, p. 23.

6. J. M. Stewart and J. D. Young, "Solid Phase Peptide Synthesis", W. H. Freeman, San Francisco, 1969.

7. R. B. Merrifield, J. M. Stewart and N. Jernberg, Anal. Chem., 38, 1905 (1966).

8. F. W. Dunn and J. M. Stewart, J. Med. Chem., 14, 779 (1971).

ION BINDING BY MACROCYCLIC DEPSIPEPTIDE ANTIBIOTICS: MUTUAL REPLACEABILITY OF AMIDE AND ESTER GROUPS

S.V. TALEKAR and K. SUNDARAM

Quantum Biology Group, Department of Biophysics, All-India Institute of Medical Sciences, New Delhi-16, India

ABSTRACT

Macrocyclic depsipeptide antibiotic molecules have alternating sequence of amide and ester linkages constituting the backbone structure. In the current paper, cylindrical polar atomic coordinates of the conformations proposed by Shemyakin <u>et al</u>. of macrocyclic depsipeptides antibiotics e.g., Valinomycin and Enniatin B were generated by building, scale models. The coordinates were further refined by simulation of these models in a digital computer. The π-electronic charge densities on parts of the molecules containing double bonds, e.g., amide and ester groups, were calculated using the variable electro-negativity - Pariser, Parr, Pople (VE-PPP) method. The σ-electronic charge densities on the molecules, (including the conjugated parts of the molecules), were evaluated using the method of Del Re. Ion multipole interaction energy for Valinomycin - K^+ - ion complex turns out to be - 107.83 K cals/mole and for Enniatin B - K^+ -ion complex -85.01 K cals/mole. Mutual replaceability of amide and ester groups in these depsipeptides has been discussed on the basis of π and σ charge distributions. The calculations show that the amide and ester groups are mutually replaceable provided this replaceability does not grossly alter the secondary molecular structure of depsipeptides, although replacement of amide group by ester in case of valinomycin will lead to the loss of formation of intramolecular hydrogen bonds, when such replaceability by ester group involves even one amide residue.

INTRODUCTION

The macrocyclic depsipeptide antibiotics, Valinomycin and Enniatin B enter into ion-multipole interaction preferentially with K^+-ion. The carbonyl oxygens of the backbone of these depsipeptides ligand with the K^+-ion. The backbones of these depsipeptides are formed by alternating amide and ester groups. The Valinomycin molecule backbone has six each of amide and ester groups; Enniatin B molecule backbone has three each of amide and ester groups (Fig. 1a & b). In the case of Valinomycin only the ester carbonyl oxygen's enter into a ligand interaction with K^+-ion while in Enniatin B both the ester as well as amide carbonyl oxygens enter into ligand interactions. The amide carbonyls of Valinomycin form six intramolecular hydrogen bonds while no intramolecular hydrogen bonds are formed in the case of Enniatin B for its amide nitrogen is methylated. Shemyakin, et al[1] have synthesised a series of linear as well as cyclic analogs of the two depsipeptides. The cyclic analogs were synthesized by replacing all or some amide groups by ester groups and vice versa. The cyclic analogs, wherein, amide groups were replaced by ester groups were found to be wholly or partially inactive in the K^+-ion complexing, transportational and antibacterial properties. On the other hand, cyclic analogs synthesized by replacing ester groups by amide groups lead to an equal or increased amount of the above mentioned activities, as compared to the naturally occuring depsipeptides.

In our laboratory we undertook to look at these structures more quantitatively to find out the reason for the above mentioned findings. A prerequisite for the theoretical study of any molecular activity (e.g., K^+-ion complexing property) is 1) a precise description of the molecular geometry and 2) electronic charge distribution over the said molecules. Similarities and differences are closely examined at the electronic levels between the amide and ester groups. Charge distribution and spatial geometry of these groups are taken as two standard parameters in judging the extent of K^+-ion complexing by the two depsipeptides. In turn, the K^+-ion complexing property is taken as the index for biological activity.

METHODS

For calculations of molecular geometry and charge distribution refer to our earlier works [2,3 & 4]. A short description is given here. The cylindrical polar atomic coordinates of the proposed models of the two depsipeptides in the K^+-ion complexed form described by Shemyakin et al[5], were generated by building scale models. The standard bond lengths (l) and bond angles (α) were taken from Pauling[6]. Following criteria were imposed for the structure solution a) K^+.... O bond length was fixed equal to the sum of covalent radii of the two atoms i.e. 2.13A°, b) all amide and ester linkages were considered trans and planer, c) perfect ring closure was sought because of the cyclic nature of the molecules, d) the side chain configuration was fixed as dictated by L or D residues and hydrogens on C(γ) carbons were put in stagger form instead of eclipse form, e) in the case of Valinomycin the hydrogen bond angle was fixed at 130°, i.e. angle C = O...H-N. The coordinates were further refined by simulation of these models in a digital computer. The charges on the backbone atoms (including the polar groups) of the two molecules were obtained by using molecular orbital techniques. The parts of the two molecules containing double bonds (e.g., the amide and ester groups) were treated by Variable electronegitivity - Pariser, Parr, Pople (VE-PPP) method (7,8,9 & 10) and the parts of the molecules containing single bonds (including the conjugated parts of the molecules) were treated by using the method of Del Re[11]. Charge interactions or ion multipole interactions in the two molecules with K^+-ion were estimated on the basis of above mentioned calculations.

RESULTS

Molecular Conformations and Ion Multipole Interactions

Figure I (a & b) show the structural formulae of Valinomycin and Enniatin B respectively. Tables I and II show (a) the internal

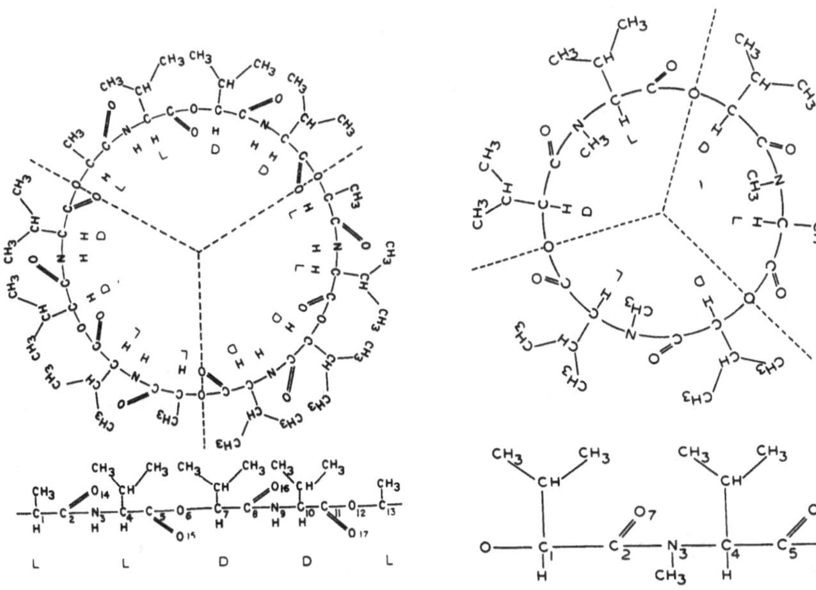

(a) (b)

Figure I: Chemical structure of (a) Valinomycin and (b) Enniatin B showing their trisymmetry. One repeat unit is shown below with numbering of atoms of the backbone. (After Brockmann et al, 1963 and Plattner et al, 1963).

coordinates; (i.e. bond length (l), bondangle (α) and dihedral angle (β); (b) cylindrical polar coordinates (ρ, φ, Z) and the radial distance (R) with reference to common coordinate frame; (c) using the π and σ charges on the backbone of the molecules and a unit positive charge on the K^+-ion interaction for all atoms with central K^+-ion are calculated. The origin of coordinates is taken at the site of K^+-ion. The Z-axis is the upward vertical axis through the K^+-ion and the X-axis is the line joining K^+-ion to the L-Lactate α-carbon in the case of Valinomycin and L-Valine α-carbon in the case of Enniatin B. Figure II (a & b) show projections of the Valinomycin molecules in the two directions respectively. Figure III shows the vertical projection of the Enniatin B. The side chain configurations have been omitted for simplicity in both molecules.

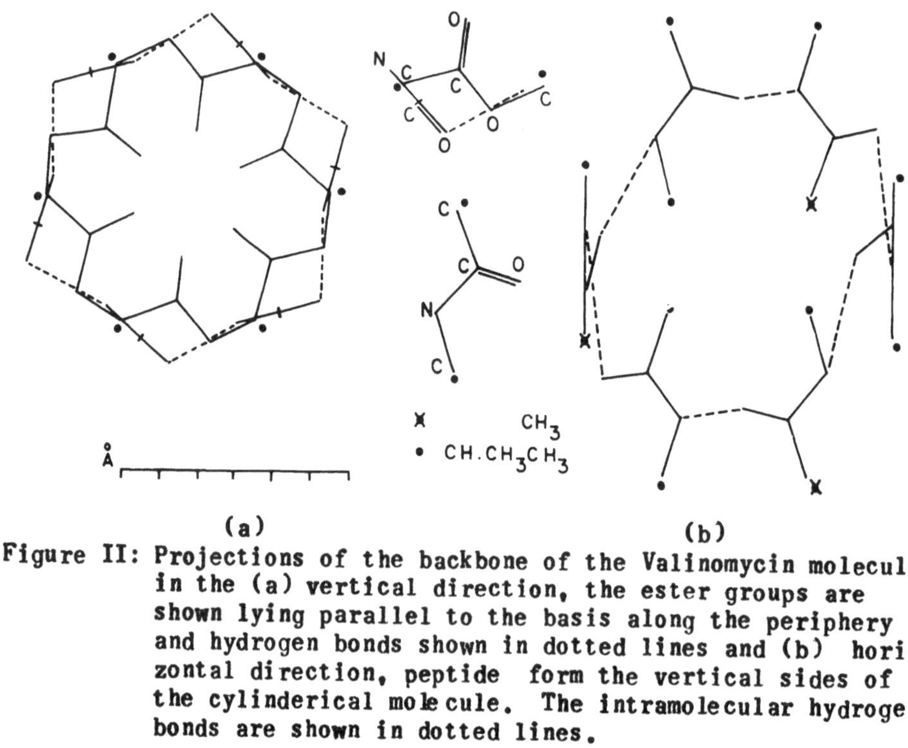

Figure II: Projections of the backbone of the Valinomycin molecule in the (a) vertical direction, the ester groups are shown lying parallel to the basis along the periphery and hydrogen bonds shown in dotted lines and (b) horizontal direction, peptide form the vertical sides of the cylinderical molecule. The intramolecular hydrogen bonds are shown in dotted lines.

The ion multipole interaction energy in Valinomycin K^+-ion complex is - 107.84 Kcals/mole and Enniatin B - K^+-ion complex is - 85.01 Kcals/mole. Only the coordinates of backbone atoms are presented here for economy of space. (Total number of atoms in a repeat unit, including the side chains for Valinomycin are fifty-five and for Enniatin B thirtyone). Also in the calculations of

		(a)			(b)				(c)			
Atom No.	Back bone atom types	l(A°)	α(°)	β(°)	ρ(A°)	φ(°)	Z (A°)	R (A°)	π-charge	σ-charge	Total charge	$q_A \cdot q_{K^+}/R_A$
1.	C	-	-	-	3.7	0	-2.05	4.232	-	+0.116	+0.116	+0.0274
2.	C	1.53	-	-	4.2	17	-0.60	4.238	+0.072	+0.263	+0.335	+0.0791
3.	N	1.52	117.0	-	3.6	22	+0.60	3.641	+0.323	-0.222	+0.001	+0.0003
4.	C	1.47	120.0	+180	3.7	0	+2.05	4.232	-	+0.071	+0.071	+0.0168
5.	C	1.53	110.0	- 45	2.6	22	+2.05	3.313	+0.140	+0.320	+0.460	+0.1388
6.	O	1.43	117.0	+135	3.8	38	+2.05	4.314	+0.180	-0.278	-0.098	-0.0227
7.	C	1.36	105.0	+180	3.7	61	+2.05	4.232	-	+0.114	+0.114	+0.0269
8.	C	1.53	110.0	+ 90	3.5	73	+2.05	4.051	+0.072	+0.263	+0.335	+0.0827
9.	N	1.52	117.0	0	4.2	38	+0.60	4.238	+0.323	-0.322	+0.001	+0.0002
10.	C	1.47	120.0	+180	3.7	61	-0.60	3.750	-	+0.071	+0.071	+0.0189
11.	C	1.53	110.0	+ 45	2.6	82	-2.05	3.313	+0.140	+0.230	+0.460	+0.1388
12.	O	1.43	117.0	-135	3.8	100	-2.05	4.314	+0.180	-0.278	-0.098	-0.0227
13.	C	1.36	105.0	+180	4.2	120	-2.05	4.670	-	+0.116	+0.116	+0.0248
14.	O Amide	1.24	121.0	+180	4.5	22	+0.30	4.504	-0.395	-0.136	-0.531	-0.1179
15.	O Ester	1.24	121.0	- 60	1.9	22	+2.05	2.796	-0.320	-0.130	-0.450	-0.1909
16.	O Amide	1.24	121.0	+180	4.5	82	-0.30	4.504	-0.395	-0.136	-0.531	-0.1179
17.	O Ester	1.24	121.0	+ 60	1.9	82	-2.05	2.796	-0.320	-0.130	-0.450	-0.1909

Table I: For Valinomycin: (a) showing the internal coordinates, (b) cylindrical polar coordinates, (c) charge interaction based on $\pi + \sigma$ calculations.

For further explanations see text.

Atom No.	Back bone atom types	(a) l (A°)	α(°)	β(°)	(b) ρ(A°)	φ(°)	Z (A°)	R (A°)	(c) π-charge	σ-charge	Total charge	$q_A \cdot q_K + / R_A$
1.	C	-	-	-	3.8	0	0	3.800	-	+0.166	+0.166	+0.0305
2.	C	1.53	-	-	3.0	21.00	+0.50	3.391	+0.072	+0.263	+0.335	+0.0987
3.	N	1.32	114	-	3.8	41.50	0	3.800	+0.323	-0.322	+0.001	+0.0002
4.	C	1.47	123	+180	3.2	63.00	-0.50	3.570	-	+0.071	+0.071	+0.0109
5.	C	1.53	114	-45	3.0	88.50	-0.20	3.066	+0.140	+0.320	+0.460	+0.1500
6.	O	1.43	114	+180	3.8	106.00	+0.20	3.852	+0.180	-0.278	+0.098	-0.0234
7.	C	1.36	105	+180	3.8	120.00	0.	3.800	-	+0.166	+0.166	+0.0305
8.	O Amide	1.24	121	0	2.6	21.00	+1.00	2.786	-0.395	-0.136	-0.531	-0.1907
9.	O Ester	1.24	121	0	2.6	88.50	-1.00	2.786	-0.320	-0.130	-0.450	-0.1615

Table II: For Enniatin B:
(a) showing the internal coordinates, (b) cylindrical polar coordinates,
(c) charge interaction calculations based on $\pi + \sigma$ calculations.

For further explanations see text.

the interaction energies only the backbone atoms (which represent the polar parts of the molecules) have been included. The coordinates of the rest of the molecule can be obtained by simple stepwise rotations of 120°, since the two molecules have trisymmetry.

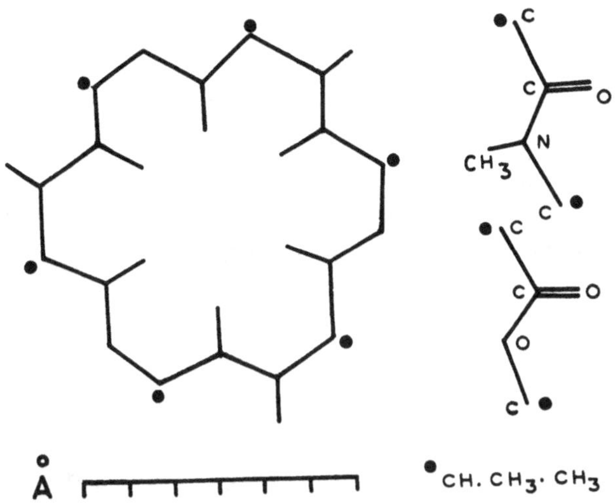

Figure III: Projection of the backbone of the Enniatin B molecule in vertical direction. The amide carbonyls are above the plane and ester carbonyls below the plane.

Total Electronic Charge Distribution on Amide and Ester Groups

As mentioned earlier, the π-electronic charge distributions were obtained by the VE-PPP method[7,8,9 & 10] and the σ-electronic charge distribution was obtained by the method of Del Re[11]. The basic results obtained from such calculations are molecular orbitals representing the distribution of electrons (in the ground and the excited states) and their energy levels. These are given in Table III for the amide and the ester groups. A diagrametic representation of the molecular orbitals is also given in Figure IV.

A number of physical parameters can be calculated from the molecular orbitals and their energies. The ionization potential turns out to be the negative of the energy of Highest Occupied Molecular Orbitals (HOMO) (Koopman's theorem[12]). Thus, we have: I (amide) = 8.620 e.V. and I (ester) = 10.528 e.V. The calculated net electronic charges ($\pi+\sigma$) with reference to neutral groups are given in Figure V.

Table III: Molecular orbitals representing the distribution of π electrons in ground and excited states with their corresponding energy levels for amide and ester groups.

		Ground state		Excited state
	O	+0.379	-0.744	-0.550
	C	+0.648	-0.211	+0.732
	N	+0.661	+0.634	-0.402
Energy levels (e.V)		-13.578	-8.620	+0.391
	O	+0.451	-0.675	-0.583
	C	+0.608	-0.246	+0.755
	O	+0.653	+0.695	-0.300
Energy levels (e.V)		-13.928	-10.528	-1.320

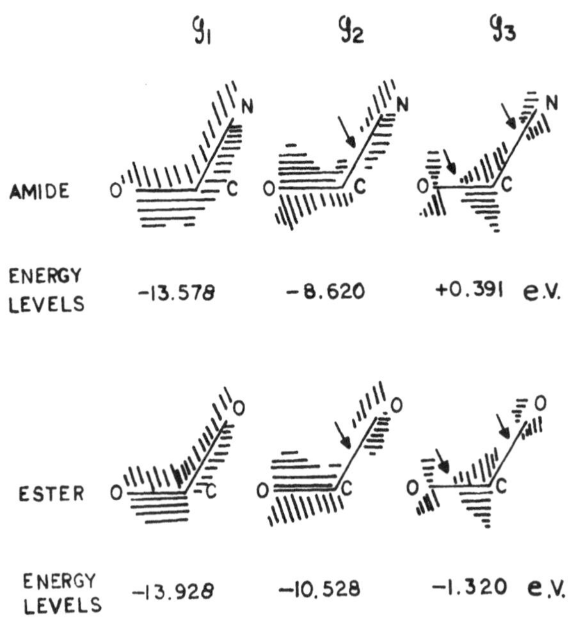

Figure IV: Schematic diagram of molecular orbitals of amide and ester groups, obtained by VE-PPP method . Vertically shaded area is negative and the horizontally shaded area is positive. Arrows indicate the nodal surfaces corresponding energy levels are shown in e.V.

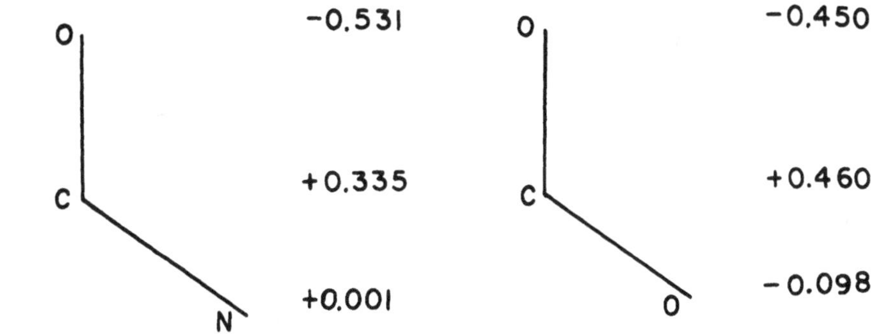

Figure V: Showing net $\pi + \sigma$ electronic changes on atoms of amide and ester groups as obtained by VE-PPP and method of Del Re.

The total ($\pi + \sigma$) dipole moment magnitude and the angle of dipole moment vector can be calculated from these charges and the positions of the atoms relative to any arbitrary coordinate system (13) as follows:

$$\mu x = x_1q_1 + x_2q_2 + x_3q_3 \quad .... (1)$$

$$\mu y = y_1q_1 + y_2q_2 + y_3q_3 \quad .... (2)$$

where μx and μy are components of the total electronic dipole vector ($\overrightarrow{\mu_{\pi+\sigma}}$) in the x and y directions (since the two groups under consideration here are planar the dipole moment vector can be expressed in terms of two components). The magnitude of $\overrightarrow{\mu_{\pi+\sigma}}$ is given by*

$$\overrightarrow{\mu_{\pi+\sigma}} = \sqrt{\mu x^2 + \mu y^2} \quad ..... (3)$$

and the angle θ, of dipole moment vector from the x axis is given by

$$\theta = \tan^{-1} \mu y/\mu x \quad ...... (4)$$

The dipole moment vector is represented graphically showing its orientation with reference to amide and ester frame works, see Figure VI.

---

* If the distance in the equations (1) and (2) are expressed in A^0 units and the charges in the electronic units the value obtained from equation (3) has to be multiplied by a factor of 4.8 to obtain the value in Debye Units.

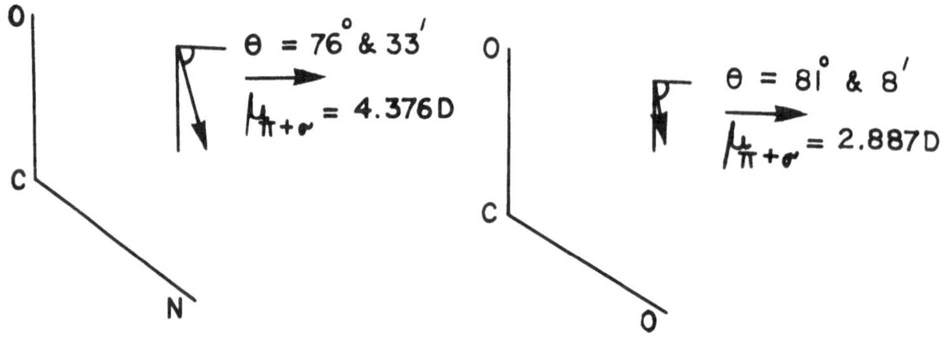

Figure VI: Showing net $\pi + \sigma$ electronic dipole moment vectors and their magnitudes of amide and ester groups.

DISCUSSION

Shemyakin et al. [(1 & 14)*] have discussed the biological activity of the two depsipeptides and their analogs. The naturally occuring depsipeptides Valinomycin has alternating six amide groups and six ester groups while Enniatin B has alternating three ester and three amide groups. The analogs of the two despspeptides were prepared by changing their topochemistry. Out of many analogs synthesized (both linear and cyclic), the cyclic analogs were prepared by replacing one or all ester groups by amide groups and vice versa. It was found by them that cyclic analogs prepared by substituting amide groups for ester groups lead to an increase in antibacterial activity and K^+-ion complexing property while the reverse replacement reduced these activities.

These experimental observations are in total agreement with our theoretical expectations. Comparing the results for amide and ester groups we find that among the basic parameters, the molecular orbital shapes are essentially similar. This can be seen from Table III and Figure IV. Qualitatively, the molecular orbital coefficients have some differences which is reflected in the electronic charge distributions (see Figure IV) and the dipole moment vector (see Figure VI). The directions of the two dipole moment vectors are nearly the same but the magnitude differs considerably. The energy levels of the π -electrons also make interesting comparison. The lowest orbitals have nearly equal energies while the next two differ significantly. This is partly reflected in the

* Refer to compound numbers VI to XXIII and especially compound number XXXIV of reference (1).

values for the ionization potentials which demonstrate that amide is a more potent electron donar than ester. These similarities and differences between the amide and the ester groups are summarized in the Table IV.

Similarities	Differences
Geometry of the two groups i.e. bond lengths and bond angles.*	& electronic charge distribution.
General shape of the Molecular orbitals.	Magnitude of the dipole moment vector (amide is more polar).
Direction of the dipole moment vector.	Ionization potentials (amide more potent electron donar).
Resonance energy*.	

Table VI: Summary of similarities and difference between amide and ester groups as brought out by VE-PPP and LCAO-MO Del Re's methods.* Refer to reference 6.

Thus it follows that in interactions depending on the general topochemistry of a depsipeptide, ester and amide groups are mutually replaceable (θester groups = 81° 8'; θ amide groups = 76° 33'). Differences are however expected in interactions involving the electrons of depsipeptides, whether they be in the form of an electrons of depsipeptides, whether they be in the form of an electrostatic binding (because of the differences in magnitude of dipole moments: $\overrightarrow{\mu_{\pi+\sigma}}$ for amide = 4.376 D; $\overrightarrow{\mu_{\pi+\sigma}}$ for ester = 2.887 D) or electron exchange interactions (because of the differences in the electron donating abilities, Ionization potentials: amide = 8.620 e.V.; ester = 10.528 e.V.). In other words, if the ester groups were replaced by amide groups and vice-versa the molecular geometry of depsipeptides, is not expected to change dramatically (since dipole moment angles θ, are similar for the two groups). But on the other hand the replacement of ester groups amide (either one or all) would lead to enhanced charge interaction with K^+-ion (due to the differences in the magnitude of dipole moments and more electron donating capacity of the amide groups and vice versa). However it is emphisized that this replaceability (i.e. amide to ester) in the case of Valinomycin will lead to loss of intramolecular hydrogen bonds, hence the "secondary" structure expected in the case of ester analogs of Valinomycin will be entirely different. In molecules, whose secondary structure does not involve hydrogen bonds (e.g., Enniatin B) this mutual replaceability is then governed by just two factors a) if the molecule is involved in steric interaction or b) electron exchange interaction. On the basis of the above study it can be summarized that K^+-ion depsipeptide multipole interactions will be higher in magnitude in the amide analogs than in the ester analogs provided no hydrogen bonds are involved in the depsipeptide's "secondary structure". These calculations have attempted to explain why amide analogs of depsipeptides are more active biologically and the ester analogs poorer. This study may have similar bearing on the biological activity of polypeptides, polypeptide hormones and their analogs.

ACKNOWLEDGEMENT

The authors wish to thank Prof. R.K. Mishra, M.B.B.S.,D.Sc., for having provided facilities in his department. One of us (S.V.T.) acknowledges the financial help of D.G.H.S., Ministry of Health, Family Planning and Urban Development, Government of India.

REFERENCES

1. M.M. Shemyakin, E.I. Vinogradova, M. Yu. Feigina, N.A. Aldanova, N.F. Loginova, I.D. Ryabova and I.A. Pavlenko; Separatum Experintia, 21, (1965), 548.

2. S.V. Talekar: (Thesis), "Quantum Chemical Studies of the Electronic Structure of Macrocyclic Antibiotic Molecules" (1970), A.I.I.M.S., New Delhi, India.

3. S.V. Talekar and K. Sundaram; In Proceeding of 2nd International Convention of Biochemists, (1970), Baroda, India . (In press).

4. S.V. Talekar and K. Sundaram; In Proceeding of First European Biophysical Congress, (1971) Vienna, Austria, Vol. 6, p.69.

5. M.M. Shemyakin, Yu. A. Ovchinnikov, V.T. Ivanov, V.K. Antonov, E.I. Vinogradova, A.M. Shakrob, G.G. Malenkov, A.V. Evstratov, I.A. Laine, E.I. Melnik and I.D. Ryabova; J. Mem. Biol. 1, (1969), 402.

6. L. Pauling: In "The Nature of Chemical Bond" III Ed., Oxford, I.B.H. Pub. Co., (1961), p. 275 & 281.

7. R. Pariser and R.G. Parr; J. Chem. Phys., 21, (1953), 466.

8. J.A. Pople; Trans. Faraday Soc., 49, (1953), 1375.

9. R.D. Brown and A. Penfold: Trans. Faraday Soc., 53, (1957),397.

10. R.D. Brown and M.L. Heffernan: Trans. Faraday Soc., 54, (1958), 757.

11. G. Del Re; J. Chem. Soc., (1958), 4031.

12. T.A. Koopmans; Physica, 1, (1933), 104.

13. R. Daudel, R. Lefebure and C. Moser; In "Quantum Chemistry Methods & Applications" Int. Sc. Pub. Inc. N.Y. (1959), p.204.

14. M.M. Shemyakin, L.A. Shchukina, E.I. Vinogradova, G.A. Ravdel, and Yu. A. Ovchinnikov; Separatum Experientia, 22, (1966),535.

15. H. Brockmann, M. Springorum, G. Traxler, I. Hofer; Natururss-enschaften, 50, (1963), 689.

16. P.A. Plattner, K. Vogler, R.O. Studer, P. Quitt, W. Kellerschierlein; Helv. Chim. Acta, 46, (1963), 927.

RADIOIMMUNOASSAY OF CIRCULATING BLOOD KININ LEVELS

M.L. Mashford and M.L. Roberts (with the technical assistance of G.W. Johnston)
Departments of Pharmacology and Medicine, (Austin Hospital), University of Melbourne, Parkville, 3052, Australia

Previous estimates of the level of circulating kinins in man have ranged widely from undetectable, (1) to a mean level of about a quarter of a ng/ml (2) to levels in the µg/ml (3) range. Some workers have questioned that there is a normal level of circulating kinin and view the small amounts found as being artefacts produced by activation of plasma kininogenase during collection. Another view is that kinins are produced extravascularly and that blood levels are likely to be irrelevant to normal or abnormal function except in such rare situations as the carcinoid syndrome in which kallikrein is released by the tumour masses and produces kinins in the blood itself. Certainly the central role of tissues and particularly the lung, (4) in clearance argues against the arterial blood kinin level being of great functional significance. Nevertheless, a reliable method of measuring blood levels would be welcome since they may give an indication, admittedly considerably damped, of what was proceeding in the tissues whence the blood drained.

Radioimmunoassay (RIA) provides an attractive means to measure the low levels present because it can provide sensitivity, specificity and the ability to handle reasonable sized batches of samples in the one assay procedure.

We have produced reasonable titres of antibodies in two rabbits by immunization with BK coupled in one case to ovalbumin and in the other to human gamma globulin.

These antibodies bind ^{14}C labelled BK and displacement curves can be constructed. However, the ^{14}C label does not have sufficiently high specific activity to give the required sensitivity and we use as antigen an analogue of BK with tyrosine in the 8-position (5) which can be iodinated with ^{125}I.

Using the antibody containing plasma diluted 1 in 6400 and this antigen a displacement curve can be constructed as shown in Fig. 1. This gives a significant displacement when 25 pg of unlabelled BK is added to the incubate, thus this assay has the sensitivity necessary to measure levels likely to be present in relatively small blood samples.

The value of such a procedure depends not only on sensitivity but also on its specificity.

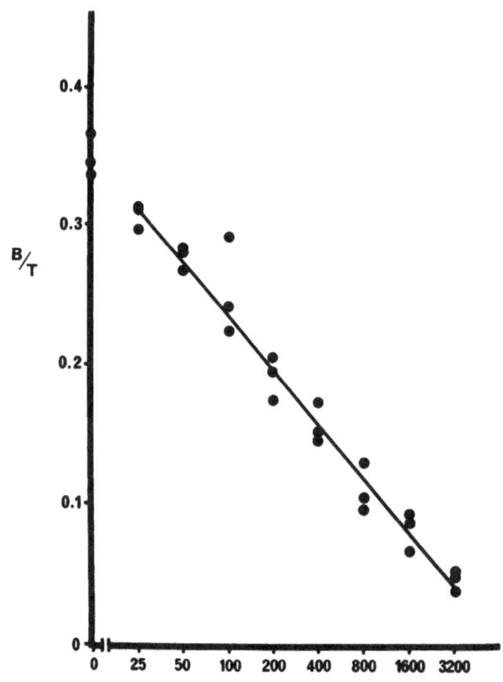

Fig. 1. Progressive displacement of (^{125}I-Tyr) BK from antibody by increasing amounts of added BK. B/T is the ratio of counts in the supernatant to total counts added.

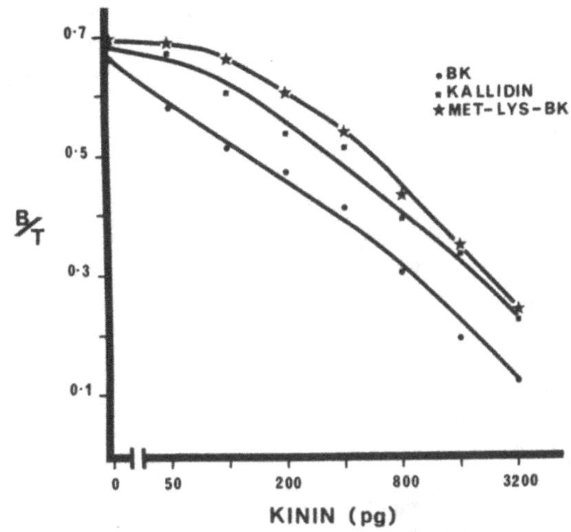

Fig. 2. Displacement of (^{125}I-Tyr) BK from antibody by BK, kallidin and met-lys-BK.

Fig. 2. shows that displacement with one of the antibodies (R18) occurs with all three kinins BK, kallidin and methionyl-lysyl-BK. With the other antibody (R12) there is virtually equal reactivity with the three peptides. On the other hand peptides such as eledoisin, angiotensin and the BK potentiating pentapeptide (6) showed no displacing ability. The two antibodies used, do show some cross-reactivity with peptide fragments as shown in Table 1. There is not a complete set but it can be seen that both antibodies bind BK but they differ in their reactivity with the fragments. For R12 the C-terminus appears to be critical and reaction occurs only if it is intact. With the other antibody (R18) the N-terminus also appears to be important since it reacts with N-terminal partial sequences as small as the hexapeptide. This fits with the data presented in Fig. 2. showing differences in reactivity of this antibody to BK, kallidin and methionyl-lysyl-BK. The other antibody R12 is not reactive to the N-terminal partial sequences and it does not distinguish between the 3 kinins which differ only at the N-terminus.

The application of the RIA to blood requires that the kinin-forming and destroying systems be inactivated

	R12	R18
Arg-Pro-Pro-Gly-Phe-Ser-Pro-Phe-Arg (BK)	+	+
Arg-Pro-Pro-Gly-Phe-Ser		
Arg-Pro-Pro-Gly		+
Arg-Pro		
Pro-Pro-Gly-Phe-Ser-Pro-Phe		
Pro-Pro-Gly-Phe-Ser		
Pro-Pro-Gly-Phe		
Pro-Pro		
Pro-Gly-Phe-Ser-Pro-Phe-Arg	+	+
Pro-Gly-Phe-Ser-Pro-Phe		
Pro-Gly-Phe-Ser		
Pro-Gly-Phe		
Pro-Gly		
Gly-Phe-Ser-Pro-Phe-Arg	+	+
Gly-Phe-Ser-Pro-Phe		
Gly-Phe-Ser		
Gly-Phe		
Phe-Ser-Pro-Phe-Arg	+	+
Phe-Ser-Pro-Phe		
Phe-Ser-Pro		
Phe-Ser		
Ser-Pro-Phe-Arg		+
Ser-Pro-Phe		
Ser-Pro		
Pro-Phe-Arg		
Pro-Phe		
Phe-Arg		

Table 1. Cross reactivity of various partial sequences of bradykinin with antibodies R12 and R18.

as soon as possible. The method of producing a suitable extract is summarized in Fig. 3.

Blood is taken into syringes kept on dry ice and containing 50 µl 0.5M phenanthroline HCl in 25% ethanol as a kininase inhibitor. As will be seen later this procedure gives virtually complete protection of added BK. Approximately 5 mls. of blood are used and injected immediately into 20 ml. of absolute ethanol at bench temperature. This is stored cold until centrifuged. The supernatant and washings of the precipitate are then reduced in volume to about 2 ml. by evaporation under vacuum. This small volume is acidified with HCl and extracted twice with ether to leave the peptide in the aqueous phase which is reduced to dryness and taken up in the diluent buffer for RIA. The assay is performed at

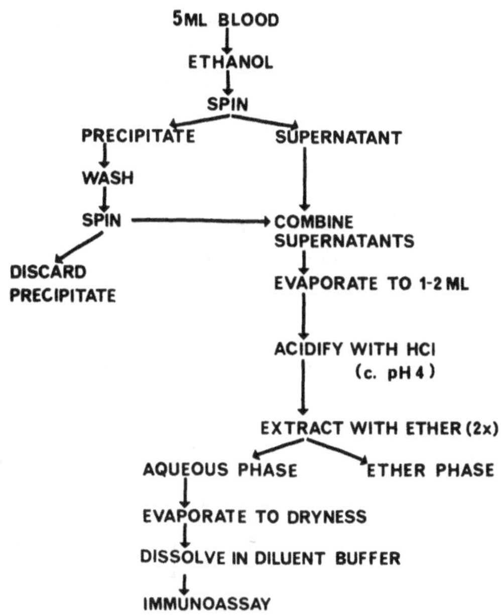

Fig. 3. Flow sheet of blood assay method.

two levels - 0.2 and 0.5 ml. respectively being added to different assay tubes and at each level, triplicate estimations are performed.

RIA is performed by a standard method using 24 hr. incubations of antibody with antigens and using dextran-coated charcoal (7) to separate bound from free label.

In any assay of blood kinins the first questions will be "Is the level found simply an artefact?" This may occur, as with any assay due to non-specificity of the assay, but in the case of kinins especially because of the possibility of generation or destruction of endogenous kinin during collection of the blood. We feel that both of these questions have been answered for this method. First the matter of specificity. As already stated binding of label is unaffected by a variety of non-kinin peptides. All assays on blood have been performed with R12 antibody which as was shown needs the C-terminus intact to react with BK fragments. The N-terminal octapeptide was not included in our survey as none was available, but incubation of BK and of extract with carboxypeptidase B* abolishes immunoreactivity. In addition immunoreactivity of both BK and of the blood extract is destroyed by chymotrypsin but unaffected by trypsin.

In radioimmunoassays it is important to show not only that the unknown and standard both react with antibody but that they follow the same dilution or dose-response curve. Blood extract prepared as described does, as can be seen in Fig. 4.

These data suggest that the material in the extract which is being measured is in fact BK or one of the kinins.

The second question that of whether the kinin detected is generated during collection has also been examined and the results of one of the experiments are shown in Table 2. Four sets of four syringes were prepared (sixteen in all). The first set of four contained only phenanthroline, i.e., they were the usual set-up for collection. These four syringes were filled in rapid succession with blood from a catheter in the forearm vein of a human subject. The first syringe was injected immediately on collection into the ethanol and the others after carefully timed delays of 10 sec, 30 sec, and 2 minutes. As can be seen there is no evidence of progressive generation of kinin. The second set of syringes contained in addition 5 mg. S.B.T.I. and the presence of this kallikrein inhibitor did not change the results. Although the initial level was lower there was no change with time. When 200 mg. of glass ballotini are added there is marked generation but it takes some time to occur. Finally these results are not due to destruction of formed kinins since recovery of standard from the 4th set of syringes shows no loss. Thus it appears that collection artefacts are not responsible for the level found.

The levels which have been formed are strikingly similar to those which were reported by Zacest and Mashford (2) using completely different methods.

The mean level in forearm venous blood of 15 subjects in the present study, was 0.2 ng/ml. (S.D. 0.20 ng/ml.) In 8 dogs anaesthetized with pentobarbital the mean level was higher,0.68 ng/ml. (S.D. 0.6 ng/ml.) In three of

* Worthington Biochemical Corporation.

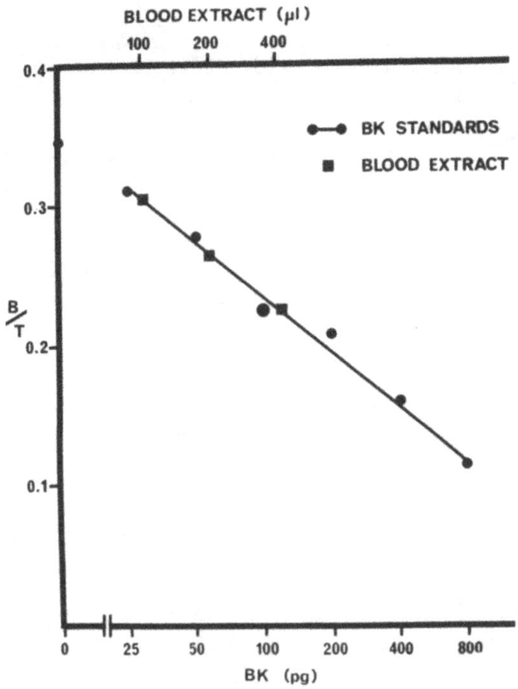

Fig. 4. Displacement curve of blood extract is congruent with the standard BK displacement curve.

Syringe Contents	Incubation Time (seconds)			
	0	10	30	120
PHENANTHROLINE	0.15	0.12	0.05	0.12
PHENANTHROLINE & S.B.T.I.	0.05	0.05	0.04	0.08
PHENANTHROLINE & GLASS	0.16	–	1.80	15.54
PHENANTHROLINE & BK	117 %	–	132 %	112 %

Table 2. Blood bradykinin levels (ng/ml) in subject M.H. following incubation of blood at room temperature for varying periods.

these experiments on dogs both arterial and venous blood was taken and there was in all of them a positive arteriovenous difference. In one dog a catheter was passed into the lower aorta through which BK was infused. Blood was collected from a catheter in the unoccluded L. femoral artery and flow in that vessel measured with a flowmeter. Infusion rates of 0.8 and 1.6 μg/minute caused increases in femoral flow and associated with these haemodynamic changes there was an increase in the level of BK in the femoral artery. (Fig. 5.)

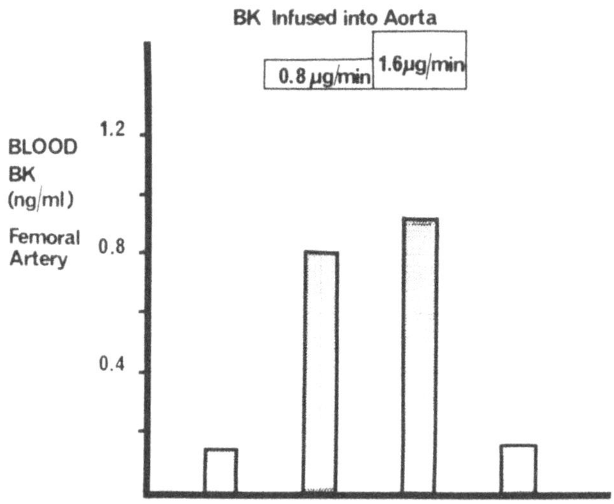

Fig. 5. Increase in femoral arterial BK levels caused by infusion of BK into the aorta.

This method has not been applied to many situations as yet but one interesting clinical application was in a 54 year old woman who had an abdominal aneurysm resected and replaced by a dacron graft from just below the renal arteries to the iliacs. Unfortunately this became infected, probably with Staph. albus and over the course of 18 months she developed florid hypertrophic osteo-arthropathy of the legs with marked pain and soft tissue swelling. BK levels in simultaneously sampled brachial and femoral arterial blood showed a persistently high femoral level. (Fig. 6.) It is suggested that generation of kinins at the infected graft or perhaps in the blood after passing the infected site may have been the cause of her disability.

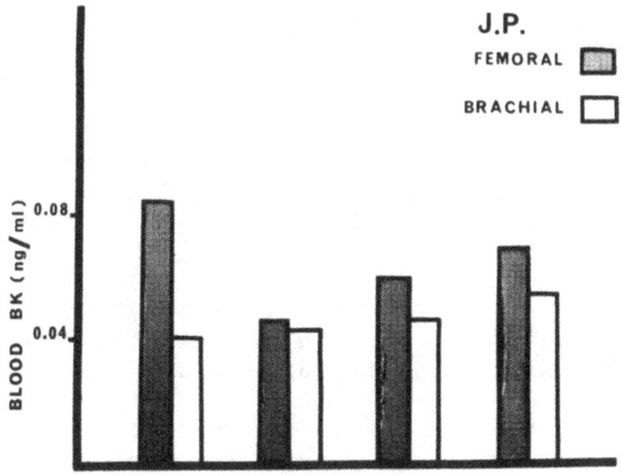

Fig. 6. Concentration of bradykinin in simultaneous samples of femoral and brachial arterial blood of a patient with an infected aortic graft.

SUMMARY

A radioimmunoassay has been developed, which can detect BK down to 25 pg. This is sensitive enough to measure blood levels in all subjects so far studied. There is good evidence that it is BK or a very similar kinin which is being measured and that what is found is not an artefact due to collection. Blood levels found in man agree with earlier estimates using bioassay procedures.

BIBLIOGRAPHY

1. Abe, K., Watanabe, N., Kumagai, N., Mouri, T., Seki, T., and Yoshinaga, K.
Estimation of kinin in peripheral blood in man.
Tohoku J. exp. Med. 89, 103 - 112.

2. Zacest, R., and Mashford, M.L. (1967)
Blood bradykinin levels in the human.
Aust. J. exp. Biol. med. Sci. 45, 89 - 95.

3. Zeitlin, I.J. and Smith, A.W. (1966)
5-Hydroxyindoles and kinins in the carcinoid and dumping syndrome.
Lancet 2. 986 - 991.

4. Ferreira, S.H., and Vane, J.R. (1967)
The disappearance of bradykinin and eledoisin in the circulation and vascular beds of the cat.
Br. J. Pharmac. 30, 417 - 424.

5. Mashford, M.L., and Roberts, M.L. (1971)
Determination of human urinary kinin levels by radioimmunoassay using a tyrosine analogue of bradykinin.
Biochem. Pharmacol. 20, 969 - 973.

6. Greene, L.J., Stewart, J.M. and Ferreira, S.H. (1970)
Bradykinin - potentiating peptides from the venom of Bothrops jararaca.
In Bradykinin and Related Kinins, eds. Sicuteri, F., Rocha e Silva, M., and Back, N.
Plenum Press, New York. pp. 81 - 87.

7. Herbert, V., Lau, K.S., Gottlieb, C.W., and Bleicher, S.J. (1965)
Coated charcoal immunoassay of insulin.
J. clin. Endocr. 25, 1375 - 1384.

SOME PROPERTIES OF BOVINE HIGH MOLECULAR WEIGHT KININOGEN

M. Komiya, H. Kato, M. Yano and T. Suzuki

Institute for Protein Research, Osaka University

Osaka, Japan

The presence of two kininogens has been shown in human, dog, guinea pig, rabbit and rat plasma by Jacobsen (1-3), in horse plasma by Henriques _et al_.(4), and in bovine plasma by Yano _et al_. (5). Pierce and Webster (6) isolated two low molecular weight (LMW) human kininogens with the same molecular weight of about 50,000 but with their kinin sequences carboxyl-terminal in one case and internal in the other. After that, two kininogens of different molecular weight were isolated from human plasma by Jacobsen and Kritz (7). LMW bovine kininogen (kininogen-II) has been highly purified by the groups of Habermann (8) and of Suzuki (9), and it was found by us that the kinin segment, which was thought to lie in the central part of LMW kininogen molecule, was released by trypsin, snake venom kininogenase or hog pancreatic kallikrein but not by glass- or casein-activated bovine plasma kallikrein. High molecular weight (HMW) kininogen (kininogen-I), a sensitive substrate of bovine plasma kallikrein, was isolated from bovine plasma by Yano _et al_.(10), and it was found that there is a methionyl-lysyl-bradykinin sequence located both at the carboxyl terminus and inside of the polypeptide chain of bovine HMW kininogen (11).

Moreover, Yano and Suzuki experienced that from the carboxypeptidase B (CPase B) treated HMW bovine kininogen plasma kallikrein liberated kinin more slowly when it was compared with the case of untreated bovine HMW kininogen (Fig. 1, unpublished data). Trypsin liberated kinin from CPase B-treated HMW kininogen as rapidly as from untreated HMW kininogen. So they tentatively supposed that two kinin moieties were located in a HMW kininogen molecule and by some conformational changes of the kininogen resulting from the treatment with CPase B, kinin segment

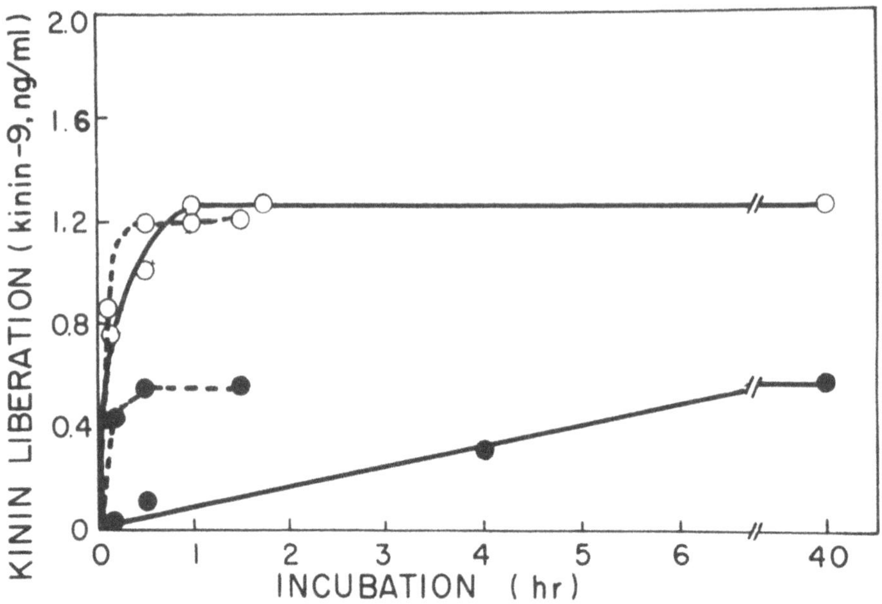

Fig. 1 Kinin liberation from carboxypeptidase B treated HMW kininogen. Experimental consitions are the same as reported previously (11). Solid lines and dotted lines indicate the amounts of kinin liberated from kininogen by plasma kallikrein activated by casein-treatment and by trypsin, respectively.
o ; untreated HMW kininogen
● ; carboxypeptidase B treated HMW kininogen

located at the inner part of the kininogen changed its reactivity with plasma kallikrein to liberate kinin.

However, to make up a conceptions for the gross-structure of HMW bovine kininogen, many doubts must be resolved. Therefore, the authors attempted to isolate HMW bovine kininogen by the improved method avoiding denaturation during the purification procedures. In this paper, we will present the possibility that HMW kininogen consists of two kininogens, one with kinin moiety at carboxyl terminus and another at inside; both are sensitive substrates for plasma kallikrein.

Purification of HMW Kininogen

Because bovine HMW kininogen is susceptible to plasma kallikrein, to inactivate the spontaneously activated kallikrein, heat treatment was previously employed, but this caused low yield of kininogen. In the improved method, heat treatment was avoided and hexadimethrine bromide (polybrene) was used as an inhibitor

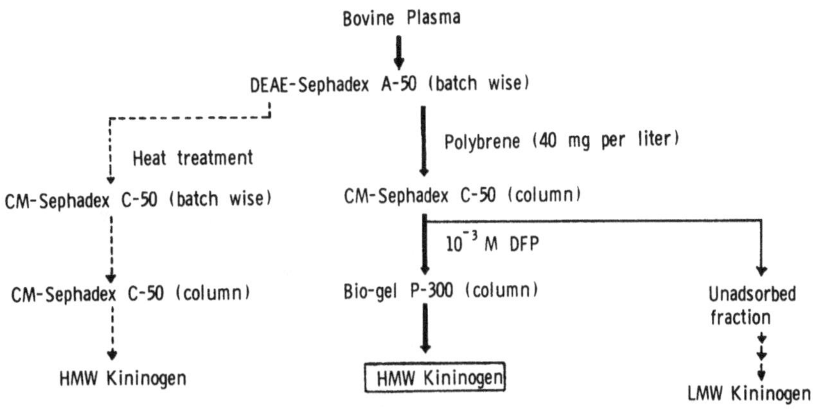

Fig. 2 Purification procedures of bovine HMW kininogen. Dotted line shows the previous method reported by Yano *et al.*(10).

of prekallikrein activator, and DFP was used to destroy the kallikrein activity.

The previous and improved purification procedures of HMW kininogen can be seen in Fig. 2. Bovine plasma was mixed with DEAE-Sephadex A-50 and after the DEAE-Sephadex A-50 has sedimented, the supernatant was discarded; the residual DEAE-Sephadex was poured into a glass column, and was washed with 2 liters of 0.02 M Tris buffer, pH 7.5, containing 0.1 M NaCl and then eluted with 2.5 liters of 0.2 M acetate buffer, pH 6.5, containing 0.6 M NaCl. The eluates from DEAE-Sephadex A-50 column were combined and dialysed against deionized water, in which polybrene was added. The solution was mixed with CM-Sephadex C-50 and after stirring in cold room, the suspension was poured into a column. LMW kininogen can be isolated from the unadsorbed fraction of CM-Sephadex C-50 column chromatography. After the bulk of proteins were washed out from a CM-Sephadex C-50 column, linear gradient elution was performed. Kinin-yielding ability was eluted as a single peak from the column (Fig. 3), however, when each fraction of the eluate was subjected to disc electrophoresis, it was found that kininogen fraction, in tubes number of 50-60, was contaminated with other protein which may be HMW kinin-free-kininogen. HMW kininogen fractions indicated by solid bar in Fig. 3 (in tubes number of 70-90) were combined and lyophilized. However, even in this fraction, a trace amount of kininogenase was contaminated and when an aliquot of the fraction was incubated for 24 hr at 37°C, liberation of kinin was detected. So, to the above lyophilized fraction 10^{-3} M DFP was added as kininogenase inhibitor, and to remove a trace amounts of aggregated proteins formed gel-filtration on a column of Bio-gel P-300 was performed. The yields of HMW kininogen thus highly purified was about 500 mg from 30 liters of blood. And from the same lot of the bovine plasma 400 mg of LMW

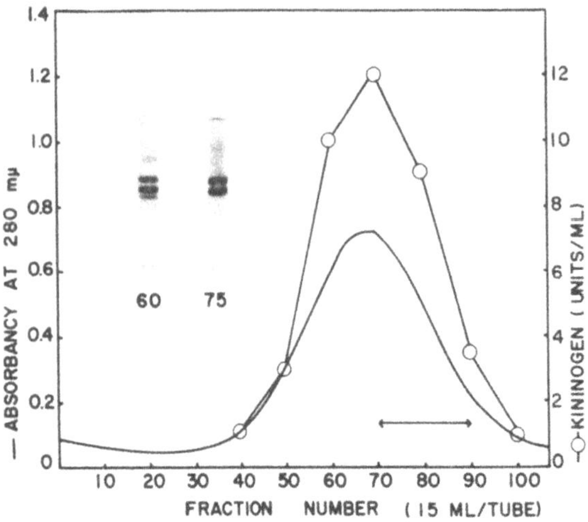

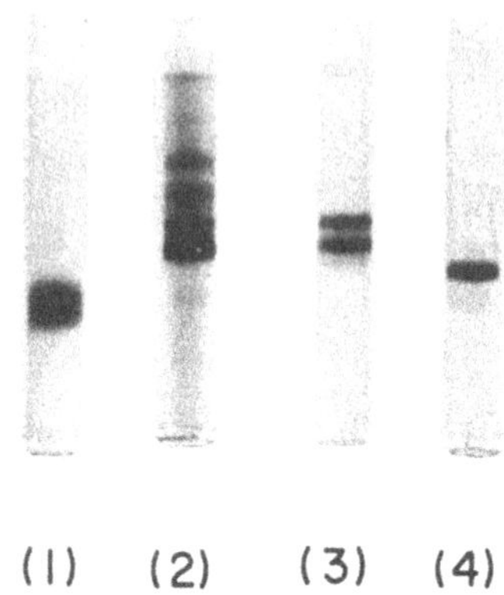

Fig. 3 Fig. 4

Fig. 3 CM-Sephadex C-50 column chromatography of HMW kininogen. CM-Sephadex C-50, which was mixed with the eluates from DEAE-Sephadex A-50, was poured into a column (5 x 35 cm) and washed with 2 liters of 0.05 M acetate buffer, pH 6.2, containing 0.25 M NaCl and polybrene (40 mg per liter). Linear gradient elution was performed from one liter of 0.05 M acetate buffer, pH 6.2, containing 0.8 M NaCl. Polybrene was added to each buffers with the concentration of 40 mg per liter. Disc electrophoretic patterns of fraction number 60 and 75 are also shown in the figure.

Fig. 4 Polyacrylamide gel disc electrophoresis of bovine kininogens.

(1) LMW kininogen
(2) HMW kininogen obtained by the heat treatment method as previously reported (11).
(3) HMW kininogen obtained by the method as described in this paper.
(4) HMW kinin-free-kininogen

Polyacrylamide gel disc electrophoresis was performed with 400 volts for one hr at pH 8.3. The gel concentration used was 5.6 %. The proteins were stained with coomassie brilliant blue.

kininogen were obtained.

When HMW kininogen preparation thus obtained was applied on disc electrophoresis, two protein components, which have the kinin moiety susceptible to plasma kallikrein, were found (Fig. 4 (3)). The HMW kininogen purified by the previous procedures in which heat-treatment was inserted to destroy the kininogenase activity was found to be contaminated with aggregated proteins (Fig. 4(2)). When HMW kininogen preparation was incubated with casein-activated

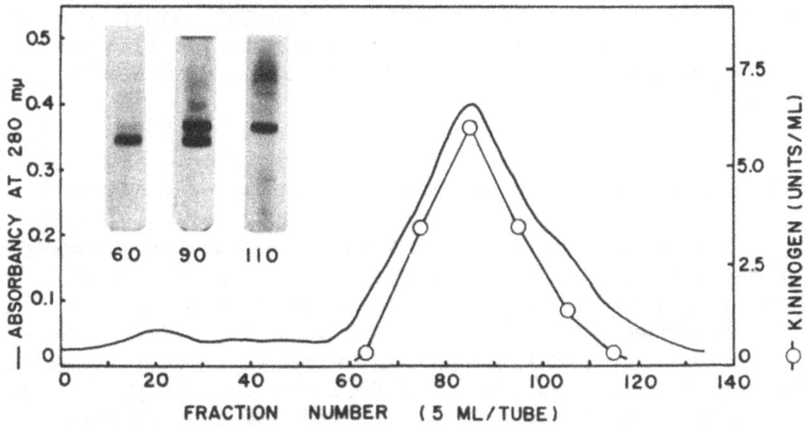

Fig. 5 Separation of HMW kininogen-a and -b by CM-Sephadex C-50 column chromatography. To a column (2 x 84 cm) of CM-Sephadex C-50 equilibrated with 0.05 M acetate buffer, pH 6.2, containing 0.3 M NaCl, 130 mg of HMW kininogen was applied. Linear gradient elution was performed from 500 ml of 0.05 M acetate buffer, pH 6.2, containing 0.3 M NaCl to 500 ml of the same buffer containing 0.6 M NaCl. Disc electrophoretic patterns of fraction number 60, 90 and 110 are shown in the same figure.

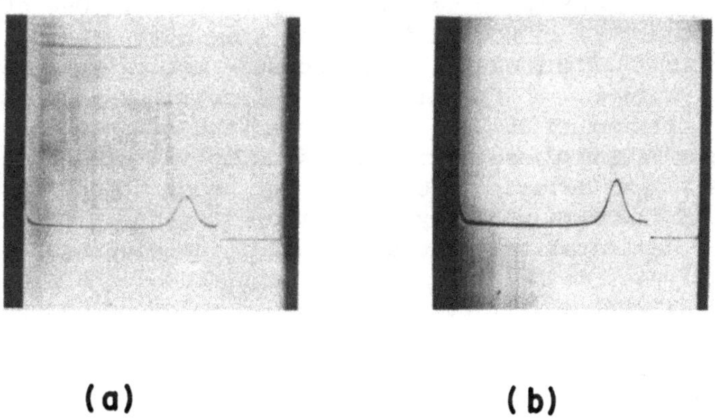

Fig. 6 Ultracentrifugal patterns of HMW kininogen and kinin-free-HMW kininogen.

(a) HMW kininogen (after 44 min)

(b) HMW kinin-free-kininogen (after 46 min)

Ultracentrifugation was performed at a concentration of 0.78 % at 50243 r.p.m. in the case of HMW kininogen; and 1.06 % at 52724 r.p.m. in the case of HMW kinin-free-kininogen. Solvent used was 0.1 M acetate buffer, pH 6.2, containing 0.2 M NaCl.

plasma kallikrein, two bands of HMW kininogen disappeared and the faster band, which corresponds to HMW kinin-free-kininogen, was newly detected (Fig. 4 (4)). LMW kininogen moved faster than HMW kininogen (Fig. 4 (1)).

To separate the two HMW kininogen components each other, rechromatography on CM-Sephadex C-50 column was performed. Although the complete separation was not achieved, as shown in Fig. 5, kininogen with faster mobility on disc electrophoresis, HMW kininogen-b, was first eluted and then the other, HMW kininogen-a, was eluted. Each of the fractions in tubes No. 60-80 (HMW kininogen-b) and No. 105-117 (HMW kininogen-a) were pooled and lyophilized. HMW kininogen-b was further purified by re-chromatography on a column of CM-Sephadex C-50. By the action of casein-activated plasma kallikrein, both of the kininogen's-a and -b produced kinin-free-kininogens with the same mobility. HMW kininogen preparation obtained from Bio-gel P-300 column and HMW kinin-free-kininogen derived from it, showed homogeneous schlieren patterns on ultracentrifugation (Fig. 6). These results suggest that HMW kininogen-a and -b may differ with their microhetrogeneities.

Each of the HMW kininogens-a and -b and the mixture of them (HMW kininogen-ab), showed a single band on SDS (sodium dodecyl sulfate) disc electrophoresis. However, when kininogen-ab was reduced by β-mercaptoethanol and subjected to SDS disc electrophoresis, two main bands were detected (Fig. 7). One of them had the same mobility as HMW kininogen-ab and another which moved faster than HMW kininogen-ab had the same mobility as that of reduced kininogen-a. HMW kininogen-b fraction was contaminated with a small amount of HMW kininogen-a, and after the reduction, the contaminating protein with the same mobility as reduced HMW kininogen-a was detected. On the other hand, reduced kininogen-b moved with the same mobility as that of original kininogen-ab. From this we tentatively speculated that the kininogen-a consists of two polypeptide chain bridged by disulfide linkage and reduced kininogen-a yielded a lower molecular weight peptide, which was not detected on SDS disc electrophoresis, probably due to its low molecular weight. On the other hand, kininogen-b consists of a single polypeptide chain, and even after it was reduced, only a polypeptide chain was detected.

HMW kinin-free-kininogen-a, -b and -ab and each of the reduced HMW kinin-free-kininogen's-a, -b and -ab showed a single band respectively on SDS disc electrophoresis (Fig. 8), and these three of the reduced HMW kinin-free-kininogens were very much the same as reduced HMW kininogen-a (Fig. 7 (3)) in their mobilities. Concerning to the faint band with faster mobility of HMW kinin-free-kininogen-b (Fig. 8 (2)), nothing is yet known. From these results, it is speculated that one chain of HMW kininogen-a may be identical with one of the two peptides obtained from the reduced HMW kinin-free-kininogen-b. Previously, kallidin and the kinin-yielding peptide were separated by cyanogen bromide treatment of HMW kininogen by Yano et al.(11). Therefore, from these results above, HMW kininogen-a seems to have kinin moiety at its C-terminus and HMW kininogen-b at inside. However, it was observed that the amounts of kallidin obtained by cyanogen bromide treatment of HMW kininogen, which correspond to the content of HMW kininogen with kinin moiety at its C-terminus, varied by the preparation used. The explanation on this problem awaits further investigation.

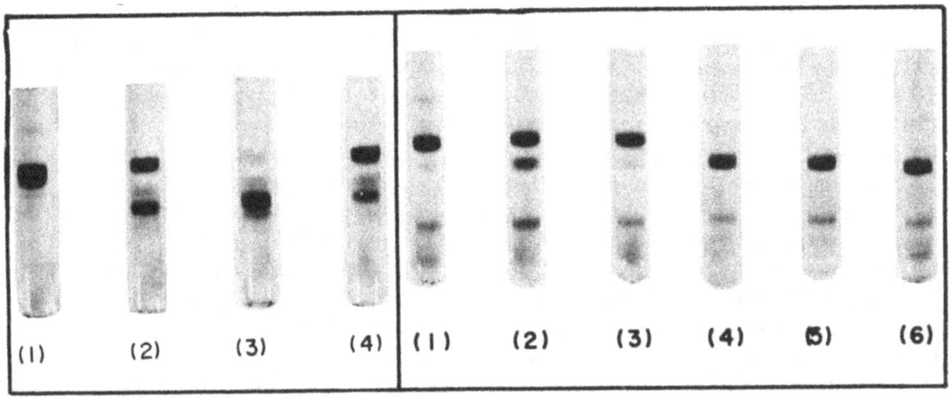

Fig. 7 Fig. 8

Fig. 7 Electrophoretic mobilities of HMW kininogens on SDS disc electrophoresis.

(1) HMW kininogen-ab
(2) Reduced HMW kininogen-ab
(3) Reduced HMW kininogen-a
(4) Reduced HMW kininogen-b

HMW kininogen and the reduced HMW kininogen with -mercaptoethanol were electrophoresed in 0.1 M phosphate buffer, pH 7.5, containing 0.1 % SDS at 7 mA per tube for 5 hr.

Fig. 8 Electrophoretic mobilities of HMW kinin-free-kininogens on SDS disc electrophoresis.

(1) HMW kinin-free-kininogen-ab
(2) HMW kinin-free-kininogen-b
(3) HMW kinin-free-kininogen-a
(4) HMW kinin-free-kininogen-ab after reduction
(5) HMW kinin-free-kininogen-b after reduction
(6) HMW kinin-free-kininogen-a after reduction

HMW kinin-free-kininogens and their reduced products were subjected to disc electrophoresis in 0.1 M phosphate buffer, pH 7.5, containing 0.1 % SDS at 7 mA per tube for 5 hr.

Properties of HMW Kininogen Comparing with LMW Kininogen

In this paper we separated HMW and LMW kininogens as distinct molecules even when the denaturation was carefully avoided during purification procedures. From the experimental results so far obtained, we suppose that some chemical interrelations may exist between LMW and HMW kininogens and also between two HMW kininogen's-a and -b molecules, and recently we observed antibody against HMW kininogen reacted with LMW kininogen. To elucidate the interrelations of chemical natures of HMW and LMW kininogens in future, some properties of HMW kininogens were examined comparing with LMW kininogen. The properties of HMW and LMW kininogens are summarized in Table I. Molecular weight of HMW kininogen was measured by Archibald method and calculated to be 70,000. Previously, the molecular weight of HMW kininogen (kininogen-I) was estimated to be about 300,000 by gel-filtration on Bio-gel P-300 by Yano <u>et al</u>.(10), but they considered that further studies are necessary on the molecular weight of HMW kininogen. We corrected the value, using sedimentation constant and stokes radius, from the equation, $M = 6\pi\eta NrS/(1-v\rho)$ and calculated to be 60,000 to 70,000. The abnormal behaviour of HMW kininogen on gel-filtration was also observed by Jacobsen and Kritz (7), and it may be due to its extended shape as we experienced in the case of LMW kininogen. Location of one of the kinin moiety of HMW kininogen differs from that of LMW kininogen.

Table I
Some properties of HMW kininogen and LMW kininogen

	HMW kininogen-a & b (kininogen-I)	LMW kininogen (kininogen-II)
Kinin moiety	C-terminal, internal	internal
Rate of kinin liberation by Plasma kallikrein	very quick	very slow
Mobility on polyacrylamide gel electrophoresis		
at pH 8.3	slow (two bands, a& b)	fast (one band)
in SDS	slow	fast
Chromatography on CM-Sephadex C-50	adsorption	unadsorption
Solubility to salt free water	insoluble	soluble
Physical properties		
extinction coefficient $E^{1\%}_{1cm}$ (280 mμ)	7.4	6.7
sedimentation coefficient $S^{\circ}_{20,w}$	3.80	3.66
partial specific volume $\bar{v}$ (ml/g)	0.718	0.688
molecular weight	70.000 ± 5,000 (Archibald)	49,500 - 47,000

From LMW kininogen, plasma kallikrein liberate kinin very slowly. HMW and LMW kininogens can be separated each other on CM-Sephadex C-50 column. LMW kininogen, and not HMW kininogen, was soluble in salt-free water. And, LMW and HMW kininogens have different molecular weight, (LMW kininogen 49,500 to 47,000) each other.

REFERENCES

(1) S. Jacobsen, Nature, 210, 98 (1966).
(2) S. Jacobsen, Brit. J. Pharmacol., 26, 403 (1966).
(3) S. Jacobsen, Brit. J. Pharmacol., 28, 64 (1966).
(4) O.B. Henriques, N. Kauritcheva, V. Kuznetsove and M. Astrakan, Nature, 215, 1200 (1967).
(5) M. Yano, S. Nagasawa, K. Horiuchi and T. Suzuki, J. Biochem., 62, 504 (1967).
(6) J.V. Pierce and M.E. Webster, "Hypotensive Peptide", ed. by E.G. Erdös *et al*., Springer-Verlag (N.Y.), p.130 (1966).
(7) S. Jacobsen and M. Kritz, Brit. J. Pharmacol., 29, 25 (1967).
(8) E. Habermann, W. Klett and G. Rosenbusch, Z. Physiol. Chem., 332, 121 (1963).
(9) T. Suzuki, Y. Mizushima, T. Sato and S. Iwanaga, J. Biochem., 57, 14 (1965).
(10) M. Yano, S. Nagasawa and T. Suzuki, J. Biochem., 69, 471 (1971).
(11) M. Yano, S. Nagasawa and T. Suzuki, Biochem. Biophys. Res. Communs., 40, 914 (1970).

BRADYKININOGEN SYNTHESIS BY LIVER

F. T. Bryan, J. W. Ryan and R. S. Niemeyer

Departments of Medicine and Surgery, University of Miami School of Medicine and the Howard Hughes Medical Institute, Miami, Florida, 33152, U.S.A.

INTRODUCTION

Kininogen disappears from peripheral blood after hepatectomy or intoxication with carbon tetrachloride (1), a point that suggests that kininogen is synthesized by the liver. In the present study we have tested this possibility by perfusing the isolated liver with Krebs-Henseleit solution containing ^{14}C-arginine and ^{14}C-proline. The radioactive protein formed by the isolated liver reacts with a specific kininogenase (2) to produce a radioactive peptide indistinguishable from bradykinin. We interpret our data to mean that kininogen is synthesized by the liver.

A preliminary report of our work has been published (3). Therefore we present a brief description of the successful isolation of radioactive bradykinin along with a description of some unsuccessful experiments, the results of which bear on the precision and accuracy of methods for assaying bradykininogen.

EXPERIMENTAL

All glassware was coated with silicone (Siliclad, Clay-Adams). Synthetic bradykinin was purchased from Schwarz BioResearch, Inc. Rat pseudoglobulin was prepared by the method of Skeggs, et al. (4), using plasma obtained from rats 48 h after bilateral nephrectomy. The kininogenase of the pseudoglobulin fraction was activated as described previously (2).

Female Sprague-Dawley rats, weighing 0.2-0.3 kg, were injected subcutaneously with diethylstilbestrol (1 ml of 1 mg/ml). The next day the rats were anesthetized with ether and underwent bilateral nephrectomy. After operation the rats received a second injection of stilbestrol. Stilbestrol was used as a stimulus for hepatic protein synthesis. Nephrectomies were performed because of the report (5) that blood levels of kininogen may triple within 24 h of removing the kidneys. Over the next 24 h, the rats received water but no food. Then the rats were anesthetized with chloral hydrate. The liver was isolated and was perfused with Krebs-Henseleit solution containing ^{14}C-arginine (1 mCi, 310 mCi/mmole), ^{14}C-proline (1 mCi, 157 mCi/mmole) and the amino acids recommended by Jefferson and Korner (6).

arginine	2178 μM	phenylalanine	693 μM
asparagine	380 μM	proline	2396 μM
isoleucine	891 μM	threonine	2871 μM
leucine	1683 μM	tryptophan	683 μM
lysine	4752 μM	valine	1980 μM
methionine	693 μM		

The perfusion solution was aerated with O_2 + CO_2 (95:5) and heated at 38°C. Circulation proceeded for 60 min, and then the Krebs-Henseleit solution was replaced with fresh solution containing all of the amino acids except for ^{14}C-arginine and ^{14}C-proline. Circulation was continued for one more h. Solutions collected at the end of each circulation period were treated separately.

Each perfusate was ultrafiltered to 1/5th of its original volume. The concentrate was dialyzed against water and then against 3 mM EDTA in 0.9 g% NaCl.

RESULTS

A. Reaction of Liver Perfusate with Plasma Kininogenase

Recently, Ryan and Rocha e Silva (2) have shown that the pseudoglobulin fraction of plasma contains a kininogenase enzyme activated by acidification and heating. The enzyme releases bradykinin but not lysyl-bradykinin nor methionyl-lysyl-bradykinin.

The ultrafiltered, dialyzed liver perfusate was mixed with lyophillized pseudoglobulin, 20 mg/ml. The resulting solution was acidified to pH 2.5, heated in a boiling water bath (5 min), cooled and then neutralized to pH 7.3 using 3.5 N NaOH. One g of pulverized glass was added to facilitate the activation of the kininogenase. The reaction mixture was incubated at 37°C for 4 h. Trichloroacetic acid was added to a final concentration of 7%. The precipitate was removed by centrifugation and the supernatant was saved. The precipitate was washed with 10% trichloroacetic acid, and the wash supernatant was added to the first. The supernatant was extracted eight times with two volumes of ether. The aqueous layer was evaporated to dryness. Its residue was dissolved in 2 ml of 1 M pyridine and was chromatographed on Bio-Gel P-2 (1 x 100 cm). Fractions corresponding to the known elution volume of bradykinin (0.9-1.2 of the void volume) were combined, 1 mg of bradykinin was added as carrier, and the solution was evaporated to dryness. The residue was dissolved in 1% trifluoracetic acid previously saturated with butan-1-ol and was used as the lower phase of the first tube of a 10-transfer countercurrent distribution. Bradykinin was located by bioassay (rat uterus) and the contents of the three tubes having the highest biological activity were combined and dried. The residue was dissolved in pH 5 pyridine-acetic acid buffer (0.1 M in respect to acetic acid) and was submitted to electrophoresis (CM-82 paper, 18 v/cm, 90 min, pH 5) in one dimension and to chromatography (CM-82 paper, pH 5) in a second dimension. Carrier bradykinin was located by reaction with ninhydrin and was the only ninhydrin-reactive material on the chromatogram. The ninhydrin-reactive spot was cut into three pieces, which were eluted with 70% alcohol. Aliquots were assayed for biological activity (Guinea Pig ileum) and radioactivity. The apparent specific activity of bradykinin was 0.8 μCi/μmole.

B. Tryptic Hydrolysis of Liver Perfusate

In fact, the successful isolation of radioactive bradykinin (see above) was accomplished after a succession of failures of attempts to isolate bradykinin from tryptic hydrolysates. The technical problems contributing to our lack of success bear on the interpretation of results of assays in common use for measuring kininogen.

In our first experiments, the non-diffusable portion of liver perfusates was mixed with acetic acid, 0.2% in final concentration. The solution was heated at 97°C for 30 min, cooled and then adjusted to pH 7.5 by adding NaOH. One-fifth volume of 0.2 M tris, containing trypsin (pretreated with TCPK to eliminate chymotrypsin), 1 mg/ml, was added and the solution was incubated at 37°C for 30 min. Then the reaction mixture was heated in a boiling water bath. This procedure is essentially that developed by Rocha e Silva and Ryan (1960, unpublished) following a previous study of kininogen by Van Arman. The full procedure was subsequently published by Diniz, et al. (7) and by Diniz and Carvalho (1).

In this study, aliquots were taken for bioassay (Guinea Pig ileum and rat uterus) and bradykinin, 1 mg, was added as carrier. After adding carrier, another aliquot was taken for bioassay. Although the reaction mixture itself contained no activity measurable by bioassay, the reaction mixture plus carrier contained bradykinin-like activity equivalent to >3 mg of bradykinin. Thus it appears that tryptic hydrolysates, as has been indicated by Hamberg, et al. (8), have a great capacity for potentiating the effects of bradykinin.

The heat-inactivated reaction mixture plus carrier was evaporated to dryness. The residue was washed four times with water and then with trichloroacetic acid. The combined washes were extracted with ether and the aqueous layer was evaporated to dryness. The residue was dissolved in 4% trifluoroacetic acid, which was then extracted eight times with butan-1-ol. The butanol extract was dried. Part of the residue was soluble in 6% acetic acid and the remainder in 1 M pyridine. These solutions were chromatographed separately on Bio-Gel P-2 (1 x 100 cm). Fractions corresponding to 0.9-1.2 of the void volume were evaporated to dryness. These residues were dissolved in pH 5 pyridine-acetic acid buffer and were chromatographed using Sephadex G-25 superfine (0.9 x 70 cm). Fractions corresponding to 0.55-0.75 of the column volume (the known elution volume of bradykinin) were dried. This residue was processed through a 10-transfer countercurrent distribution as described in Section A. Aliquots of the transfer tubes were reacted with ninhydrin and bioassayed. Bradykinin was far from being pure. Tubes corresponding to K 1.0-2.3 were combined and dried. The residue was then submitted to electrophoresis and chromatography on CM-82 paper as described in Section A. The final chromatogram contained a variety of radioactive, ninhydrin-reactive spots and bradykinin was far from being pure. This attempt to isolate bradykinin was abandoned because of the large variety of peptides left and because loss of carrier bradykinin exceeded 99%.

In the early steps of the purification, losses of radioactivity and bradykinin-like activity were nearly parallel, allowing only a two-fold purification up to the point of chromatography on Bio-Gel P-2. The major losses occurred during evaporation steps, 60% of the radioactivity being lost in attempts to dissolve the residue of the original reaction mixture. Silicone-coated glassware was used throughout, but adsorption on glass still appeared to be the major source of loss. Radioactivity lost by adsorption could be recovered with very alkaline detergents, but the detergents did not favor the recovery of bradykinin, possibly because of adverse alkaline effects on the serine residue.

Yet, the overwhelming problem was that of the very large numbers of peptides formed by trypsin. In terms of solubility, molecular size, partition coefficients and charge, trypsin appears to form a number of peptides whose molecular differences from bradykinin must be minor.

DISCUSSION

We have found that the isolated liver incorporates ^{14}C-arginine and ^{14}C-proline into a non-dialyzable substance. Incubation of the non-dialyzable substance with a plasma kininogenase results in the formation of a radioactive bradykinin-like polypeptide. All available evidence indicates that the polypeptide is bradykinin itself. The product is formed by a highly specific kininogenase (2). The radioactive polypeptide behaves like bradykinin on chromatography using Bio-Gel P-2, countercurrent distribution, electrophoresis and chromatography on paper substituted with carboxymethyl groups. The distribution of radioactivity matches the distribution of biological activity of carrier bradykinin. Our results indicate that bradykininogen is synthesized by the liver.

Bradykinin formed on incubation of the non-dialyzable portion of liver perfusate with a specific kininogenase was isolated without difficulty. In contrast, tryptic hydrolysis of liver perfusate caused the formation of a large number of peptides having solubility, size and charge characteristics like those of bradykinin. Some of the tryptic peptides were capable of potentiating the actions of bradykinin on the guinea pig ileum and rat uterus. These peptides caused a three-fold overestimation of bradykinin content. While the similarities of liver perfusate and plasma are not known, assays in common use for the measurement of kininogen in plasma are based on quantities of kinin released by trypsin and therefore should be re-investigated to rule out spurious effects of potentiating peptides.

REFERENCES

1. C. R. Diniz and I. F. Carvalho (1963). Ann. N. Y. Acad. Sci., 104, 77.
2. J. W. Ryan and M. Rocha e Silva (1971). Biochem. Pharmac., 20, 459.
3. J. W. Ryan, F. T. Bryan and R. S. Niemeyer (1971). Ciênc. Cult., 23, No. 4.
4. L. T. Skeggs, K. E. Lentz, J. R. Kahn, F. E. Dorer and M. Levine (1969). Circ. Res., 25, 451.
5. E. Werle, G. Leysath and A. Schmal (1968). Z. Physiol. Chem., 349, 107.
6. L. S. Jefferson and A. Korner (1969). Biochem. J., 111, 703.
7. C. R. Diniz, I. F. Carvalho, J. W. Ryan and M. Rocha e Silva (1961). Nature, 192, 1194.
8. U. Hamberg, P. Elg and P. Stelwagen (1969). Scand. J. Clin. Lab. Invest., 24, Suppl. 107, 21.
9. This research was supported in part by grants from the Florida Heart Association and the U.S. Public Health Service (HE 12843 and HE 14360.

CHARACTERISTICS OF SPONTANEOUS FORMATION OF VASOPRESSOR PRINCIPLE (VA) AND OF PLASMA KININ AS REVEALED BY TEMPERATURE CHANGES **

Désirée Armstrong, H. Croxatto and J. Roblero

Department of Pharmacology, The Middlesex Hospital Medical School, University of London, England*, and

Laboratorio de Fisiologia, Universidad Catolica de Chile, Santiago, Chile

Normal human serum in polyethylene vessels between 37° and 0° C spontaneously produces at least three vasomotor substances. They include a vasopressor substance called VA, a large molecular depressor substance, here called TPD kinin, and the better known plasma kinin.

VA is a large molecular weight pressor substance formed at 37°, as reported by Kenner and Waldhäusl (1964) and Croxatto, Roblero and Ossio (1965). It relaxes the rat's isolated duodenum and is inactivated by chymotrypsin, but it does not contract the rat's isolated uterus and is not inactivated by carboxypeptidase B. The molecule, which is larger than bradykinin, is stable in serum (Roblero and Croxatto, 1967).

TPD kinin, so called because it can be digested by trypsin also by pepsin, is obtained from serum along with VA. It is a large molecular substance resembling bradykinin in many respects (see Roblero and Croxatto, 1967). TPD kinin depresses the blood pressure of the normal rat; it contracts the

*Present address of Dr. Désirée Armstrong is:
Duhring Laboratories - Department of Dermatology
Hospital of the University of Pennsylvania, Phila., Pa. 19104 USA

** The full text of this communication is being published elsewhere.

rat's isolated uterus; it relaxes the rat's isolated duodenum; and it is digested by both chymotrypsin and carboxypeptidase B. However TPD, too, is larger than bradykinin since it cannot be dialysed through a membrane. Moreover, it is digested by trypsin, also by pepsin, and thus it produces bradykinin-like polypeptides.

Freshly collected normal human serum does not form plasma kinin spontaneously at 37°, but it does so at room temperature (eg. 20°) or, better, 0° (8, 9, 10). Following storage at 37° for several days, however, serum spontaneously forms kinin at 37°, 20°, or 0°. The kininogen levels in the cold sera fall concurrently. The kininogen level of the warm, 37°, serum is maintained for many hours or several days unless and until a sudden eruption of plasma kinin takes place. By heating the fresh serum to 43-50° for several hours, then reducing the temperature to 37°, 20°, or 0° rapid kinin formation can be induced in it (10). Once the kinin forming system has been fully activated, the rate of kinin formation falls with fall in temperature. The diagram, Fig. 1, illustrates the appearance of all three agents, viz. VA, TPD kinin, and plasma kinin, at 37°.

We considered that there might be some sequence of events starting from the vasoactive proteins and leading to the vasoactive polypeptides of plasma kinin type. Moreover, it has been suggested that these vasopressor and vasodepressor systems may modify the human blood pressure. We therefore designed experiments to investigate together and simultaneously,whether the productions of VA and of plasma kinin in stored human serum appeared related. Nine normal human subjects provided the blood for the studies. Our methods were as follows. Blood was collected from the ante cubital vein into siliconed or polyethylene apparatus. It was allowed to clot at 37° and the serum was divided into aliquots which were placed at fixed temperatures, 37°, 20°, 0°. The aliquots were then tested for kinin production on the isolated rat uterus and for VA production on the blood pressure of the rat pretreated with pentobarbitone and pentamethonium. Kininogen was detected by kinin formation on exposure to glass, to dilution, and to human siliva. Some aliquots were exposed to 37°, 50°, 56° or 60° for 1 hour before incubation and formation of vasoactive principles at 37° or 0°. Other aliquots were invited to manufacture kinin at 0° after prolonged exposure to 37°. Still others were asked to produce VA at 37° after prolonged exposure to 0°, which exposure had caused the

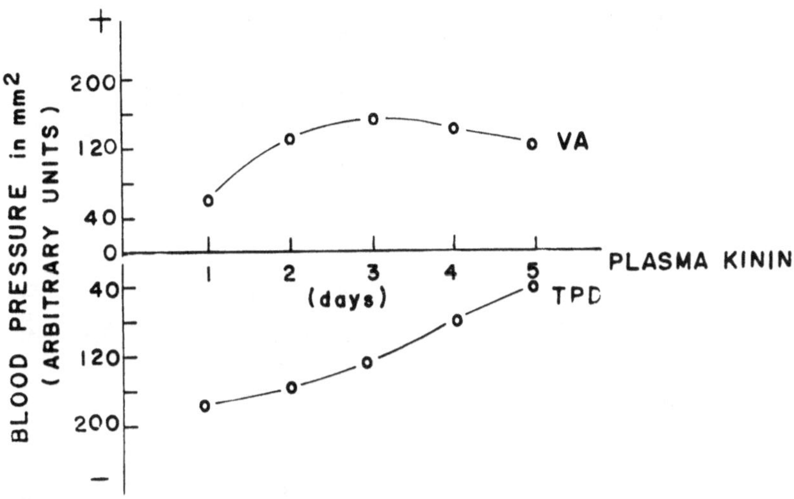

Fig. 1.

Diagram to show production of VA, plasma kinin and TPD kinin in normal human serum stored in polyethylene vessels at 37° C.

Ordinate: blood pressure (in arbitrary units) of a normal rat following intravenous injection of the active principles of the stored serum.

Abscissa: duration of storage (days)

Table 1

	TPD kinin	VA	Plasma kinin
1. (a)	Arises before VA and declines whilst VA is still increasing (15).	Readily detected in serum kept at 37° for 5-10 hr. or more.	An abnormal constituent of normal human serum at 37°, unless in trace amounts. Difficult to detect even in the presence of ε—amino caproic acid to inhibit kininase (4, 8, 9, 10).
(b)	Can be detected during a few days at 37° (15).	Generation is progressive with time and the agent can be detected during several days at 37°.	Generation may flare after several days at 37°. Agent can be detected only during a few hours. Thereafter the serum is kininogen depleted.
2.		Does not form at 0°	Forms spontaneously at 0°
3.		SBTI poorly inhibits the formation.	SBTI inhibits, apparently completely (see also 6 & 7.)

4.	HB does not inhibit the formation.	HB inhibits. The kinin forming reaction induced by cold is Hageman Factor dependent (11, 1, 3, 4).
5.	Pre-incubation of the serum at 0° for many hours does not inhibit the subsequent VA formation at 37°.	Pre-incubation at 0° depletes the serum of kininogen. It readily removes the kinin forming ability in response to cold dilution and glass. The response to human saliva trypsin or crotalus atrox venom is reduced or removed. (see also 8, 9, 10).
6.	Pre-incubation at 50° for 1 hr. does not inhibit VA formation at 37°.	As for VA. (see also 2 & 5).
	Pre-incubation at 60° slows, but does not arrest, VA formation 37°.	Pre-incubation at 60° (or even 56°) for 1 hr. completely inhibits the kinin formation which occurs on cooling (see also 2 & 5).

serum to be depleted of kininogen. We also studied the inhibitory effects of the soya bean trypsin inhibitor (SBTI; 2 mg/ml.) and of hexadimethrine bromide (HB; 50-200μg/ml.) on the productions of VA and of plasma kinin. The results are shown in Table 1.

We have concluded that the characteristics of the spontaneous production of VA in human serum stored in polyethylene vessels differ significantly from those of the plasma kinin formation. The productions(i) utilize different substrates (ii) are permanently inactivated at different temperatures (iii) are not susceptible to the same inhibitors (iv) have completely different optimal temperatures for their activity. The VA system and the cold-activated kinin forming system thus appear to be independent.

Acknowledgements

This investigation has been made possible by the generosity of the British Council and the Catholic University of Chile. We are also indebted for their support to colleagues at The Middlesex Hospital Medical School and to Dr. C. G. Van Arman, Merck Institute for Therapeutic Research, West Point, Pennsylvania, for advice and criticism.

References

1. Armstrong, D. (1968). The effects of temperature on the response of human plasma kinin-forming system to promoting and inhibiting agents. Br. J. Pharmac. 34, 670P.

2. Armstrong, D. 1969a). Actions on the human cold-activated plasma kinin-forming system of pre-heating at 56° and 60°C. Pharm. Res. Comm. 1: 30-35.

3. Armstrong, D. (1969b). Hageman Factor and the spontaneous human plasma-kinin generation occurring in homogeneous media. Activation and inactivation. In: Inflammation, Biochemistry and Drug Interaction. Ed. Bertelli, A. and Houck, C. pp 153-169. Excerpta Medica Foundation, Amsterdam.

4. Armstrong, D. (1970). Pain. In: Heffter's Handbuch der experimentellen Pharmakologie: Bradykinin:

Kallidin: Kallikrein. Ed: Erdös, E. G. pp 434-481 New York: Springer Verlag.

5. Armstrong, D. and Dias Da Silva, W. (1970). Kinin formation in human blood serum induced by cooling and by heat aggregated human gamma globulin preparations. In: Bradykinin and Related Kinins. Eds. Sicuteri, F., Back, N. and Rocha e Silva, M. pp 31-37. New York: Plenum Press.

6. Armstrong, D., Jepson, J. B., Keele, C. A., and Stewart, J. W. (1955). Activation by glass of pharmacologically active agents in blood of various species. J. Physiol. 129, 80-81P.

7. Armstrong, D., Jepson, J. B., Keele, C. A. and Stewart, J. W. (1957). Pain-producing substance in human inflammatory exudates and plasma. J. Physiol. 135, 350-370.

8. Armstrong, D. and Mills, G. L. (1965). The reversible cold-induced activation of the human plasma kinin-forming system between 37° and 0°C. J. Physiol. 179, 89-90P.

9. Armstrong, D., Mills, G. L. and Sicuteri, F. (1966). Physiological influence on the liberation of human plasma kinin at low temperatures. In: Symposium on Hypotensive Peptides. Ed: Erdös, E. G., Back, N. and Sicuteri, F. pp. 139-148. New York: Springer Verlag.

10. Armstrong, D., Mills, G. L. and Stewart, J. W. (1967). Thermally induced effects on the kinin-forming system of native human plasma, 37° - 0°C. and 37° - 50°C. In: International Symposium on Vasoactive Polypeptides; Bradykinin and Related Kinins. Ribeirão Prêto, August 1966; Eds: Rocha e Silva, M. and Rothschild, H. A. pp 167-176. Sao Paulo: Soc: Bras. Farmacologia e Terapêutica Experimental.

11. Armstrong, D. and Stewart, J. W. (1962). Antiheparin agents as inhibitors of plasma kinin formation. Nature (Lond.) 194, 689.

12. Croxatto, H. and Cruzatt, H. (1969). Ineffectiveness of carboxypeptidase B in inhibiting the pressor effect of incubated human plasma. Experientia. 25: 59-60.

13. Croxatto, H., Roblero, J. and Ossio, M. (1965) Vasopressor substances formed in the blood serum by incubation at 38°C. at normal pH. Arch. med. Exp. (Chile) 2: 55-62.

14. Kenner, Th. and Waldhäusl, W. (1964). A non-dialysable, heat-stabile, hypertensive substance in the serum of rats. Nature 204, 581-582.

15. Roblero, J. and Croxatto, H. (1967). Vasoactive substances formed in blood serum by incubation at 37°C. at normal pH. In: International Symposium on Vasoactive Polypeptides; Bradykinin and Related Kinins. Ribeirão Prêto, August 1966; Eds. Rocha e Silva, M. and Rothchild, H. A. pp 119-125. Sao Paulo: Soc: Bras. Farmacologia e Terapêutica Experimental.

PLASMA KININOGEN ASSAY. IMPORTANCE OF MAINTAINING BLOOD AT BODY TEMPERATURE UNTIL ASSAY

P. Periti and F. Sicuteri

Clinica Ostetrica e Ginecologica dell'Università di Firenze

Firenze (Italy)

We know that when *in vitro* blood is cooled there is a decrease in kininogen (Armstrong, 1965). A decrease of barely 10°C causes a significant kininogen depletion (Melmon, 1968).

Since the methods used by us from 1962 to 1968 for plasma kininogen assay (Diniz, 1963) did not entail the maintenance of blood samples at body temperature (that is, 38°C) from the moment the blood was drawn until it was used in laboratory, further tests were carried out. The blood was drawn with a warmed syringe and immediately transferred to test tubes placed in a 38°C bath. Samples of the same blood were kept at room temperature as controls.

Blood samples from 33 women were analyzed. The women were hospitalized in the Clinica Ostetrica e Ginecologica of Florence University and were in the third trimester of normal (12 cases) and toxemic (21 cases) pregnancies. The figure shows the values of the venous and arterial plasma kininogen, following withdrawal and maintenance of blood either at room temperature or at 38°C.

Studies were carried out on 86 blood samples (33 of arterial and 53 of venous blood), drawn with a warm syringe and kept at 38°C, and 64 samples (20 of arterial and 44 of venous blood), obtained with a non-prewarmed syringe and kept at room temperature for 15-120 minutes after withdrawal.

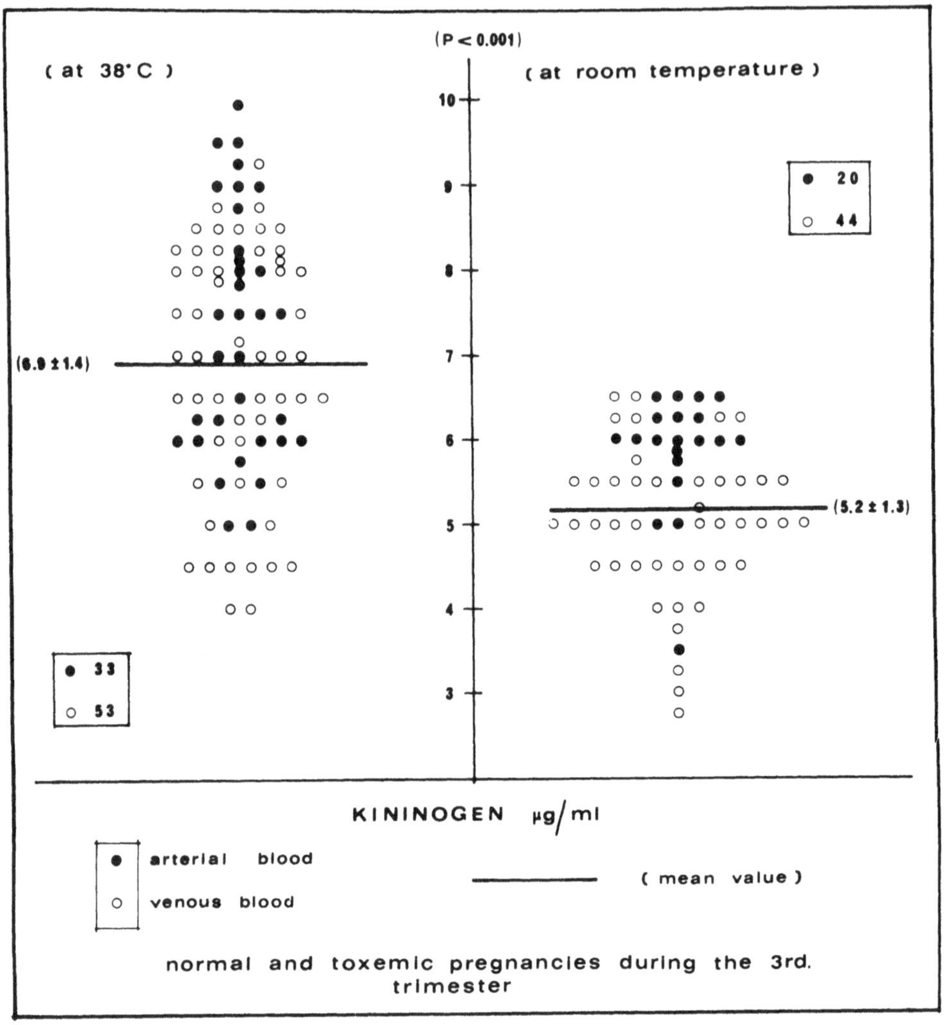

For the venous and arterial samples kept at 38°C, the mean kininogen value per ml of plasma was 6.9 ± 1.4 μg. For those kept at room temperature (12°-22°C), the value decreased to 5.2 ± 1.3 μg. The difference between these two values is significant ($P < 0.05$, according to Student's t).

The kininogen depletion, caused by the cooling of the blood samples at room temperature, is the same both in arterial and in venous blood and mostly occurs during the first 15 minutes after drawing the blood.

The conclusion we have reached is that for plasma kininogen assay the blood should be drawn and kept at a temperature of 38°C until plasma is used. A period of 15 minutes (which is often the time needed for taking the blood to the laboratory, centrifugating it and waiting for the separation of plasma) at room temperature, and in any case below 38°C, is enough to cause a depletion of kininogen of about 30%.

REFERENCES

Armstrong D. and Mills G.L.: J. Physiol. 170, 89P, 1965.
Diniz C.R. and Carvalho I.F.: Ann. N.Y. Acad. Sci. 104, 77, 1963.
Melmon K.L. et al.: J. Clin. Invest. 47, 1295, 1968.

GUINEA PIG PREKALLIKREIN ACTIVATOR*

Mary P. Treloar, ** Helen A. Pyle, Penelope J. Fuller and Henry Z. Movat

Division of Experimental Pathology and Institute of Immunology

University of Toronto, Toronto, Ontario, Canada

Davies and Lowe (2, 3) demonstrated that guinea pig serum contains a permeability factor distinct from anaphylatoxin after incubation with immune aggregates. It has also been shown that such treatment activates the kinin-forming system (5, 6). In the course of investigations into the components of the guinea pig intrinsic plasma kinin-forming system, a kinin-forming component has been differentiated from kallikrein by its inability to generate kinin from plasma heated to 61° for 60 minutes (8). This activator can be adsorbed to antigen-antibody precipitates and subsequently eluted (10). The isolated material possesses potent vascular permeability enhancing activity and the ability to contract the isolated rat uterus when incubated with fresh, non-glass contacted plasma. These properties have been used in the further isolation, purification and characterization of this material as an activator of plasma prekallikrein.

Isolation and Purification. Non-glass contacted guinea pig serum was activated by washed antigen-antibody precipitates as described previously (7). Following Sephadex G-200 gel filtration of part of this serum, three areas of vascular permeability enhancing and kinin-forming activity were observed (Fig. 1). The fractions eluting after the main albumin peak, while forming kinin when added to fresh plasma, were inactive when incubated with plasma heated to 61° for 60 minutes. Upon rechromatography on QAE Sephadex A-50 this low molecular size fraction eluted with 0.2 M NaCl

* Supported by the Atkinson Charitable Foundation.

** Present Address: Pharmakologisches Institut, Justus Liebig-Universität, Giessen, Germany.

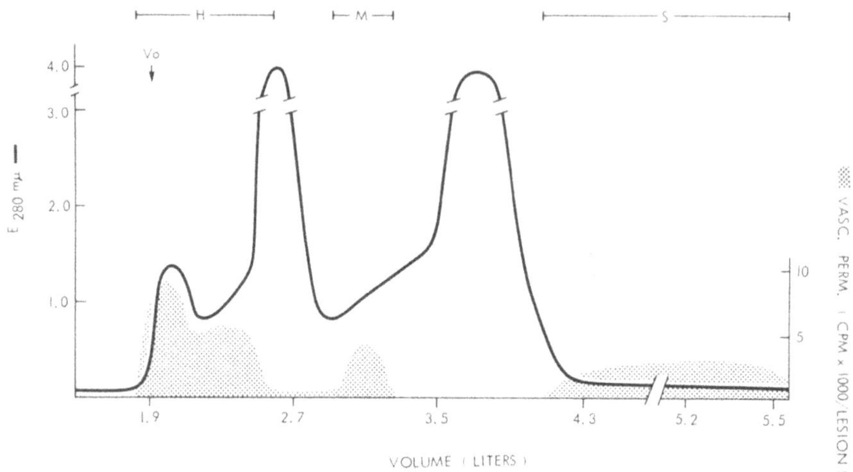

Fig. 1. Chromatography of 150.0 ml of activated guinea pig serum on two interconnected (10 x 90 cm) columns of Sephadex G-200 superfine. Elution was performed with 0.1 M Tris-HCl containing 0.2 M NaCl at 4°C. All fractions enhancing vascular permeability also formed kinin from fresh guinea pig plasma. Vascular permeability tests were done as described in detail in reference 16.

and after preparative polyacrylamide electrophoresis the bulk of the active material was eluted from prealbumin fractions. In contrast to the small amount of material recovered from these low molecular weight Sephadex G-200 fractions, over 85% of the activator present in the applied serum was isolated from the large molecular size region (designated H, Fig. 1). When applied to QAE Sephadex (equilibrated with 0.1 M Tris-HCl, pH 8.0) kallikrein appeared in the excluded peak, whereas fractions forming kinin from fresh, but not heated plasma, appeared in a broad irregular peak which eluted between 0.15 and 0.3 M NaCl. When these anionic fractions were pooled and again chromatographed on Sephadex G-200, the bulk of the activity, in contrast to the initial G-200 run, now had an elution volume greater than the marker protein human serum albumin (Fig. 2). On rechromatography on Sephadex G-75, the peak of activity appeared with an elution volume intermediate between egg albumin and SBTI and was estimated to have a molecular size of 38,000 to 40,000. Preparative disc gel electrophoresis was carried out and the active material corresponding to the prealbumin band was subjected to amino acid analysis.

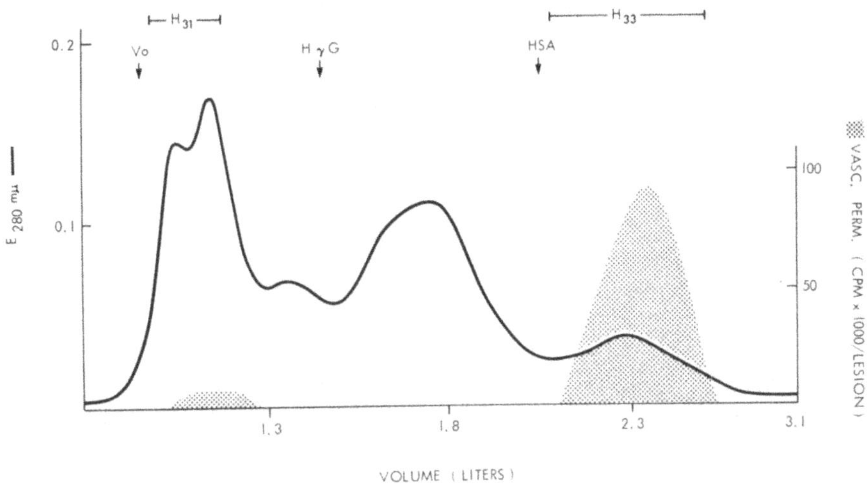

Fig. 2. Sephadex G-200 fine gel filtration (two 5 x 90 cm columns). Fraction H shown in Fig. 1 was chromatographed on QAE Sephadex A-50 and the most anionic fractions rechromatographed by gel filtration.

A portion of the same batch of activated guinea pig serum was chromatographed initially on QAE Sephadex A-50. As shown in Fig. 3, a series of anionic peaks (C and D) with activator activity were eluted. The isolation and properties of two prekallikrein activators from the less anionic peak C have previously been reported (Treloar and Movat, 1970). Although possessing similar properties, one of the activators had a molecular size of about 60,000 and the other approximately 40,000. Subsequent rechromatography of the 60,000 molecular size component resulted in the recovery of activity with a molecular size of approximately 36,000.

When the most anionic pool D was chromatographed on Sephadex G-200, a small molecular weight activator was eluted and subsequently purified; but prekallikrein activator also appeared in the macroglobulin fractions. These fractions also contained a small amount of kallikrein but formed about 100 times more kinin from fresh than from heated plasma. It appears that although some prekallikrein activator exists in free form, the bulk of the active material is bound to a macroglobulin carrier protein and can be dissociated from it during anion exchange chromatography.

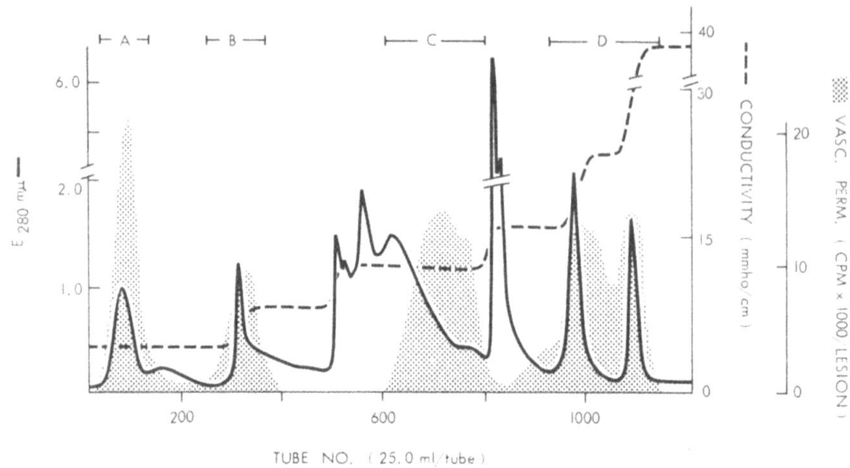

Fig. 3. Anion exchange chromatography of 620.0 ml of the activated guinea pig serum (same batch as shown in Fig. 1) on QAE Sephadex A-50 (5 x 50 cm column). A stepwise gradient of NaCl in 0.1 M Tris-HCl, pH 8.0 was used for elution: the salt concentration is expressed as conductivity. Prekallikrein activator was purified further from Fractions C and D.

Prekallikrein activator was also isolated and purified from celite eluates of guinea pig plasma. Celite (Hyflo Supercel 150 mg/ml plasma) was incubated with fresh, non-glass contacted, $Al(OH)_3$ adsorbed guinea pig plasma for 10 minutes. The celite was separated from the plasma by centrifugation and after repeated washings with 0.001 M Tris-HCl (pH 8.0) the protein adsorbed to it was eluted with 2.0 M NaCl in 0.1 M Tris-HCl. The dialysed washings were applied to QAE Sephadex A-50. Potent vascular permeability enhancing and kinin-forming activity was eluted between 0.13 and 0.3 M NaCl. Polyacrylamide disc gel electrophoresis of this material revealed a prealbumin band just like the material obtained by chromatographing antigen-antibody activated serum (Fig. 4). Gels were cut at 0.25 cm intervals and eluted with 1.0 M NaCl, dialysed, concentrated and tested for BAEe hydrolysing activity, induction of esterase activity on incubation with prekallikrein and clot-promoting activity. The eluted prealbumin band was rerun and revealed a homogeneous protein (Fig. 4). The amino acid analysis carried out on this material is shown in Table 1.

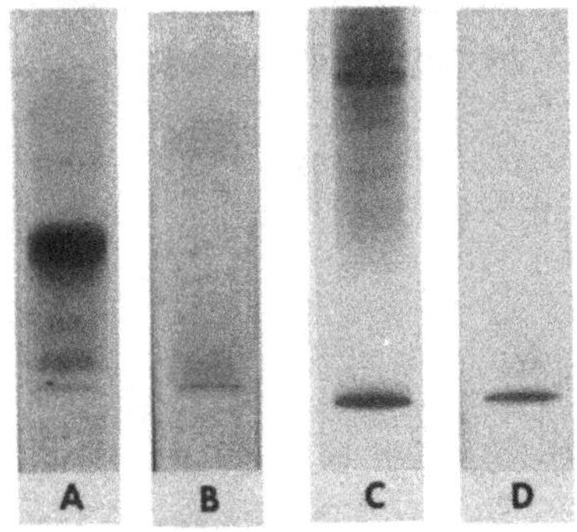

Fig. 4. Polyacrylamide electrophoresis (pH 8.3) of PKA obtained by repeated chromatography of activated serum (A), and of PKA obtained from celite eluates (C); (B) represents electrophoresis of the fractional cut corresponding to the prealbumin band in (A) and (D) that eluted from (C).

TABLE 1. AMINO ACID COMPOSITION OF PREKALLIKREIN ACTIVATOR OBTAINED BY ELUTION FROM POLYACRYLAMIDE GELS (Fig. 4)
(Moles/100 moles recovered)

Lysine	5.3	Alanine	5.7
Histidine	1.6	Valine	4.9
Arginine	2.4	Isoleucine	3.3
Aspartic Acid	2.5	Leucine	7.6
Threonine	2.5	Tyrosine	10.0
Serine	9.1	Phenylalanine	3.2
Glutamic Acid	15.1	Cystinic Acid	5.4
Proline	6.4	Methionine Sulfone	3.5
		Total	99.9

Properties. All of the activators isolated by the various procedures possessed identical properties. Prekallikrein activator migrates as a prealbumin on polyacrylamide electrophoresis at pH 8.3 and has a pI of 4.0 to 4.6 as determined by isoelectric focusing in ampholines. Amino acid analysis confirmed the acidic nature of the protein. A sedimentation coefficient of 2.8 to 3.2 was determined by sucrose density gradient ultracentrifugation. Gel filtration studies indicated a molecular weight between 36,000 and 40,000, whereas the molecular weight calculated from the amino acid analysis was approximately 34,000.

TABLE 2. EFFECT OF PREKALLIKREIN ACTIVATOR (PKA) ON PREKALLIKREIN

Sample	BAEe (n-moles) hydrolysed/min/ml	ng-Kinin released from 1.0 ml heated substrate*
Prekallikrein + Buffer	3.0	~ 30
PKA #1 + Buffer	1.2	0
Prekallikrein + PKA #1	14.0	1600
PKA #2 + Buffer	1.8	0
Prekallikrein + PKA #2	12.0	1500

Activation of guinea pig prekallikrein by its activator (preparations PKA #1 or #2): 0.2 ml of prekallikrein was incubated at room temperature for 5 minutes with 0.2 ml of PKA, followed by incubation with either BAEe or plasma heated to 61° for 60 minutes. In the control runs the activator or prekallikrein were incubated with 0.01 M phosphate buffered saline.

*Tested on the estrus rat uterus.

When the activator was incubated with guinea pig prekallikrein, kallikrein activity was generated as evidenced by the appearance of arginine esterase and kinin-forming activity in the incubation mixture. As shown in Table 2, the induction of esterase activity was paralleled by the ability of the incubation mixture to release kinin from plasma heated to 61° for 60 minutes. The rate of conversion of prekallikrein to kallikrein increased with the amount of activator added. The amount of generated kallikrein depended on the amount of prekallikrein present in the reaction mixture.

The activator, in contrast to kallikrein, is resistant to heating to 56° for 30 minutes. The most purified preparations contained no TAMe or BAEe esterase activity. In vascular permeability tests and assays for kinin release, the activator was partially inhibited by 10^{-4} M diisopropylfluorophosphate.

Lima bean trypsin inhibitor (50 μg/ml), which has little or no effect on kallikrein, induced 50% inhibition of the activator.

All preparations with prekallikrein activating-activity shortened the partial thromboplastin time of Factor XII-deficient human plasma, while not shortening that of plasmas deficient in Factors XI, IX or VIII. The slower migrating bands appearing in the ß-region on disc gel electrophoresis of the celite eluates of guinea pig plasma (Fig. 4) also shortened the partial thromboplastin time of Factor XII-deficient plasma. The electrophoretic behaviour, of these fractions correspond to properties described for bovine Factor XII (12), which has been reported to activate bovine prekallikrein (11, 14).

A partially purified preparation of guinea pig Factor XIIa did not induce prekallikrein-activation unless it was first treated with trypsin or adsorbed to and eluted from celite.

Concluding Remarks: The observations reported here correlate well with data obtained with human (4, 9, 13) and rabbit (1) plasmas, leading to the conclusion that prekallikrein activator is derived from Factor XII.

The mode of action of contact activators, such as immune precipitates or celite, is as yet uncertain. Preliminary data in our laboratory indicate that mere contact does not convert Factor XII to PKA in highly purified preparations, suggesting the requirement of other factors.

References

(1) Cochrane, C. G. and Wuepper, K. D.: Isolation in precursor form of the first component of the kinin-forming system. Fed. Proc., 30: 451, 1971.

(2) Davies, G. E., and Lowe, J. S.: A permeability factor released from guinea pig serum by antigen-antibody precipitates. Brit. J. Exp. Path., 41: 335, 1960.

(3) Davies, G. E., and Lowe, J. S.: Further studies on a permeability factor released from guinea pig serum by antigen-antibody precipitates. Int. Arch. Allergy, 20: 235, 1962.

(4) Kaplan, A. P., and Austen, K. F.: A prealbumin activator of prekallikrein. J. Immunol., 105: 802, 1970.

(5) Movat, H.Z.: Activation of the kinin-system by antigen-antibody complexes. In Vasoactive Polypeptides: Bradykinin and Related Kinins. Rocha e Silva, M., and Rothschild, H.A., eds., Soc. Bras. Farmacol., São Paulo, Brazil, 1967.

(6) Movat, H.Z., DiLorenzo, N.L., Mustard, J.F., and Helmel, G.: Activation of Hageman and a vascular permeability factor in serum by Ag-Ab precipitates. Fed. Proc., 25: 682, 1966.

(7) Movat, H.Z., and DiLorenzo, N.L.: Activation of the plasma kinin-system by antigen-antibody aggregates. I. Generation of permeability factor in guinea pig serum. Lab. Invest., 19: 187, 1968.

(8) Movat, H.Z., DiLorenzo, N.L., and Treloar, M.P.: Activation of the plasma kinin-system by antigen-antibody aggregates. II. Isolation of permeability-enhancing and kinin-releasing fractions from activated guinea pig serum. Lab. Invest., 19: 201, 1968.

(9) Movat, H.Z., Poon, M.-C., and Takeuchi, Y.: The kinin-system of human plasma. I. Isolation of a low molecular weight activator of prekallikrein. Int. Arch. Allergy, 40: 89, 1971.

(10) Movat, H.Z., Treloar, M.P., and Takeuchi, Y.: A small molecular weight permeability factor in guinea pig serum: adsorption to antigen-antibody aggregates. J. Immunol., 4: 875, 1969.

(11) Nagasawa, S., Takahashi, H., Koida, M., Suzuki, T., and Schoenmakers, J.G.G.: Partial purification of bovine plasma kallikreinogen, its activation by Hageman factor. Biochem. Biophys. Res. Com., 32: 644, 1968.

(12) Schoenmakers, J.G.G., Matze, R., Haanen, C., and Zilliken, F.: Hageman factor, a novel sialoglycoprotein with esterase activity. Biochim. Biophys. Acta, 101: 166, 1965.

(13) Soltay, M.J., Movat, H.Z., and Özge Anwar, A.H.: The kinin-system of human plasma. V. The probable derivation of prekallikrein activator from activated Hageman factor (XIIa). Proc. Soc. Exp. Biol. Med., 138: 1971, in press.

(14) Temme, H., Jahreiss, R., Habermann, E., and Zilliken, F.: Aktivierung von Gerinnungs-und Kinin-system durch eine Plasmaesterase (Hageman-Factor). Reinigung und Wirkungsbedingungen. Hoppe-Seyler's Z. Physiol. Chem., 350: 519, 1969.

(15) Treloar, M.P., and Movat, H.Z.: Isolation of two small molecular weight activators of the plasma kinin-system in the guinea pig. Fed. Proc., 29: 576, 1970.

(16) Udaka, K., Takeuchi, Y., and Movat, H.Z.: Simple method for quantitation of enhanced vascular permeability. Proc. Soc. Exp. Biol. Med., 133: 1384, 1970.

THE KALLIKREIN-LIKE ACTIVITY PRESENT IN HUMAN EUGLOBULIN FRACTION

W. Dias da Silva*

Rocival L. Araujo** and Maria N.R. Vieira***

The euglobulin fraction was prepared essentially as described by Lepow et al. (1963). One volume of fresh human serum was adding to 8 volumes of pH 5.5 acetate buffer, ionic strength 0.02, the mixture was allowed to stand overnight and the precipitate was resuspended in 0.5M NaCl to one-tenth the original volume of serum. The preparation was centrifuged at 29.000 RPM for 60 minutes, the particles of lipid were removed by filtration in glass-wool and the resultin opalescent solution containing the euglobulins was then dialyzed vs. pH 7.5 phosphate buffer ionic strength 0.15. Human euglobulin prepared in this manner (C_1 F_1) contracts rat uterus only after

*From the Department of Biochemistry and** Department of Physiology Instituto de Ciências Biológicas da UFMG. Brasil

***Recipient of fellowship support from Conselho Nacional de Pesquisas (CNPq) Brasil.

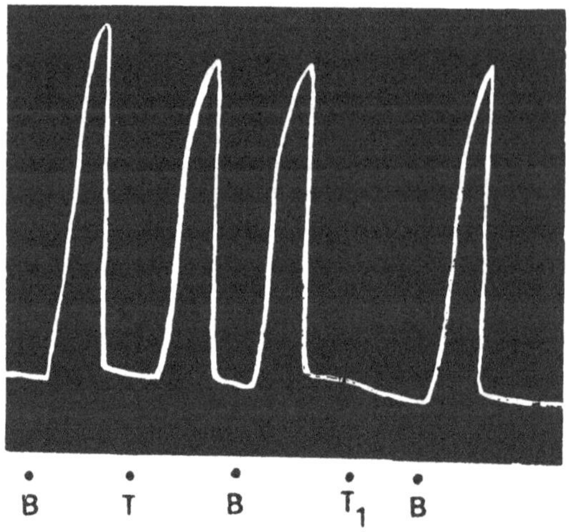

Fig. 1. Contractions of the rat uterus produced by human euglobulins preparation. B = 1 ng Bradykinin; T = 0.1 ml C_1 F_1 at 0ºC; T_1 = 0.1 ml C_1 F_1 at 97ºC during 20 min.

a prolonged contact of 2 to 3 minutes (Fig. 1), fails to contract guinea-pig ileum, enhances vascular permeability in guinea--pig skin and forms kinin when incubated or either diluted human plasma or with partialy purified preparations of bradykininogen. The activity only appears after dialysis of the euglobulin preparation against pH 7.5 phosphate buffer ionic strength 0.15 (fig. 2).

In order to see if the kallikrein-like activity present in the euglobulin preparation was a contamination with adsorbed unwashed pseudoeuglobulin, a preparation of euglobulins was washed 10 times with 0.5M NaCl before the dialysis against the pH 7.5 phosphate buffer, ionic strength 0.15. The activity present in this preparation was roughly the same as in the control sample.

Applied to a Sephadex G-100 columns, the kallikrein-like activity was found along the descending portion of the first protein peak rather than with the peak itself, suggesting a molecular weight less than 100.000 (fig. 3). Its activity is destroyed by heating at 97ºC and is unstable a pH lower than 4.0 and over 9.5. Further purification of the kallikrein-like activity was performed on DEAE cellulose chromatography. As is showed in fig. 4 the activity appears in the first protein peack together with the unadsorved proteins.

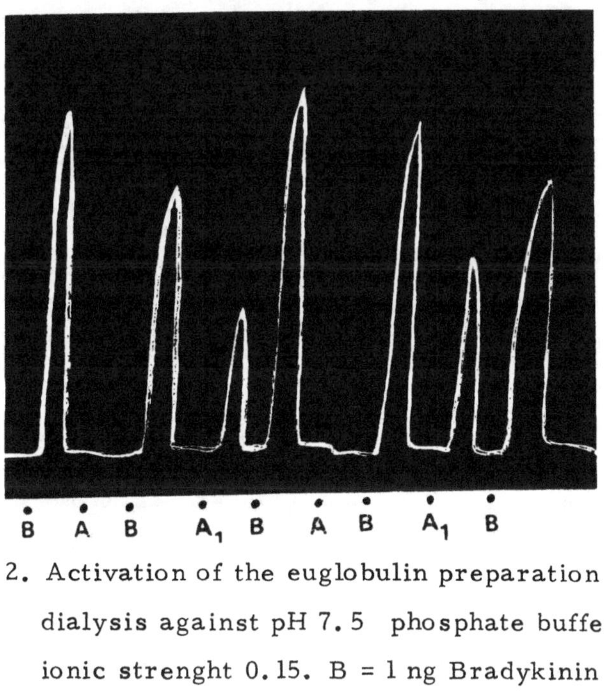

Fig. 2. Activation of the euglobulin preparation by dialysis against pH 7.5 phosphate buffer, ionic strenght 0.15. B = 1 ng Bradykinin A = 0.1 ml C_1 F_1 before dialysis; A = 0.1 ml C_1 F_1 after dialysis.

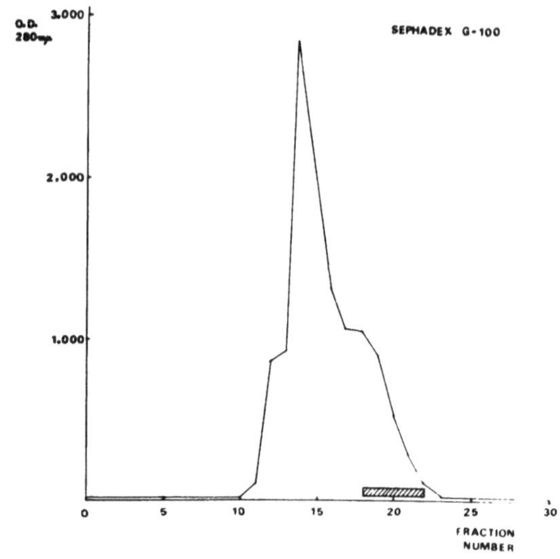

Fig. 3. Elution pattern of human euglobulins resulting from filtration through a Sephadex G-100 columns.

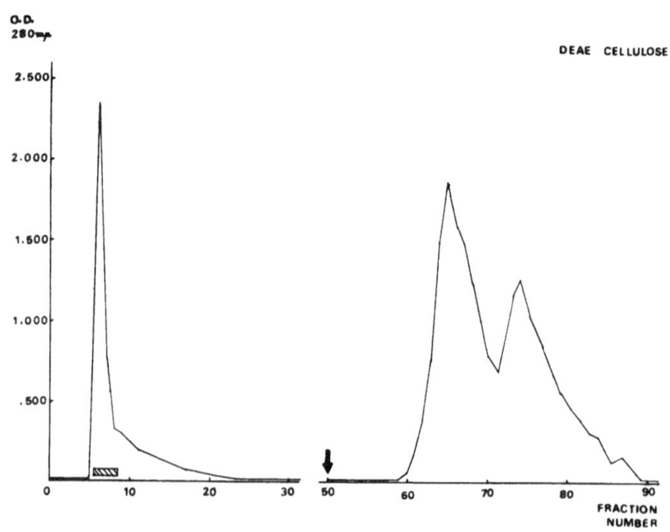

Fig. 4. Chromatographic resolution of the oxytocic activity on DEAE cellulose.

The oxytocic activity of this material was inhibited by SBTI (soy bean trypsin inhibitor) and by trasylol. It was also observed that the activated euglobulins potentiate the bradykinin activity on rat uterus (Fig. 5).

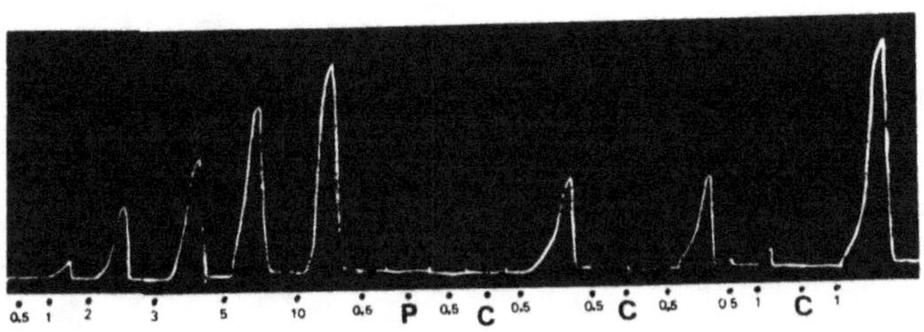

Fig. 5. Potentiation of the bradykinin activity in rat uterus produced by the activated human euglobulin preparations. Numbers represent additions of Bradykinin (ng). P = 0.1 ml Phosphate buffer pH 7.5 u = 0.15 C = 0.1 ml C_1 F_1

The evaluation of its possible relationship either the clot or with the complement system is presently under investigation.

PROTEIN COMPONENTS WHICH RELATE TO THE KININ RELEASING SYSTEM IN BOVINE PLASMA

T. Suzuki, H. Takahashi, M. Komiya, K. Horiuchi and S. Nagasawa

Institute for Protein Research, Osaka University

Osaka, Japan

Blood plasma contains several discrete systems consisting of the sequential conversion of proenzymes into active enzymes, in which their activations lead to a hypotension due to the liberation of vasoactive polypeptide termed kinin, and to thrombosis and hemorrhage due to the formation and lysis of fibrin clot. From physiological point of view, it is of interest how is the kallikrein-kinin system to be concerted with the other enzyme systems, such as blood coagulation and fibrinolysis. Recently, it was suggested that Hageman factor, which is known to have triggering action for the intrinsic coagulation and fibrinolysis systems, also pull the trigger for the activation of kallikrein-kinin system. To clarify the above relations, it is extremely important to isolate the pure components, which concern to the liberation of kinin, and to elucidate their interrelations, reconstructing the kinin liberating systems *in vitro*.

In this paper we attempted to present the purification procedures of prekallikrein, prekallikrein activator and plasminogen in bovine plasma, and to discuss their activating mechanisms.

Hageman Factor (Prekallikrein Activator)

Previously, we have demonstrated that bovine prekallikrein is activated directly by bovine Hageman factor (1). After that, Dr. Nagasawa and Komiya isolated a protein component, and after the treatment with ellagic acid, it had the activity to convert prekallikrein to active kallikrein. The purification procedures of it were essentially similar to that of bovine Hageman factor reported by Dr. Schoenmakers, except that adsorption on glass

Table I

Summary of Purification of Prekallikrein Activator

Procedures	Total protein (A_{280})	Total activity (units) Activator[1]	Esterase[2]	Specific activity (units/A_{280}) Activator	Esterase	Ratio of Act. : Est.
Defibrinated euglobulin (from 15 liters plasma)	96,000					
CM-Sephadex C-50	364	7470	138	20.5	0.38	54.1
DEAE-Sephadex A-50	28.4	2220	38	77.1	1.33	58.4
CM-Sephadex C-50	1.1	333	7.0	306.4	6.4	47.6

1) One unit of activator induces 1 unit of TAME esterase activity from prekallikrein under the standard condition.

2) One unit of TAME esterase hydrolyzes 1 μmole of TAME per min.

powder was avoided: namely column chromatographies on CM-Sephadex C-50 and DEAE-Sephadex A-50 and rechromatography on CM-Sephadex C-50 were applied to defibrinated euglobulin fraction of bovine plasma. The results can be seen in Table I. It was found that the preparation obtained had been partially transformed into active form during the purification procedures, and was fully

Table II

Properties of Prekallikrein Activator (Hageman Factor)

	Bovine		Human		Rabbit
	Komiya (1969)	Schoenmakers (1965)	Kaplan (1971)		Cochrane (1970)
Activation	Ellagic acid	Glass contact	Glass contact	Plasmin	Trypsin
Molecular weight	90,000	82,000		35,000	32,000
Ability to correct HF-deficiency	+	+	+	±	±
Prekallikrein activator activity	+	+	±	+	+
Synthetic substrates					
TAME	+	+			
BAEE	+	+			−
TLME	−				+
Inhibition by					
DFP	+	+			+
LBTI	+	+			+
SBTI	−	−			−

activated by the addition of ellagic acid. The properties of this preparation with prekallikrein activating activity were similar to that of bovine Hageman factor with respect to molecular weight, chromatographic behaviours on CM- and DEAE-Sephadex columns, functional activities to hydrolyze TAME and to activate bovine prekallikrein. In addition the prekallikrein activator corrected the coagulation defect of Hageman factor-deficient plasma, and the clot-promoting activity was associated with prekallikrein activator activity on disc gel electrophoresis. These results in Table II supported the evidence that Hageman factor was responsible for the activation of prekallikrein as previously reported(1).

Prekallikrein

Purification method of prekallikrein was improved by combining the affinity chromatography on an arginine-Sepharose 4B (Arg-Sepharose 4B) column to the previous method. By the chromatography on Arg-Sepharose 4B, the contaminant protein was well removed even from the subfraction which contains a small amount of prekallikrein (Fig. 1), and the yield was efficiently ascended by adding the prekallikrein thus obtained to the main prekallikrein fraction.

And the next gel-filtration on a Sephadex G-150 column removed the γ-globulin type protein contaminated in the eluate

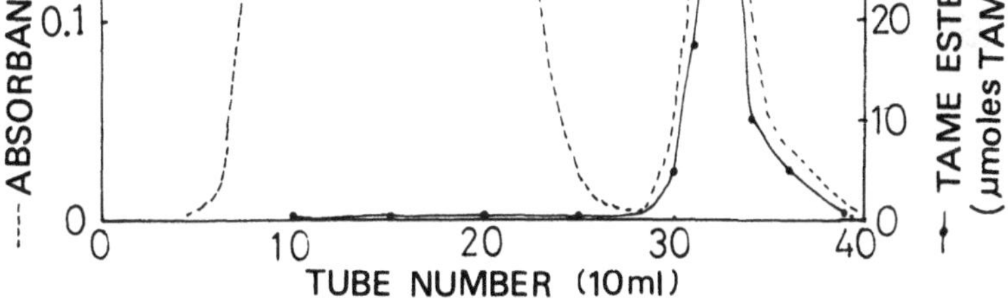

Fig. 1 Affinity chromatography of bovine plasma prekallikrein on Arg-Sepharose 4B. Prekallikrein fraction obtained from CM-Sephadex C-50 chromatography was applied on a column (2 x 10 cm) of Arg-Sepharose 4B equilibrated with 0.04 M Tris-HCl buffer, pH 8.5. After washing the column with starting buffer, prekallikrein was eluted with 0.07 M NaCl in the buffer.

Table III
Summary of Purification of Bovine Plasma Prekallikrein

Procedures	Total protein (mg)	Esterolytic activity (total units)*	Specific activity (units/mg protein)	Yields (%)
Pseudoglobulin fraction (from 7.5 liters plasma)	242.4×10^3	—		
1st DEAE-Sephadex A-50	22.608×10^3	872	0.039	100
2nd DEAE-Sephadex A-50	4.29×10^3	895.9	0.209	103
CM-Sephadex C-50	95.02	731.9	7.70	83.9
Affinity chromatography on Arg-Sepharose	17.35	509	29.34	58.3
Sephadex G-150	8.12	403.5	49.69	46.2
Re-gel filtration on Sephadex G-150	4.62	240	51.95	27.5

* One unit of enzyme, developed by the addition of prekallikrein-activator, hydrolyzes 1.0 μmole of TAME per min.

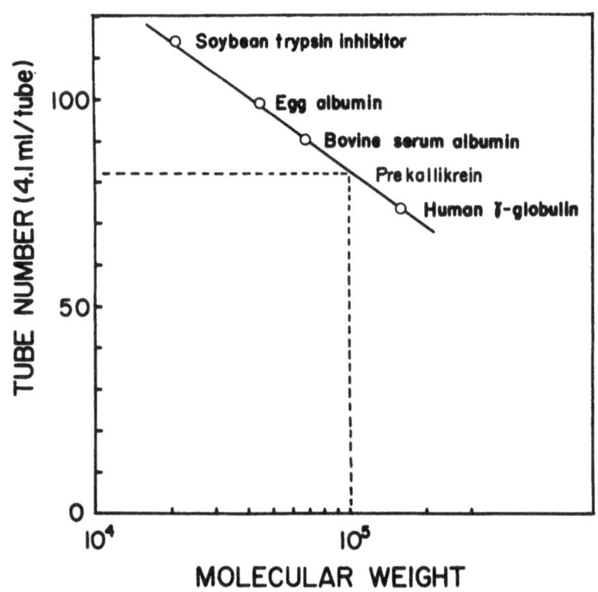

Fig. 2 Determination of molecular weight of prekallikrein by gel filtration on Sephadex G-150. The molecular weight of prekallikrein was determined from the elution volume of prekallikrein from a column (3 x 91 cm) of Sephadex G-150.

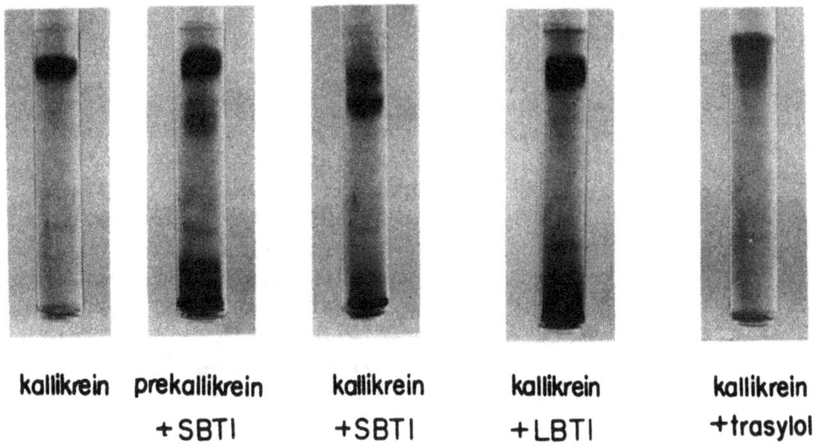

Fig. 3 Disc electrophoretograms of the reaction products of trypsin inhibitors and Hageman factor-activated kallikrein. Each 80 µg of Hageman factor-activated kallikrein was incubated with trypsin inhibitors of soy bean (SBTI, 50 µg), lima bean (LBTI 100 µg) or trasylol (100 K.I. units) at 37°C for 10 min. The reaction mixture was subjected to polyacrylamide gel disc electrophoresis at 300 V for 1.5 hr in pH 8.3. The protein was stained with Coomassie brilliant blue.

from Arg-Sepharose column, and the highly purified prekallikrein was obtained. The average yield of prekallikrein was about 5 mg from 7.5 liters of bovine plasma (Table III). The purified protein was homogeneous on ultracentrifugation and isoelectric focusing, and was found to give a single band on polyacrylamide gel disc electrophoresis. There was no evidence to be contaminated with prekallikrein activator (Hageman factor), plasminogen, plasmin, thrombin and plasma kininase. Thus, the material should be functionally pure.

The molecular weight of prekallikrein was estimated to be about 80,000 by sucrose density gradient ultracentrifugation, 90,000 by SDS-polyacrylamide gel disc electrophoresis, 90,000 by the Archibald method and 100,000 by gel filtration on a Sephadex G-150 column (Fig. 2). Thus, its molecular size was almost the same as that of casein activated kallikrein previously purified from bovine plasma (3). From this, it may be tentatively suggested that in bovine plasma, Hageman factor will be activated by the treatment with casein, and the active Hageman factor reacts to prekallikrein.

The activation of prekallikrein by Hageman factor did not result in the changed electrophoretic behaviour on SDS-polyacrylamide gel and no change of molecular size was observed by the

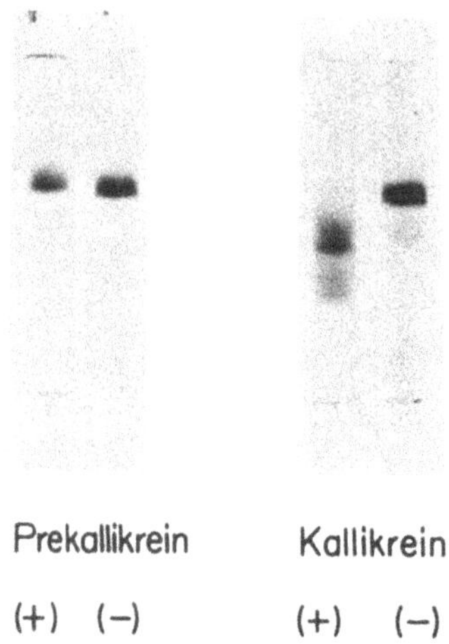

Fig. 4 SDS-polyacrylamide gel disc electriphoreses of prekallikrein and Hageman factor-activated kallikrein. Prekallikrein or kallikrein treated (+) or not treated (-) with 0.1 M β-mercaptoethanol was subjected to electrophoresis in 0.1 M phosphate buffer, pH 7.5, containing 0.1 % SDS at 7mA/tube for 4 hr.

activation. Although the prekallikrein did not react with soybean trypsin inhibitor (SBTI) or Trasylol, activated kallikrein reacted with them and the electrophoretic behaviours of the reacted product in polyacrylamide gel were changed from that of kallikrein (Fig. 3). Lima bean trypsin inhibitor (LBTI) did not react with plasma kallikrein and even after 35 μg of kallikrein and 100 μg of

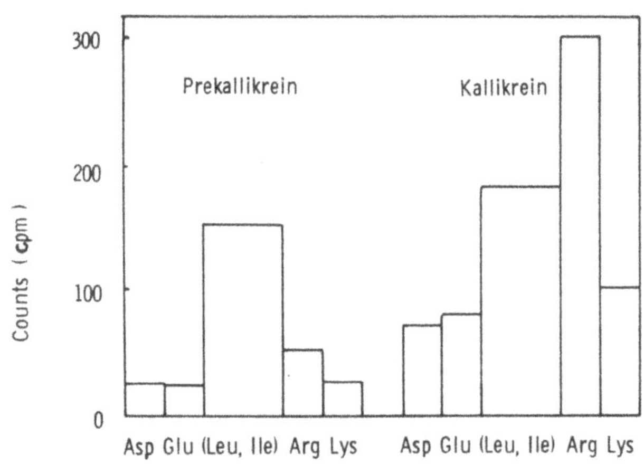

Fig. 5 Incorporation of tritium into C-terminus of prekallikrein and of Hageman factor-activated kallikrein. Each 1 mg of prekallikrein and kallikrein was treated with 50 mC of tritium water in pyridine-acetic anhydride mixture. After 12 hr, the reaction mixtures were dialyzed against water and hydrolyzed by 6 N HCl at 105°C for 24 hr. Incorporation of tritium into amino acids was determined after the application of high voltage paper electrophoresis to the hydrolysates.

LBTI were incubated at 37°C for 30 min, TAME esterase activity of kallikrein was fully recovered. When prekallikrein was treated with β-mercaptoethanol, only a single band was found on SDS-disc gel electrophoresis, however, kallikrein was cleaved into protein pieces after the reduction (Fig. 4). These results suggest that activation of prekallikrein associates with the limited proteolysis of the peptide chain located between two cysteine molecules which form the disulfide linkage of prekallikrein; and one of the newly appeared amino acids in kallikrein molecule reacted with SBTI to make kallikrein-inhibitor complex. Then, we attempted to identify the newly appeared C-terminus of active kallikrein by the tritium labelling method, and identified arginine as the C-terminus, in addition to the C-terminal leucine (isoleucine) of prekallikrein (Fig. 5).

Plasminogen

Purification method of plasminogen was also improved by combining the affinity chromatography on lysine-Sepharose 4B to the previous method (4) (Table IV). By this, the plasminogen preparation with a symmetrical single peak on isoelectric focusing was obtained (Fig. 6). Activation of plasminogen by urokinase seems to associate with the limited hydrolysis of the peptide chain in its molecule. However, autodigestion of plasmin was observed when the plasminogen and urokinase was incubated. Therefore, to avoid the autodigestion of plasmin, plasminogen was incubated with urokinase in the presence of SBTI, which inhibits the activity of plasmin and not the activity of urokinase.

Table IV

Purification of Bovine Plasminogen

Bovine plasma euglobulin fraction

↓ defibrination by addition of $CaCl_2$ solution to final concentration of 0.1 %.

Defibrinated euglobulin solution

↓ Sat. $(NH_4)_2SO_4$ solution was added to final concentration of 33 %: the precipitate was dissolved in water and dialyzed against 0.05 M phosphate buffer, pH 7.5.

Lysine-Sepharose 4B affinity chromatography

↓ plasminogen was eluted from the column with 0.1 M ε-aminocaproic acid and was precipitated by dialysis against sat. $(NH_4)_2SO_4$. The precipitate was dissolved in 0.04 M Tris-HCl buffer, (pH 7.5)-0.1 M NaCl.

Sephadex G-100 gel filtration

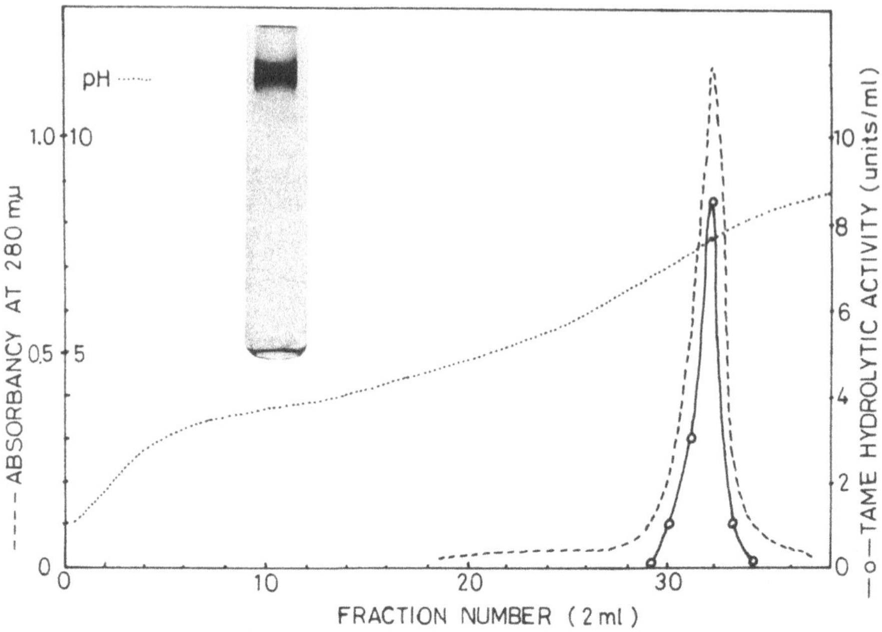

Fig. 6 Isoelectric focusing of bovine plasminogen. 10 mg of plasminogen was applied to the electrophoresis under the gradient of pH 3-10 at 1000 V for 16 hr.

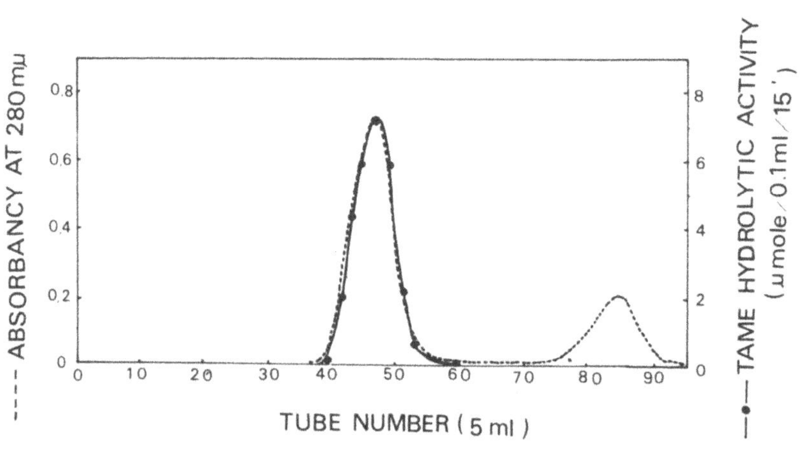

Fig. 7 Gel filtration of bovine plasminogen activated by urokinase in the presence of SBTI. 11 mg of plasminogen was activated by 500 units of urokinase in the presence of 15 mg of SBTI in 5 ml of 0.04 M Tris-0.02 M lysine buffer, pH 8.5, at 37°C for 30 min. The mixture was directly applied on a column (2.5 x 125 cm) of Sephadex G-100 equilibrated with 0.1 M acetic acid.

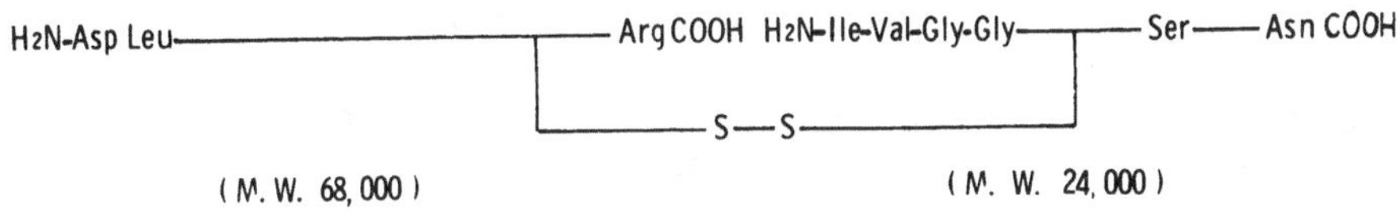

Fig. 8 Polypeptide chains of bovine plasmin

Plasmin-SBTI complex thus formed was completely dissociated in acidic condition and one symmetrical peak with plasmin activity was obtained after gel filtration on a Sephadex G-100 column equilibrated with 0.1 M acetic acid (Fig. 7). Plasminogen is composed of a single polypeptide chain, but plasmin thus obtained was dissociated into two polypeptide chains after the reduction with β-mercaptoethanol. The N-terminal amino acid of plasmin was found to be aspartic acid and isoleucine by Edman's procedure, and the C-terminal amino acid of it was found to be asparagine and arginine by the tritium labelling method and by the digestion method with carboxypeptidases A and B. Since aspartic acid and asparagine were identified as N- and C-terminal amino acids of original plasminogen, isoleucine and arginine were deduced to be newly appeared N- and C-terminal amino acid in active plasmin. Gross-structure of plasmin can be seen in Fig. 8.

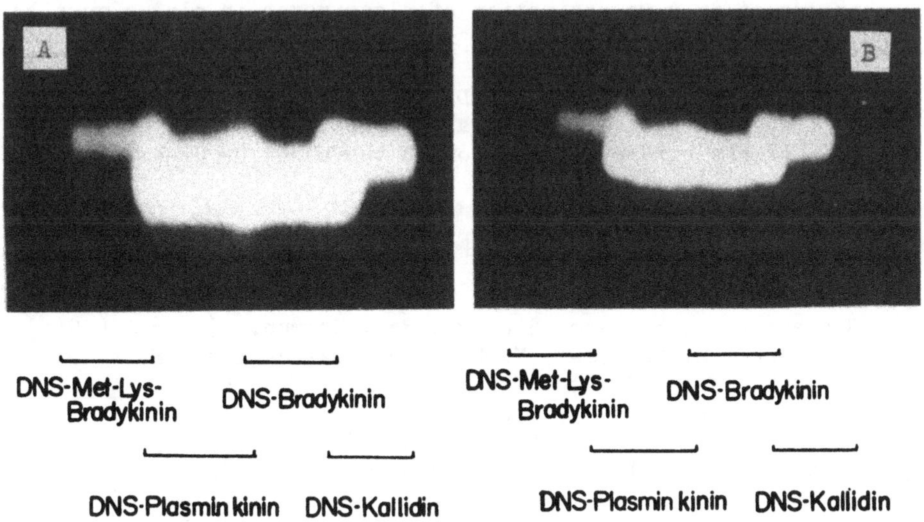

Fig. 9 Thin layer chromatograms of kinins released from bovine HMW and LMW kininogens by bovine plasmin. Kinins liberated were treated with 0.5 % dansyl chloride-acetone solution and subjected to thin layer chromatography with a solvent system of n-buthanol: acetic acid:water = 4:1:5. A is a chromatogram of kinin obtained from HMW kininogen, and B from LMW kininogen, respectively.

From the above results, it seems that both prekallikrein and plasminogen can be similarly activated by cleaving only an arginylpeptide bond on the peptide chain located between two cysteine molecules which form the disulfide linkage.

Previously, we observed that when bovine high molecular weight (HMW) kininogen (kininogen-I), specific substrate for bovine serum kallikrein, was incubated with bovine serum kallikrein, bradykinin was predominantly liberated; however, a minor peak of kallidin was identified by CM-Sephadex C-50 chromatography of the kinin fraction of the above reaction mixture. Existence of a small amount of kallidin in the reaction mixture was also confirmed by thin layer chromatography on a silica gel plate of dansylated kinins (5). Therefore, identification of kinins liberated by plasmin from HMW and low molecular weight (LMW) bovine kininogens (kininogen-II) were performed. In the case of plasmin, only bradykinin was found in the reaction mixtures of HMW or LMW kininogens (Fig. 9). Recently, Henriques *et al.* observed that Met-Lys-bradykinin was released from equine kininogen by human plasmin (6), and they suggested that the release of Met-Lys-bradykinin in the acid-treated bovine pseudoglobulin in the experiemnt of Elliott and Lewis (7) should be caused by the action of bovine plasmin. In this study, we activated bovine plasminogen with urokinase *in vitro*, and the mechanisms of *in vivo* activation of plasminogen in plasma must be studied in future.

We are greatly indebted to Dr. A. Kaplan for the clotting assay of Hageman factor, and we also wish to thank Dr. O. D. Ratnoff for his kind consideration on the assay of Hageman factor.

REFERENCES

(1) S. Nagasawa, H. Takahashi, M. Koida, T. Suzuki and J.G.G. Schoenmakers, Biochem. Biophys. Res. Commun., 32, 644 (1968).
(2) T. Suzuki, S. Nagasawa, M. Yano, H. Takahashi and H. Kato, "Bradykinin and Related Kinins", ed. by F. Sicuteri *et al.*, Plenum Press, p.15 (1970).
(3) M. Yano, S. Nagasawa and T. Suzuki, J. Biochem., 67, 713 (1970).
(4) S. Nagasawa and T. Suzuki, J. Biochem., 66, 273 (1969).
(5) M. Yano, S. Nagasawa and T. Suzuki, J.Biochem., 69, 471 (1971).
(6) O.B. Henriques, E. Gapanhuk, N. Kauritcheva and P. Budnitskaya, Biochem. Pharmacol., 18, 1788 (1969).
(7) D.F. Elliott and G.P. Lewis, Biochem. J., 95, 437 (1965).

THE PRESENCE OF A SMOOTH MUSCLE CONTRACTING AND BRADYKININ POTENTIATING FACTOR IN PLASMA

Antti Louhija and Risto Sipilä

The Wihuri Research Institute

Helsinki, Finland

The final proof of the activation of the kinin-system is the demonstration of the liberation of free kinin into the blood or other biological fluids. The current methods for free kinin determination include rapid denaturation of the blood proteins (including kallikreins, kininogens and kininases), purification of the extracts to eliminate active substances interfering with the assay, concentration of the extracts, and finally the assay for which smooth muscle has mostly been used.

The methods used for the purification of the crude blood extracts fall principally into two categories. Ion exchange chromatography has been used e.g. by Oates et al.(1964), Webster and Gilmore (1965), Zacest and Mashford (1967) and Melmon et al. (1968), while others, including Binia et al. (1963), Abe et al.(1966) and Brocklehurst and Zeitlin (1967) have used the extraction of the kinins with butanol at low pH.

We have been trying to find a method for blood

free bradykinin bioassay that could be applied to clinical material. Crude alcoholic extracts of whole blood and plasma (Allwood and Lewis 1964), as well as more purified extracts, were used. Great care was taken when drawing the blood samples; only siliconed needles and plastic syringes were used. Blood samples varying from 10 to 40 ml were immediately denaturated in either three volumes of ethanol, two volumes of 0.5 M perchloric acid or half volume of 0.8 N hydrochloric acid. To obtain plasma samples, heparin or EDTA blood was immediately centrifuged in a refrigerated centrifuge, and the plasma then denaturated in the same way as whole blood. The extracts were purified either by using ion exchange chromatography on Amberlite IRC-50, H^+, resin (Oates et al.1964, Melmon et al.1968) or by butanol extraction (Abe et al.1966) with subsequent ion exchange chromatography. Bioassay was mostly performed on an estrus rat uterus in de Jalon's solution. Atropine and methysergide were used as inhibitors, and the uteri were sensitized with chymotrypsin according to Edery (1964). In some experiments chymotrypsin sensitized guinea-pig ileum was used in Tyrode solution with atropine, methysergide and promethazine as inhibitors.

All the crude alcoholic extracts as well as the more purified extracts of blood and plasma showed smooth muscle contracting activity. The contractions, however, differed from those obtained by synthetic bradykinin by showing a shorter delay before contraction and a steeper dose-response line. The activity also proved to be more or less resistant to chymotrypsin thus differing from bradykinin. Subcontractile doses of the extracts remarkably potentiated the contractions elicited by synthetic bradykinin. The activity was further shown to be dialysable, thermostabile and even resistant to total acid hydrolysis in 6 N HCl at 110°C overnight.

In order to separate this smooth muscle contracting and bradykinin potentiating factor from bradykinin a gel filtration procedure was undertaken. When the eluate after ion exchange chromatography of a whole blood extract was applied on a 0.9 x 53 cm Sephadex G-25 (superfine) column and eluted with distilled water, one single uterus contracting fraction was found. This fraction also potentiated the action of synthetic bradykinin and possessed all the other characteristics of the less purified blood extracts. When synthetic bradykinin was added to the blood extracts before gel filtration , there appeared two uterus contracting fractions in the eluate. The first of them behaved like synthetic bradykinin and the second showed all the characteristics of the uterus contracting and bradykinin potentiating factor described above.

Thus in conditions prevailing in the current methods of free bradykinin bioassay there seems to be a smooth muscle contracting factor, which also remarkably potentiates the action of bradykinin. The factor can be separated from bradykinin by Sephadex gel filtration, where it moves more slowly than bradykinin (Louhija and Sipilä 1971).

We consider this factor a potential source of error in the measurement of blood free kinin activity when bioassay on smooth muscle is used. The more precise identity of this factor is unknown, but obviously it is of smaller molecular size than synthetic bradykinin and can hardly be a peptide because of its resistance towards hydrolysis.

REFERENCES

1. Abe,K., Watanabe,N., Kumagai,N., Mouri,T., Seki,T. and Yoshinaga,K.; Tohoku J.exp.Med. 89,103,1966.

2. Allwood,M.J. and Lewis,G.P.; J.Physiol.(Lond.) 170, 571,1964.

3. Binia,A., Fasciolo,J.C. and Carretero,O.A.; Acta physiol. latino-amer. 13,101,1963.

4. Brocklehurst,W.E. and Zeitlin,I.J.; J.Physiol. (Lond.) 191,417,1967.

5. Edery,H.; Brit.J.Pharmacol. 22,371,1964.

6. Louhija,A. and Sipilä,R.; Biochem.Pharmacol. In press.

7. Melmon,K.L., Cline,M.J., Hughes,T. and Nies,A.S.; J.clin.Invest. 47,1295,1968.

8. Oates,J.A., Melmon,K., Sjoerdsma,A., Gillespie,L. and Mason,D.T.; Lancet I,514,1964.

9. Webster,M.E. and Gilmore,J.P.; Biochem.Pharmacol. 14,1161,1965.

10. Zacest,R. and Mashford,M.L.; Aust.J.exp.Biol.med. Sci. 45,89,1967.

OBSERVATIONS ON THE CHYMOTRYPSIN PEPTIDE RELEASING ACTIVITY ON PLASMAS

Alba A. C. Lavras, Mina Fichman, Elisa Hiraichi, Tomoyo Tobo, Marisa A. Boucault and Paulina Schmuziger

Servico de Farmacologia de Instituto Butantan
C.P. 65, Sao Paulo - 05504, Brasil

INTRODUCTION

Previous report[1] has shown the liberation of a pharmacologically active principle by the action of Padutin on the Bothrops jararaca plasma and has demonstrated its distinction from kinins. Afterwards, experiments were carried out to clarify whether such release is due to kallikrein or to another enzyme present in Padutin as an impurity. Taking in account the chymotrypsin destructive ability on the active released substance[1], this enzyme was reconsidered as a possible Padutin active contaminant. This hypothesis was reinvestigated by some experiments herein presented. On the other hand, this paper presents the results obtained in the study of chymotrypsin action extended from Bothrops jararaca to mammal plasmas.

MATERIAL AND METHODS

I) Enzymes and drugs

The following enzymes and drugs were used in this work: Padutin[R]: Pd (obtained from Bayer, Leverkusen, Germany, through the courtesy of Dr. Otterbach, C.) ; 3 x crystallised Chymotrypsin; Cht; 2 x crystallised Trypsin: Try; Carboxypeptidase B from porcine pancreas: Cbx; Ethylenediaminetetracetic acid: EDTA and Synthetic Bradykinin triacetate: SB (Sigma Chemical Co., St. Louis, U.S.A); 3 x crystallised Pepsin: P; (General Biochemicals, Inc., Ohio, U.S.A.) ; pure pig Pancreatic Kallikrein B, 185 U/mg: pKal (kindly provided by Prof E. Werle); Valyl 5 - angiotensin II: Ang (Ciba Basel, Switzerland; kindly donated by Dr. T. Paiva from the

Escola Paulista de Medicine); one year stored at 4 C solution of 150 ug/ml Physalaemin : Phy (kindly supplied by Prof. V. Erspamer); L-adrenalin: Adr (E. Merck Ag. Darmstadt, Germany); Amberlite XE-64-IRC-50 (British Drug Houses Ldt., Poole, England); Nembutal (Abbott-Laboratorios do Brasil Ltda., Sal Paulo, Brasil); Heparin (Evans Medical Ltd., Liverpool and London, England) and Sodium citrate (Baker & Adamson, New York, U.S.A.).

II) Plasmas

1) Fresh plasmas. Human (Hp), equine (Ep), dog (Dp) and *Bothrops jararaca* (BJp) plasmas were used. Human and horse, dog and finally snake blood samples were respectively collected by vein, cardiac and arterial[1] puncture into polyethylene syringes through siliconized needles. Immediately after collection, the blood samples were centrifuged at 0-5 C and the plasmas separated.

2) Treated plasmas. *Bothrops jararaca* plasma treated by Horton's method[1]: BJH and equine plasma treated by Werle's process[2]: EW.

III) Methods

1) Plasma kininase inhibition. Dog, horse and human plasma kininases were inhibited by 1 hour incubation at 37 C with 3×10^{-3}M EDTA.

2) Partial purification of the active materials released by Cht from plasmas. The principles liberated by Cht from *Bothrops jararaca* and dog plasmas were partially purified on Amberlite as previously described[1]

3) Biologic assays.

a) Isolated preparations. The following were used: guinea pig ileum[3], rat uterus[4] and duodenum[5], rabbit colon[6] and rabbit aorta strip[7]. *Bothrops jararaca* aorta strip was employed in the same conditions referred for rabbit aorta.

b) Rat blood pressure assays were carried out as previously described[1].

c) Rabbit blood pressure assays. Rabbits were anesthetized with 40 mg/Kg Nembutal intraperitonially. Anesthesia is maintained during the experiments by additional doses of 10 mg Nembutal given intravenously. The blood pressure was recorded from the carotid artery by means of a mercury manometer; coagulation was avoided by using 10% Sodium citrate solution. Material to be tested was injected through a plastic catheter in the marginal ear vein.

RESULTS AND DISCUSSION

In the search for the Padutin component enzyme responsible for its activity on BJH[1] the occasional findings of some Pd batches with low or without activity on this substrate in spite of being active on EW as kallikrein, indicated that the principle liberated from BJH by Pd could not be ascribed to its main component enzyme. An additional support to this evidence arose from the demonstration that hog purified pancreatic kallikrein was unable to release the active principle from BJH. Actually, figure 1 presents the absence

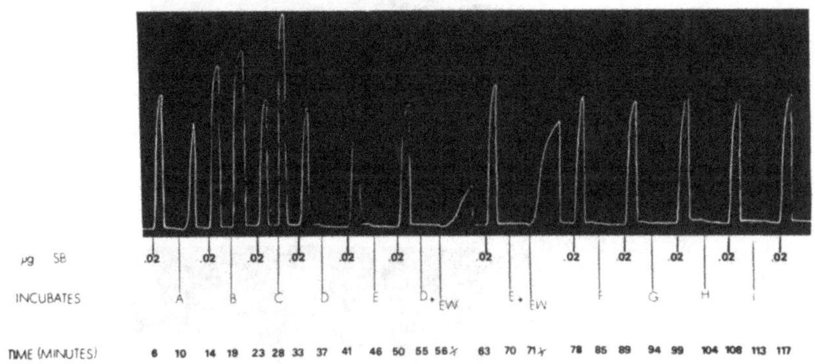

Fig. 1 Purified hog pancreatic kallikrein (pKal) failure and Padutin (Pd) ability to release, from the acid treated <u>Bothrops jararaca</u> plasma (BJH) a principle detected by its oxytocic effect (isolated rat uterus).
Incubation mixtures: 30 min. at 37 C; 0.1 ml enzyme + 0.4 ml substrate.
A: 4 U Pd + BJH; B: 4 U Pd + EW (equine treated plasma); C: 2 U pKal +EW; D: 2 U pKal +BJH; E: 10 U pkal + BJH; F, G,H and I: blanks of enzymes and substrates.
The sign + indicates that EW was added in the organ bath without washing out D or E in order to make evident the presence of active kininogenase in these incubates that were inactive on BJH.

of response of the isolated rat uterus to pKal-BJH incubation mixture compared to the effect of the kinin liberated by pKal from EW. It also shows the organ contraction induced by that incubation mixture plus EW, demonstrating that the failure of pKal activity on BJH is not due to its complete destruction or inactivation by this substrate. The next approach to the problem was to consider again chymotrypsin as the active Pd contaminant enzyme based on its ability to destroy the active product released by Pd from BJH[1] and on the demonstration that Pd, immediately after an acid treatment used to destroy chymotrypsin[8], no more exhibits its activity on BJH while its kininogenase activity on EW is preserved even after 90 minutes in the same acid conditions (Fig.2). On the other hand, incubating BJH with Cht in adequate doses and experimental conditions, in order to prevent not only the total destruction or inactivation of the enzyme by the substrate, but also of the released

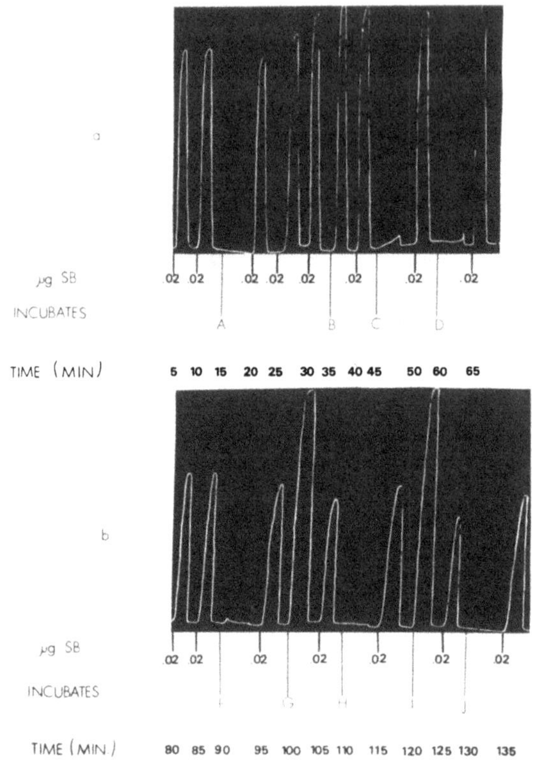

Figure 2 Destruction of Padutin activity on BJH (a) and preservation of its kallikrein content (b) by acid treatment. Assay on the isolated rat uterus.
Incubation mixtures: 30 min. at 37 C.
a) 0.2 ml enzyme solution + 0.4 ml substrate; B: 8U Pd +BJH; D: 8 U acid treated Pd + BJH; A and C: blanks of Pd and acid treated Pd.
b) 0.1 ml enzyme solution + 0.4 ml substrate; G: 4 U Pd + EW; I: 4 U acid treated Pd + EW; F, H and J: blanks of substrate and enzymes.

product by this proteolytic enzyme and/or by the substrate, it was possible to demonstrate that Cht is really active on BJH. Thus

Figure 3 shows the screening and finding of convenient enzyme

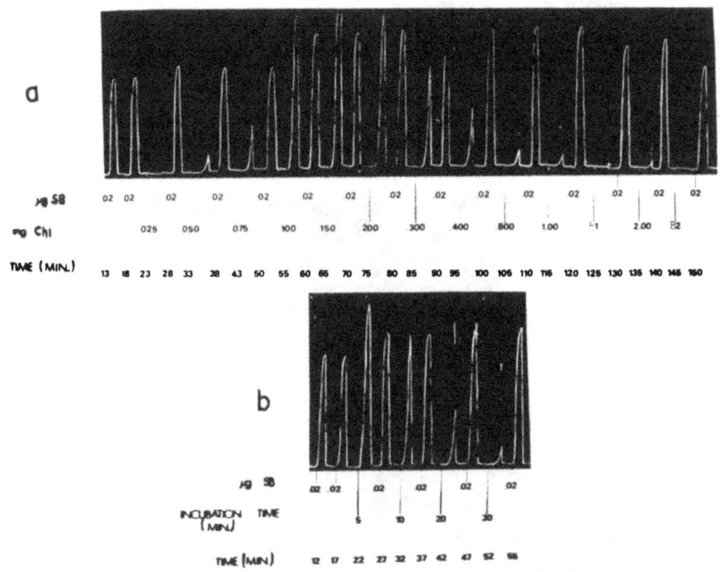

Figure 3 Chrmotrypsin ability to liberate an active substance from BJH, besides its destructive action on this released product. Assay on the isolated rat uterus.
Incubation mixtures at 37 C 0.1 ml enzyme solution + 0.4 ml substrate. Reaction interrupted by 5 min heating in a boiling water bath.
a) Organ responses to mixtures of BJH with various doses of Cht after a 5 min. incubation period; B_1 and B_2 blanks of BJH and 2 mg Cht respectively.
b) Time influence on the Cht destructive action upon the active substance released by 150 ug of this enzyme from BJH.

doses and incubation time to reveal, on the isolated rat uterus, the liberation of an active substance from BJH by Cht. In parallel it evidences the Cht destructive action on the released product. Similarly it was possible to establish suitable conditions to reach detectable levels of the active substance in the fresh BJp-Cht incubation mixtures.

Assays of this principle released by Cht from BJH, on the isolated rat duodenum, guinea pig ileum and rat blood pressure, have also demonstrated strict similar effects to those caused by the action of the material liberated by Pd from BJH[1]. This pharmacological similarity was reinforced since it was found that both principals contract the rabbit as well as the <u>Bothrops jararaca</u> aorta strips (Fig. 4). Besides their pharmacological resemblances and

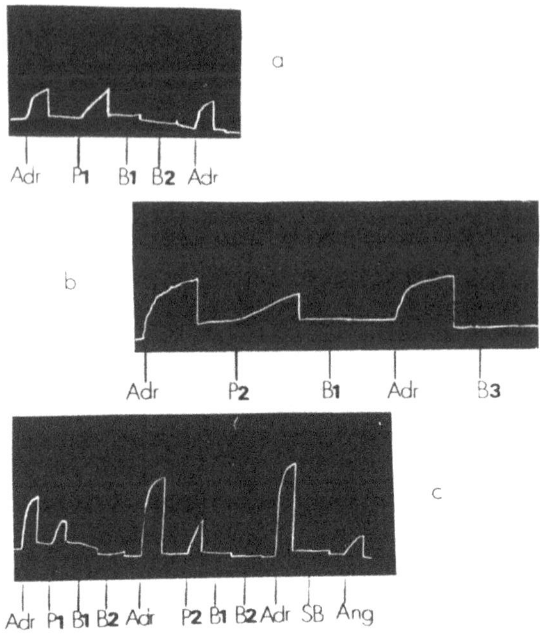

Figure 4: Effect of the principles released from BJH by Pd and Cht on *Bothrops jararaca* (a,b) and rabbit (c) aorta strips. P1 and P_2: partially purified principles released at 37 C from BJH by 120 U of Pd in 30 min. and by 4 mg of Cht in 5 min., respectively.
B_1, B_2 and B_3: blanks of BJH, Pd and Cht respectively.
Adr, SB and Ang in the aorta bath: 1 ug/ml.

common protein origin, these substances are dyalisable, equally destroyed by Cht, Try, Cbx and P and possible of being absorbed on Amberlite. Therefore it is reasonable to assume the release of the same peptide either by Cht or Pd.

After succeeding in the demonstration of an active peptide release by Cht from *Bothrops jararaca* plasma, in which there is no kinin interference[9] a study was carried out to verify whether Cht led to a similar peptide formation in mammal plasmas, which could be masked by a simultaenous kinin liberation, since Rocha e Silva and col. detected chymotrypsin kinin release activity in guinea pig fresh plasma [10] (in vitro) or dog plasma (in vivo). Taking in account the observation of changeful levels of anti-Cht activity in the various animal plasmas as well as the known kinin destructive ability of Cht it was accomplished the search of adequate experimental conditions to increase the accuracy of the kinin releasing detection in each kininase inhibited plasma-Cht incubation mixture. Thus, the demonstration of Cht kinin releasing activity on kininase inhibited human, dog and horse fresh plasmas fully succeeded since it was found for each kininase inhibited plasma sample an appropriate Cht concentration to overcome the expontaneous kinin release that occurs in the correspondent control plasma mixture and to prevent the complete destruction of the kinin released by the enzyme. In fact, figure 5 shows on the isolated rat uterus and duodenum the effect of the kinin released by Cht from dog kininase inhibited fresh plasma. In addition,

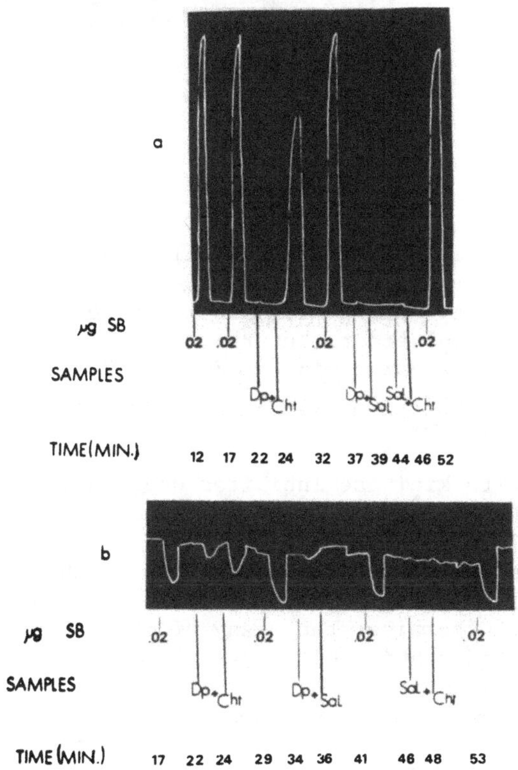

Fig. 5: Cht ability to release kinin from kininase inhibited dog fresh plasma (Dp).
Assays directly on the bath containing the isolated rat uterus (a) or duodenum (b).
The sign + indicates that 0.2 ml Cht (200 ug) or 0.2 ml saline (Sal.) was added in the organ bath without washing out Dp or saline.

figure 6 illustrates the interference of Cht kinin destructive ability on the disclosure of the kinin release by different doses

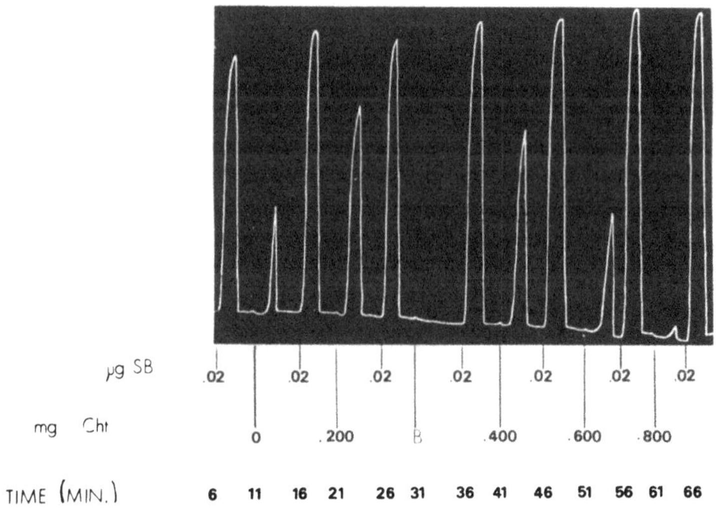

Figure 6: Interference of the kinin destructive ability of different doses of Cht with the disclosure of its kinin releasing activity on dog kininase inhibited plasma.
Assay on the isolated rat uterus.
Incubation mixtures: 3 min. at 30 C: 0.2 ml kininase inhibited Dp + Cht in increasing doses. B- blank of 0.200 mg Cht.
Final volume: 7 ml.

of this enzyme with kininase inhibited dog plasma. In the incubates of Cht with dog, horse, or human fresh plasma it was possible to remove the kinin interference and so to demonstrate another active substance release, taking advantage of the kinin destructive ability of suitable doses of Cht augmented by each plasma own kininase activity. Otherwise, the assay of the incubates was carried out on the isolated rabbit colon that presents a very low sensibility to kinins, ascertaining therefore, that any eventual kinin leftover would remain below the threshold of measurement as confirmed on the isolated rat uterus (Fig. 7). Reducing the anti-Cht level of dog, equine or human fresh plasma by 3 hours heating at 56 C, the formation of the principle active on the isolated rabbit colon was brought about with doses of Cht lower than the liberating effective ones on the correspondent fresh plasma. The peptide nature of the principle released by Cht from dog plasma was drawn, after its partial purification on Amberlite, as above described for the substance released from Bj plasma and based on its dyalisis capacity as well as on the demonstration that some proteolytic enzymes abolished its pharmacological activity on the isolated rabbit colon. This peptide, in a dose up to 7 times higher than the effective one to contract the rabbit colon, is inactive on the isolated guinea pig ileum, rat uterus and duodenum. More-

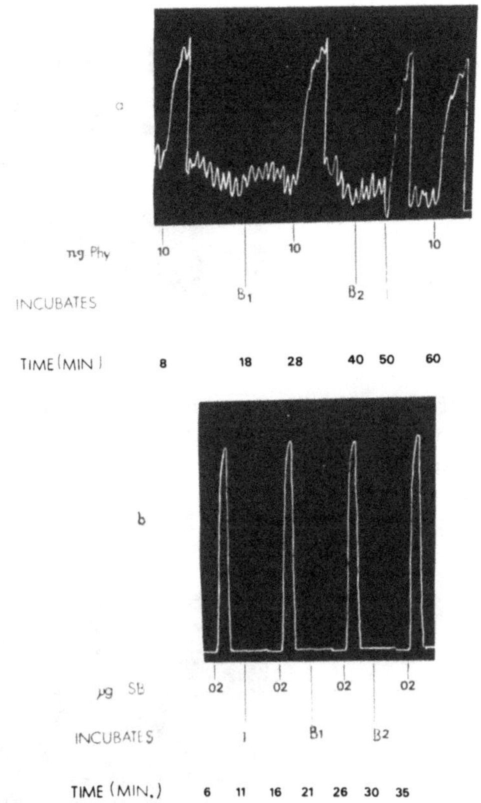

Fig. 7 a) Response of the isolated rabbit colon to a principle different from kinin relased by Cht from Dp.
b) The absence of kinin interference is shown on the isolated rat uterus.
Incubation mixtures: 30 min. at 30 C; I: 0.2 ml Dp + 0.2 ml Cht (200 ug); B_1 and B_2 blanks of Dp and Cht respectively.

over it was shown to increase the arterial blood pressure of the rat (Fig. 8) and also of the rabbit.

Since the rat serum breaks down bradykinin rapidly[12] this animal was selected to verify in vivo an eventual Cht pressor effect correspondent to the vasopressor peptide liberation demonstrated in vitro by the action of this enzyme on dog fresh plasma. Really, figure 9 shows the pressor effect of an intravenous injection of 2 mg Cht in the rat.

Attending to the above results, the observation of Rocha e Silva et al.[10] that "Cht when injected into dogs causes a drastic reduction of bradykininogen content of blood plasma without

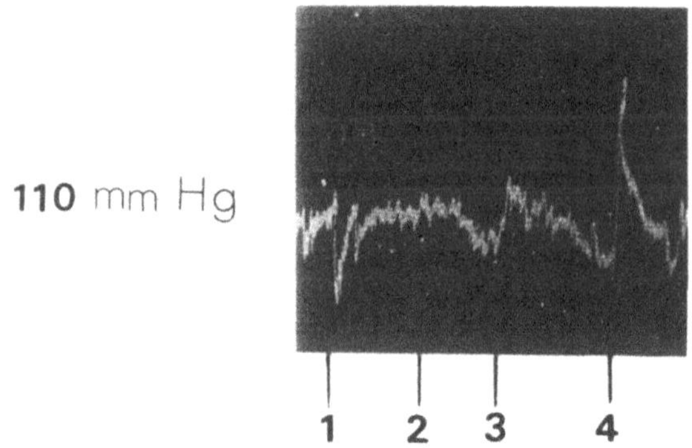

Figure 8 Effect of the peptide released by Cht from fresh dog plasma on the rat blood pressure (300 g male rat). 1-SB: 0.05 ug/100 g rat; 2-Saline; 3-Blank; 4-Partially purified peptide released by 10 mg Cht from Dp All substances were injected in a 0.4 ml volume.

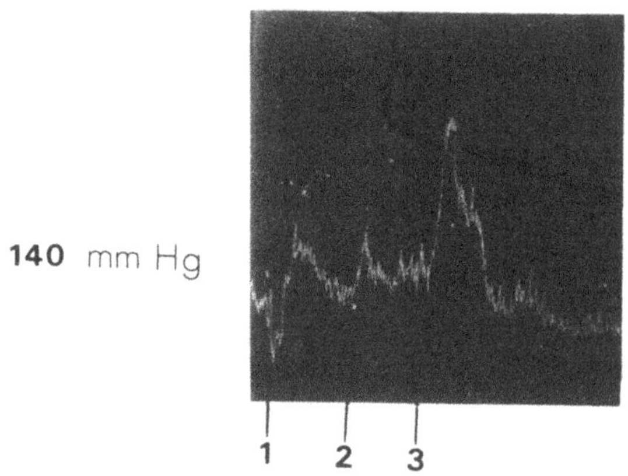

Figure 9 Cht pressor effect in the rat (400 g male rat). 1-SB: 0.05 ug/100 g rat; 2-Saline; 3-Cht: 0.5 mg/100 g rat.
All substances were injected in a 0.4 ml volume.

signs of any effect upon the blood pressure" could be explained by the simultaneous release of kinin and that pressor peptide.

As a matter of fact it is conceivable that the Cht pressor effect in the dog could be suppressed by the kinin interference since this animal presents a weak plasma kininase activity.

Further purification and characterization of the pressor peptides released by Cht from *Bothrops jararaca* and dog plasmas will be undertaken in order to permit their comparison with other pressor peptides, particularly with substance A[13] and angiotensins.

SUMMARY

1. It was demonstrated that Cht is the Padutin contaminant enzyme responsible for its ability to release an active peptide from *Bothrops jararaca* plasma.

2. It was also shown the contracting effect of the peptide released by Pd or Cht from BJH on the rabbit and *Bothrops jararaca* aorta strips.

3. In addition, it was revealed the Cht ability to release kinin from dog, human or horse kininase-inhibited fresh plasma.

4. Removing the interference of kinin produced by Cht acting on dog, human or horse plasma it was disclosed, in the isolated rabbit colon preparation, its ability to release another active principle from those plasmas.

5. After concluding upon the peptide nature of the principle released by Cht from dog plasma its activity was assayed on various isolated smooth muscle preparations and on the rat and rabbit blood pressure.

6. By intravenous injection of Cht in the rat it was observed a hypertensive effect that could be correlated to the pressor peptide liberated in vitro by the action of the same enzyme on dog fresh plasma.

Acknowledgements: Many thanks are due to Dr. Z.P. Picarelli, from the Departamento de Farmacologia e Bioquimica da Escola Paulista de Medicina, for the assistance in some of the phases of this work and helpful criticism during the writing of this paper. We are also grateful to Dr. L. Nahas from the Divisao de Patologia do Instituto Butantan for her suggestions, to Mr. D.F. Oliveira for technical assistance and to Mr. W.B. D'Avila for helping in the illustration of this paper.

REFERENCES

1. Lavras, A.A.C., Fichman, M., Hiraichi, E., Schmuziger, P., and Picarelli, Z.P.: Advances Exp. Med. Biol. 8: 89, 1970.
2. Henriques, O.B., Lavras, A.A.C., Fichman, M., and Picarelli, Z.P.: Biochem. Pharmacol. 15: 31, 1966.
3. Henriques, O.B. ,Picarelli, Z.P., and Oliveira, M.C.F.:

Biochem. Pharmacol. 11: 707, 1962.

4. Picarelli, Z.P., Henriques, O.B., and Oliveira, M.C.F.: Experientia, 18: 77, 1962.
5. Horton, E.W.: Brit. J. Pharmacol. 14: 125, 1959.
6. Erspamer, V., and Erspamer, G.F.: Brit J. Pharmacol. 19: 337, 1962.
7. Furchgott, R.F., and Bhadrakom, S.: J. Pharmac. exp. Ther. 108: 129, 1953.
8. Habermann, E., and Blennemann, G.: Arch. exp. Path. Pharmak. 249: 357, 1964.
9. Lavras, A.A.C., Fichman, M., Hiraichi, E., and Picarelli, Z.P.: Abstracts of the Fourth International Congress on Pharmacol., Basel, Switzerland, 209, 1969.
10. Rocha e Silva, M., Reis, L.M., and Ferreira, S.H.: Biochem. Pharmacol. 16: 1665, 1967.
11. Ferreira, S.H., and Rocha e Silva, M. : Brit. J. Pharmacol. 36: 611, 1969.
12. Fasciolo, J.C.: Acta physiol. lat. am. 14: 170, 1964.
13. Walaszek, E.J., and Hugghins, C.G. : J. Pharm. Exp. Therap. 126: 258, 1959.

AN ALTERNATIVE SCHEME OF KININ FORMATION DURING CONTACT ACTIVATION OF HUMAN PLASMA

W.Vogt, M.Schlaeger, G.Seidel and H.-U.Stücker

Max-Planck-Institut für experimentelle Medizin,

Dept. of Biochem. Pharmacology, 34 Göttingen,Germany

Among the various reactions that lead to kinin formation those are of particular interest which proceed by factors endogenous to plasma, as these represent the most readily available natural kinin forming system. As is well-known plasma contains simultaneously prekininogenases, activators and substrates. Armstrong, Keele, Jepson and Stewart (1954) found that the kinin forming system in plasma can be triggered by contact with foreign surfaces, e.g. collagen or glass, and Margolis (1960) demonstrated that it is Hageman Factor which is activated by absorption to these surfaces, and which in turn activates a kininogenase ("component A").

Based on these findings and additional results a scheme was presented at the first Symposium on Hypotensive Peptides, at Florence (Vogt, 1966) according to which two separate kinin forming systems exist in human plasma: Hageman Factor absorbed onto glass etc. activates kininogenase II (corresponding to component A of Margolis) which consumes kininogen II (= component B), liberating plasma kinin. Kininogenase I (equal to classical serum kallikrein) which can be activated in fresh plasma by acid or acetone and acts on kininogen I, does hardly react during contact activation of human plasma. Accordingly, kininogen I is largely left intact after even exhaustive treatment of plasma with glass. As neither acid nor acetone activate kininogenase I in post-contact plasma, it was concluded that during contact kininogenase I <u>is</u> activated simultaneously with enzyme II, but is then blocked by endogenous inhibitors at a rate sufficiently high to prevent both consumption of kininogen I and further activation subsequent to acid or acetone treatment, or new contact.

The following new findings suggest that the fate of (pre-)kininogenase I during contact of human plasma has to be described in a different manner.
1) Activation of kininogenase I by plasmin.
Treatment of human plasma with streptokinase leads to kinin formation merely by an indirect mechanism, namely by activation of kininogenase I (Seidel, Stücker and Vogt, 1971). This is evident from the facts that the kinin forming activity of streptokinase-treated plasma is directed to kininogen I, is not inhibited by lima bean inhibitor, but destroyed by heating at pH 4 to 61°, the two latter features being different from the behaviour of plasmin. The induction of kininogenase I by streptokinase is possible also in plasma pretreated once ("B-dep"; Margolis, 1960) or repeatedly ("A+B-dep"; Margolis, 1960) with glass. This means that activatable prekininogenase I must still be present in these plasma preparations although they do not develop new kininogenase I activity upon glass contact or treatment.

2) Passage of human plasma through a column of insolubilized trypsin activates kininogenase I. Again, this activity can be induced not only in fresh but also in post-contact plasma (B-dep or A+B-dep) (table 1).

3) It seemed not impossible that in glass-treated plasma trypsin (and maybe plasmin) only <u>re</u>activated kininogenase I which had been activated already during the earlier contact with glass and had decayed subsequently because of binding to inhibitors. However, Werle and Vogel (1960) found that inactivated plasma kallikrein cannot be reactivated by trypsin, and own experiments support this finding. An acetone-activated plasma kallikrein preparation was added to fresh human plasma (180 mU esterase/ml mixture). Estimation of esterolysis (BAEE; Frey, Kraut and Werle, 1968) two hours later showed a residual activity of 56 mU/ml mixture. After treatment with insolubilized trypsin the esterolytic potential of the mixture rose to 86 mU/ml, 27 mU of which were due to activation of esterase of the plasma used as source of inhibitors. Accordingly the activity of the added plasma kallikrein preparation after trypsin treatment was only 86 - 27 = 59 mU/ml, i.e. no reactivation took place.

Table 1

No.	plasma	substrate added	Kinin formation before	trypsin treatment after
1	normal	-	< 0.06	< 0.06
2	"	KI	-	1.1
3	"	61°-P1	-	0.7
4	A+B-dep	-	< 0.06	< 0.06
5	"	KI	-	1.1
6	"	61°-P1	-	0.9

Kinin forming activity induced by trypsin in human plasma. The plasma specimen were first treated with Na-p-iodo-benzoate (0.074 M) to eliminate excess of protease inhibitors. Aliquots were then stirred with 3 - 4 batches of 0.6 g/ml quartz powder (surface 3.1 m^2/g), 10 min each time, to deplete them from kininogenase II and kininogen II ("A+B-dep"). Non-contact portions were kept in siliconized vessels ("normal"). None of these samples developed detectable kinin activity when incubated for 60 min at 37°, in the presence of 1 mg/ml o-phenanthroline-HCl (to prevent kinin destruction) (Ns 1 and 4; before trypsin treatment). Other portions of normal or A+B-dep plasma were passed through a column of solid phase trypsin (Sephazyme TRY, Pharmacia). Kininogen lost during the passage was restored by adding a kininogen I preparation ("KI") or plasma heated for 30 min to 61° ("61°-P1"), and the mixtures were incubated for 60 min at 37° in the presence of o-phenanthroline. Kinin formed not only in incubates of normal but also of A+B-dep plasma (Ns 2,3; 5,6). No kinin activity was detected when substrate was omitted (Ns 1 and 4; after trypsin treatment). The substrate preparations alone did not show detectable kinin activity.

Our findings demonstrate that prekininogenase I is not wholly activated and consumed during glass contact of human plasma, but is largely left intact as preenzyme. The fact that it cannot be efficiently activated in post-contact plasma by new contact, acid, or acetone - although Hageman Factor is still present - suggests that there is an essential link in the activation chain between Hageman Factor and prekininogenase I, present in fresh plasma but rapidly consumed during glass contact. This link, an activator, is not essential for the activation of prekininogenase II by Hageman Factor, as kininogenase II can be induced several times by repeated contact (Margolis, 1960).

A preparation of PF/dil kindly obtained from Dr. E. Becker (Washington) was able to induce kinin formation in glass-treated plasma (B-dep) although it had no kininogenase activity _per se_ - it did not liberate kinin from kininogen I containing substrate preparations. This PF/dil preparation apparently contained the essential activator, the missing link between Hageman Factor and prekininogenase I, in an active form. It may not be identical with PF/dil proper, however, as another PF/dil preparation was devoid of such activity.

The new scheme of reactions resulting from our findings and considerations is presented below.

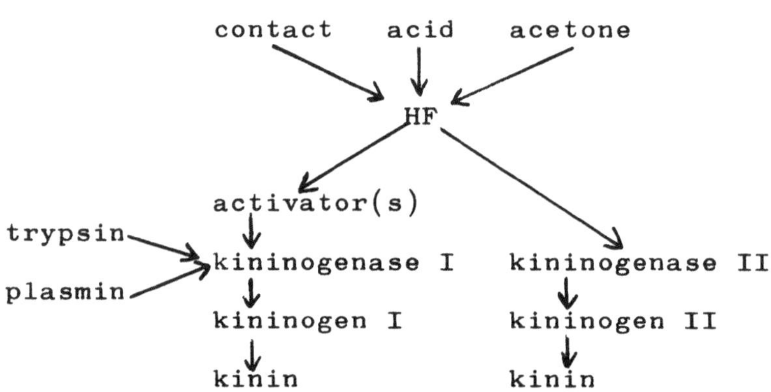

It appears to be in accordance with all results relating to this subject, except possibly one finding of Jahrreis and Habermann (1971) who found that purified Hageman Factor was able to directly activate a purified prekallikrein preparation which was assumed to correspond to prekininogenase I. Possibly there are only quantitative differences. Sufficiently large amounts of Hageman Factor and preenzyme I may create measurable activity in a pure system whereas such activity may not be attained under natural conditions, in plasma.

Summary. In human plasma kininogenase I can be activated by acid, acetone, and a fraction of it by contact with glass. After contact treatment of plasma practically no kininogenase I activity can be induced by these procedures. However, plasmin and trypsin still activate kininogenase I in post-contact plasma. This indicates that the preenzyme is still present, and that an activator essentially intercalated between Hageman Factor and prekininogenase I is missing (consumed) after glass contact. One preparation of PF/dil, but not a second one, was found to have such activator potency. A new scheme of reactions during contact of human plasma with foreign surfaces is presented.

References

Armstrong, D., Keele, C.A., Jepson, J.B., Stewart,J.W.: Development of pain-producing substance in human plasma. Nature (Lond.) 174, 791 (1954).

Frey, E.K., Kraut, H., Werle, E.: Das Kallikrein-Kinin-system und seine Inhibitoren. Enke-Verlag Stuttgart, 1968 p.12.

Jahrreis, R., Habermann, E.: On the endogenous mechanism of kinin release. I. Attempts at discriminating kallikreins. Naunyn-Schmiedebergs Arch. Pharmak. 269, 85 (1971).

Margolis, J.: The mode of action of Hageman Factor in the release of plasma kinin. J. Physiol. (Lond.) 151, 238 (1960).

Seidel, G., Stücker, H.U., Vogt, W.: Significance of direct and indirect kinin formation by plasmin in human plasma. Biochem. Pharmacol. in print, 1971.

Vogt, W.: Demonstration of the presence of two separate kinin forming systems in human and other plasma. In: Hypotensive peptides, p.185, ed. by E.G.Erdös, N.Back and F. Sicuteri. Berlin - Heidelberg - New York: Springer 1966.

Werle, E., Vogel, R.: Über die Kallikreinausscheidung im Harn nach experimenteller Nierenschädigung. Arch. int. Pharmacodyn. 126, 171 (1960).

THE RELATIONSHIP BETWEEN THE PLASMA KININ-SYSTEM AND THE CONTACT PHASE OF BLOOD-COAGULATION IN MAN*

Henry Z. Movat, Mary J. Soltay, Penelope J. Fuller
and Ayse Özge-Anwar

Division of Experimental Pathology and Institute of
Immunology
University of Toronto, Toronto, Ontario, Canada

Three mechanisms have been postulated for the activation of plasma kallikrein (Fig. 1). According to the oldest theory activated Hageman factor or Factor XIIa activates the permeability globulin Pf/dil, which then acts on prekallikrein and the latter, as kallikrein, liberates kinin from kininogen (1, 2). The second hypothesis, based on good experimental data, assumes that Factor XIIa acts directly on prekallikrein (3, 4, 5). More recently, an activator of prekallikrein distinct from Pf/dil has been isolated from guinea pig (6), rabbit (7) and human plasma (8, 9, 10, 11). Some evidence was presented that this prekallikrein activator (PKA) is a fragment of Factor XIIa (8). Accordingly, this substance is designated as XIIf in Fig. 1.

We have previously shown that PKA adsorbs to antigen-antibody precipitates (6) and to celite (12, 13) when serum or plasma is incubated with the surface active agents. When plasma is incubated with large quantities of celite little or no Factor XIIa is recovered, most of it seems to convert to PKA. It remained to ascertain what happens when a preparation of Factor XIIa is treated with celite. In the course of these studies Factor XIIa was treated also with trypsin, also giving rise to PKA.

Preparative and analytical procedures are described in detail in other publications (10, 13, 14). Non-contacted platelet poor plasma or celite eluates of plasma was fractionated on QAE Sephadex A-50, Sephadex G-200

* Supported by the Canadian Arthritis and Rheumatism Society, the Medical Research Council of Canada and the Ontario Heart Foundation.

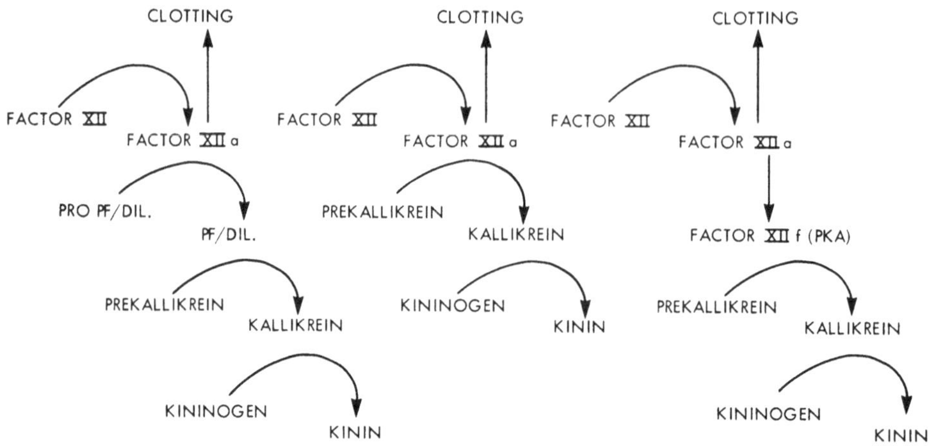

Fig. 1. Three hypotheses of activation of the kinin system.

TABLE 1. EFFECT OF FACTOR XIIa AND OF PKA ON CLOTTING OF FACTOR XII DEFICIENT PLASMA AND ON PREKALLIKREIN

Reaction Mixture	Clotting	Prekallikrein Activation (n-moles BAEe hydrol./min./ml.)
Factor XIIa + Buffer	72	0
Factor XIIa + Prekallikrein	-	2.7
PKA + Buffer	512	11.3
PKA + Prekallikrein	-	73.4
Buffer + Prekallikrein	-	0
Buffer	926	-

Factor XIIa was incubated for 10 minutes with celite (150 mg./ml. Factor XIIa). The celite was spun out and washed with 0.001 M Tris HCl (pH 8.0). The PKA was then eluted from the celite with 2.0 M NaCl in 0.1 M Tris HCl (pH 8.0) and concentrated to 1/2 the original volume of the incubated Factor XIIa. Prekallikrein activation and clotting were assayed as described in references 10, 13, 14 and 15.

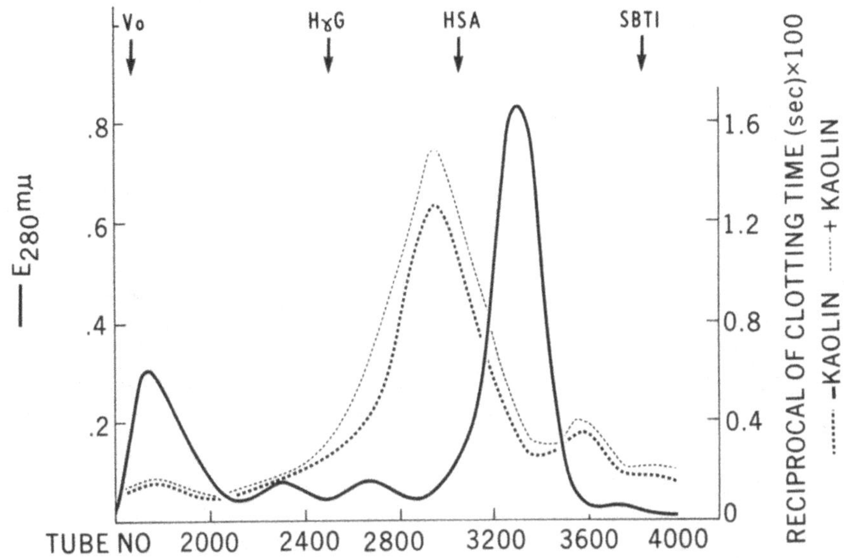

Fig. 2. Gel filtration of 23.5 ml of partially purified Factor XII on three interconnected columns (5 x 93, 5 x 89 and 5 x 86 cm.) of Sephadex G-200 fine, equilibrated with 0.1 M NaCl in 0.1 M Tris HCl (pH 8.0). Flow rate 25.0 ml/hr (10.0 ml/tube). The elution volumes of the following markers are shown by arrows: HγG (human γ-globulin), HSA (human serum albumin) and SBTI (soybean trypsin inhibitor). The void volume (Vo) was determined with blue dextran. Hageman factor activity was measured using Factor XII-deficient plasma as substrate in the presence or absence of kaolin, as described in references 13 and 14 (From Soltay, Movat and Özge Anwar - Proc.Soc.Exp.Biol.Med., 139, 1971).

and further purified and analysed by electrofocusing, preparative and analytical acrylamide electrophoresis and density gradient ultracentrifugation.

A functionally pure preparation of Factor XIIa was obtained, eluting with 0.1 M NaCl in 0.1 M Tris HCl (pH 8.0) from the anion exchanger. The bulk of the albumin was separable from this preparation by gel filtration (Fig. 2). Before gel filtration Factor XII was in an unactivated form, being activatable with kaolin. After gel filtration most of it was activated (XIIa).

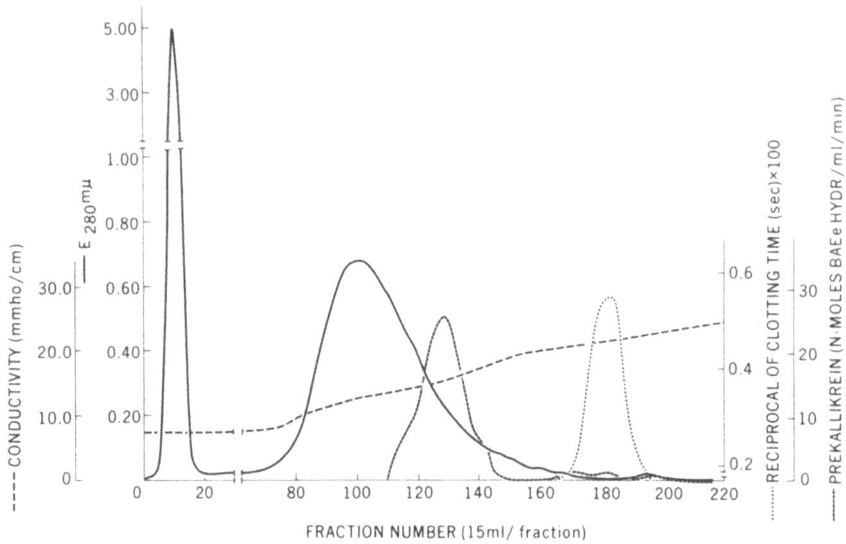

Fig. 3. Separation of prekallikrein from Factor XI and from immunoglobulin G by chromatography on SP Sephadex C-50. The excluded fraction (50.0 ml) obtained with a QAE Sephadex column was applied to a 3.5 x 30.0 cm column of SP Sephadex equilibrated with 0.01 M phosphate buffer (pH 6.0) containing 0.07 M NaCl (conductivity 7.5 mmho/cm). After passing 700 ml of starting buffer through the column a linear gradient (3500 ml) was applied to the column. The limiting buffer was 0.35 M NaCl in 0.01 M phosphate buffer (conductivity 33.8 mmho/cm). Prekallikrein (0.3 ml) was activated to kallikrein with trypsin (3.3 μg) and after inhibition of the trypsin with lima bean trypsin inhibitor (4.0 μg) the BAEe hydrolysing activity of kallikrein was measured. The tubes hydrolysing BAEe were pooled; this pool liberated kinin from plasma heated to 61°C for one hour. The prekallikrein contained traces of IgG, (tested by double diffusion), but no clot-promoting activity. Clot-promoting activity eluted in highly cationic fractions; using Factor XI-deficient plasma as substrate in partial thromboplastin assays, correction of the clotting defect was achieved in the presence of kaolin. The fractions containing Factor XI were free of prekallikrein and IgG.

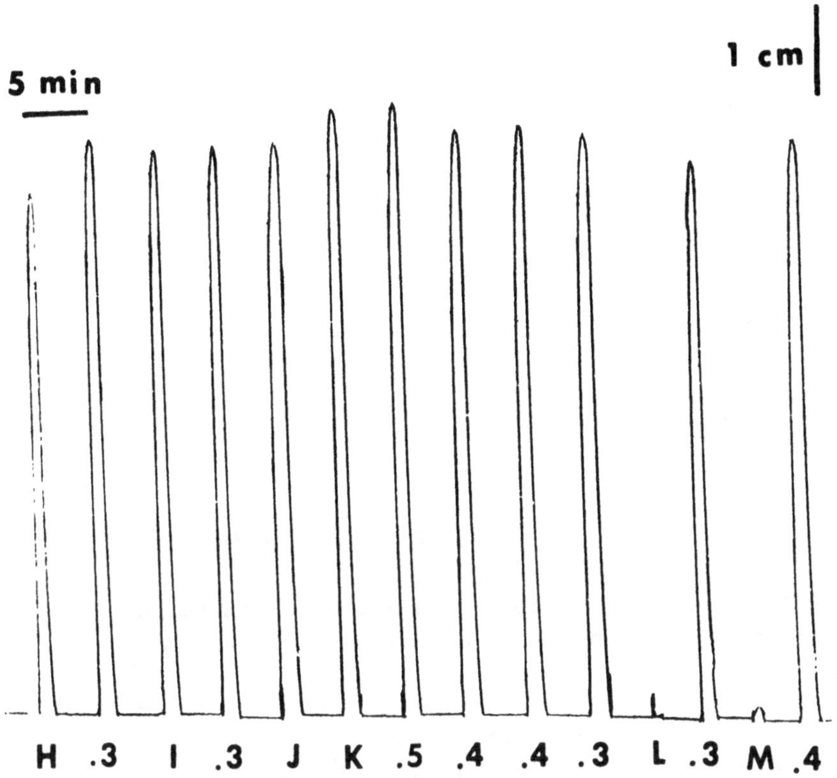

Fig. 4. Kymograph tracing. Kinin assay on the rat uterus. The numerals represent ng/ml of synthetic bradykinin added to the bath. "H" is PKA with prekallikrein 1/200; "I" and "J" are the same mixture 1/150 and "K" 1/100. "L" and "M" represent Factor XIIa and prekallikrein in dilutions of 1/100 and 1/50 respectively.

Its molecular weight was estimated to be about 120,000. The isoelectric point lay between pH 5.0 and 5.3 and by disc electrophoresis it migrated slightly faster than kallikrein in the ß-region. The preparation of Factor XIIa induced good correction of plasma congenitally deficient in this blood-clotting factor (Table 1). The preparation of Factor XII obtained by ion exchange chromatography hydrolysed BAEe, but after gel filtration arginine esterase activity was no longer present in the Factor XIIa preparation.

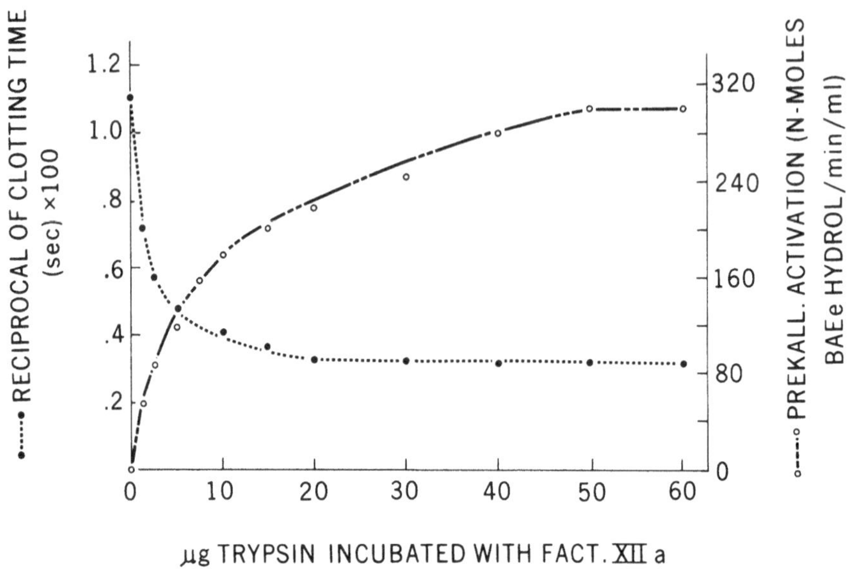

Fig. 5. Effect of trypsinization on Factor XIIa. Increasing concentrations of trypsin were incubated with Factor XIIa for 5 min., followed by neutralization of the trypsin with LBTI. Correction of the clotting defect of Factor XII-deficient plasma decreased by about 2/3 of the original Factor XIIa after adding 20 μg of trypsin, but further increase of the trypsin did not abolish this residual clot-promoting activity. PKA-activity levelled off after 50 μg of trypsin. (From Soltay, Movat and Özge Anwar - Proc. Soc. Exp. Biol. Med., 139, 1971).

Prekallikrein was present in the excluded protein peak obtained with QAE Sephadex. It was further purified by passing through Sephadex G-200 or SP Sephadex C-50. SP Sephadex readily separated the prekallikrein from the bulk of immunoglobulin and from Factor XI (Fig. 3).

Upon incubation of Factor XIIa with prekallikrein no conversion of the latter to kallikrein took place. However, when the Factor XIIa preparation was exposed to celite, the celite eluates contained PKA activity. The PKA activated the prekallikrein to kallikrein, the latter exhibiting BAEe-hydrolysing activity (Table 1). When the BAEe was replaced by kininogen (plasma heated at 61°C for 1 hour) liberation of kinin was demonstrable (Fig. 4).

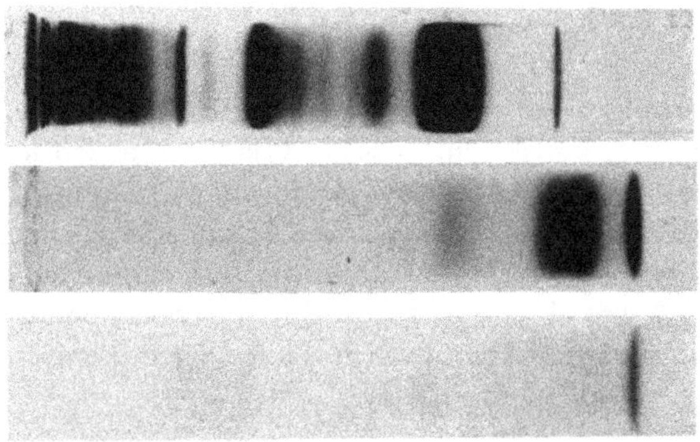

Fig. 6. Polyacrylamide disc electrophoresis of whole human plasma (upper), of chromatographed celite eluates of plasma (middle) and of chromatographed celite eluates of Factor XIIa (lower).

Factor XIIa could be converted to PKA also by incubating 1.0 ml of the XIIa preparation for 2 minutes with 25.0 μg of trypsin, followed by inactivation of the trypsin with 30 μg of lima bean trypsin inhibitor (LBTI). In this experiment, it was shown that the clotting time of Factor XIIa decreased rapidly when incubated with trypsin, but there was no complete abolition of the clot-promoting activity. This was more or less paralleled by an increase in the prekallikrein activating activity (Fig. 5). The LBTI does not inactivate kallikrein, but it has an inhibitory effect on the PKA generated from Factor XIIa. This was shown by increasing the LBTI above that required to inhibit trypsin.

PKA elutes from QAE Sephadex with 0.2–0.25 M NaCl in 0.1 M Tris HCl, has a sedimentation coefficient of about 2.8 S, a molecular weight of approximately 35,000, an isoelectric point of 4.2–4.6 and migrates in the prealbumin region by disc electrophoresis (Fig. 6).

Kallikrein, obtained by activating prekallikrein, was found to have an S-value of about 6.0, an approx. molecular weight of 95,000, migrating in the γ-region by disc electrophoresis at alkaline pH. Its isoelectric point lay between pH 8.0 and 8.5.

The molecular weight of Factor XI was estimated to be 160,000 to 170,000 and the pI of this clotting factor was 8.3-8.9.

Concluding Remarks: The data presented indicate that only PKA can activate prekallikrein to kallikrein. Factor XIIa when added to prekallikrein induces no conversion of this zymogen to the active enzyme. Upon massive contact exposure of partially purified Factor XII conversion to PKA occurs. However, recently, using a further purified Factor XII preparation, no conversion to PKA was observed, suggesting that additional factors are required for the conversion of Factor XII to PKA. Such highly purified Factor XII could be converted to PKA by trypsin.

The early data with Pf/dil are difficult to evaluate. Pf/dil has never been obtained in a reasonably pure form. On the basis of its elution pattern and electrophoretic mobility (16, 17, 18, 19) one could assume that the Pf/dil preparations contained plasmin and Factor XII. As described elsewhere (20, 21, 22) plasmin can bring about the generation of PKA from Factor XII.

References

(1) Miles, A.A.: Large molecular substances as mediators of the inflammatory reaction. Ann. N.Y. Acad. Sci., 116: 855, 1964.

(2) Ratnoff, O.D., and Miles, A.A.: The induction of permeability-increasing activity in human plasma by activated Hageman factor. Brit. J. Exp. Path., 45: 328, 1964.

(3) Nagasawa, S., Takahashi, H., Koida, M., Suzuki, T., and Schoenmakers, J.G.G.: Partial purification of bovine plasma kallikreinogen, its activation by Hageman factor. Biophys. Biochem. Res. Com., 32: 644, 1968.

(4) Temme, H., Jahreiss, R., Habermann, E., and Zilliken, F.: Aktivierung von Gerinnungs-und Kinin-system durch eine Plasmaesterase (Hageman-Factor). Reinigung und Wirkungsbedingungen. Hoppe-Seyler's Z. Physiol. Chem., 350: 519, 1969.

(5) Jahreiss, R., and Habermann, E.: In vitro investigations on some components of the kinin-system (kininogen, serum kallikrein, Hageman factor), and their interaction. In Bradykinin and Related Kinins - Cardiovascular, Biochemical and Neural Actions, Sicuteri, F., Rocha e Silva, M., and Back, N., eds., Plenum Press, New York and London, 1970.

(6) Movat, H.Z., Treloar, M.P., and Takeuchi, Y.: A small molecular weight permeability factor in guinea pig serum: adsorption to antigen-antibody aggregates. J. Immunol., 103: 875, 1969.

(7) Wuepper, K.D., Tucker, E.S. III., and Cochrane, C.G.: Plasma kinin-system: proenzyme components. J. Immunol., 105: 307, 1970.

(8) Kaplan, A.P., and Austen, K.F.: A pre-albumin activator of prekallikrein. J. Immunol., 105: 802, 1970.

(9) Movat, H.Z., Sender, H.B., Treloar, M.P., and Takeuchi, Y.: The isolation and partial characterization of kinin-forming enzymes and other active components. In Bradykinin and Related Kinins - Cardiovascular, Biochemical and Neural Actions, Sicuteri, F., Rocha e Silva, M., and Back, N., eds., Plenum Press, New York and London, 1970.

(10) Movat, H.Z., Poon, M.-C., and Takeuchi, Y.: The kinin system of human plasma. I. Isolation of a low molecular weight activator of prekallikrein. Int.Arch.Allergy, 40: 89, 1971.

(11) Kaplan, A.P., and Austen, K.F.: Immunologic activation of the bradykinin-forming system in man. In Biochemistry of the Acute Allergic Reactions, Austen, K.F., and Becker, E.L., eds., F.A. Davis Company, Philadelphia, in press.

(12) Movat, H.Z., and Özge Anwar, A.H.: The interrelationship between the kinin and blood-clotting systems. Fed. Proc., 30: 451, 1971.

(13) Özge Anwar, A.H., Movat, H.Z., and Scott, J.G.: The kinin system of human plasma. IV. The interrelationship between the contact phase of blood-coagulation and the plasma kinin system. Thromb. Diath. Haemorrh., in press.

(14) Soltay, M.J., Movat, H.Z., and Özge Anwar, A.H.: The kinin system of human plasma. V. The probable derivation of prekallikrein activator from activated Hageman factor (XIIa). Proc.Soc. Exp. Biol. Med., 139: 1971, in press.

(15) Özge Anwar, A.H., Movat, H.Z., and Scott, J.G.: The human plasma kinin system. II. Contact activation of plasma prekallikrein and Factor XI in Factor XII-deficient plasma. Proc.Soc.Exp. Biol.Med., 138: 330, 1971.

(16) Wilhelm, D.L., Mill, P.J., and Miles, A.A.: Enzyme-like globulins from serum reproducing the vascular phenomena of inflammation. III. Further observations on the permeability factor and its inhibitor in guinea pig serum. Brit.J.Exp.Path., 38: 446, 1957.

(17) Elder, J.M., and Wilhelm, D.L.: Enzyme-like globulins from serum reproducing the vascular phenomena of inflammation. V. Activable permeability factor in human serum. Brit.J.Exp.Path., 39: 335, 1958.

(18) Becker, E.L., and Kagen, L.: The permeability globulins of human serum and the biochemical mechanism of hereditary angioneurotic edema. Ann.N.Y.Acad.Sci., 116: 866, 1964.

(19) Webster, M.E.: Human plasma kallikrein, its activation and pathological role. Fed.Proc., 27: 84, 1968.

(20) Kaplan, A.P., and Austen, K.F.: A prealbumin activator of prekallikrein. II. Derivation of activators of prekallikrein from active Hageman factor by digestion with plasmin. J.Exp.Med., 133: 696, 1971.

(21) Burrowes, C.E., Movat, H.Z., and Soltay, M.J.: The kinin system of human plasma. VI. The action of plasmin. Proc.Soc.Exp.Biol. Med., 139: 1971, in press.

(22) Burrowes, C.E., Movat, H.Z., and Soltay, M.J.: The role of plasmin in the activation of the kinin system. In Vasoactive Polypeptides, Back, N., and Sicuteri, F., eds., Plenum Press, New York and London, in press.

CONNECTIONS BETWEEN KININ FORMATION AND COMPLEMENT

V. Eisen and C. Loveday

Rheumatology Research Department

Middlesex Hospital Medical School, London, W. 1

Intrinsic blood clotting, thrombolysis, plasma kinin formation and complement may be regarded as four major reactive systems which are triggered into activity when blood is exposed to noxious stimuli. The four processes do not of course rely solely on intrinsic factors, but cooperate with mechanisms available in surrounding tissues. In each of the four systems there are proteolytic enzymes, often acting in sequence, i. e. on precursors of other proteases either in their own reaction sequence or in one of the other systems. In this way, numerous functional connections inside and between systems are formed. The links between blood clotting, fibrinolysis and plasma kinin formation have been under study for many years. It is probably true to say that in vivo these reactions only rarely proceed in isolation. Only two of these connections will be mentioned here. Plasmin, a protease capable of attacking a very large number of proteins is not only responsible for the lysis of blood clots, but also intervenes in the formation of clots and kinins and in the activation of complement (Ratnoff & Naff, 1967). There is now a great deal of evidence that clotting factor XII (Hageman factor) occupies a central position as the initial step in the intrinsic sequences of blood clotting, fibrinolysis and plasma kinin formation. In recent years, evidence has been produced that factor XII or its molecular subunits may also trigger the first step in the complement sequence by catalyzing the activation of subcomponent C'1s to C'1 esterase (Donaldson, 1968a; Austen, 1971). This action of factor XII may be mediated through plasmin or plasma kallikrein.

The activators which complement shares with plasma kinin formation and thrombolysis are not the only reason why it should be mentioned in this Symposium. Donaldson et al. (1969) have reported that complement also forms a kinin-like peptide which appears to be largely responsible for the permeability effects in hereditary angio-oedema (HAO). This complement kinin raises the blood pressure and is inactivated by trypsin, properties which clearly distinguish it from bradykinin, kallidin and methionyl-kallidin.

Rothschild (1968) reported that cellulose sulphate (polymerized cellobiose sulphate) leads to exhaustive plasma kinin formation in the rat, probably as a result of activation of factor XII (Kellermeyer & Kellermeyer, 1969). We therefore examined whether cellulose sulphate reduces complement titres as well, and found that in vitro concentrations of 50-600 μg/ml did so very effectively. Indeed in human plasma, complement titres could be reduced by 80-90%, whilst kininogen fell only by 15-25% (Eisen & Loveday, 1970). Hexadimethrine bromide, an inhibitor of factor XII, prevented both effects, SBTI only the fall in plasma kininogen. The polymerized galactose sulphate carrageenin acted in a similar manner as the closely related cellulose sulphate.

In the rat, cellulose sulphate 6.5 mg/kg i.v. produced a significant fall in complement titres lasting 3-4 hours; in individual rats falls ranged from 17-65%. The kininogen depletion was rapid and rather more complete than in man. After the dose of 6.5 mg/kg i.v., rats were in good condition alert and lively.

Now the finding that cellulose sulphate (and carrageenin) reduced both kininogen and complement levels obviously bears upon the interpretation of reports that pre-treatment with these polysaccharides depressed inflammation produced by various means. The depression has been attributed either to the low kininogen (Rocha e Silva, 1970) or low complement (Willoughby et al., 1969). We produced mild inflammation of the rat paw by heat (45.5°C/min), or more severe injury of the ear by painting with xylene (Eisen & Loveday, 1971). Inflammatory oedema was expressed as the ratio $\frac{\text{fresh wet weight}}{\text{dry weight}}$ of the injured parts (Fig. 1). Both heat and xylene produced significant oedema. This oedema was not reduced in rats pre-treated with cellulose

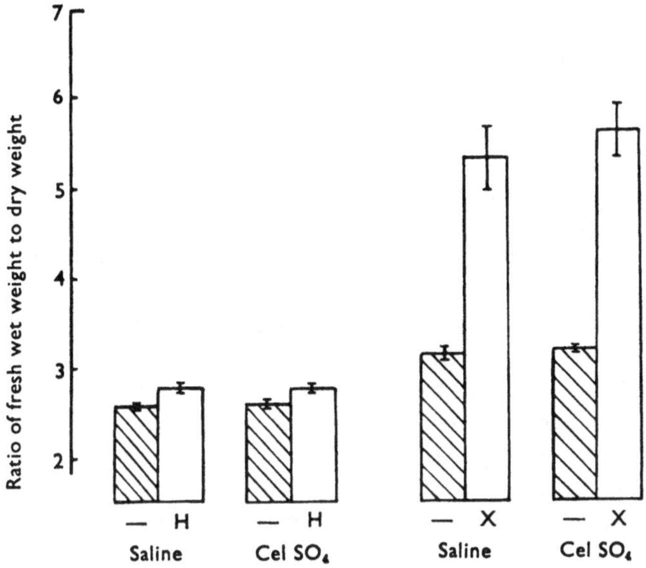

Fig. 1. Effect of cellulose sulphate (6.5 mg/kg i.v.) on inflammation. Oedema was produced in the rat paw by heating (H) (45.5°C/min) and on the ear by xylene (X) (xylene/20 min) at time 0. Cellulose sulphate was infused from -45 to -30 minutes. The clear columns show the oedema expressed as the ratio of fresh wet weight to dry weight of the injured parts. The shaded columns give the ratio of fresh wet weight to dry weight of uninjured parts. The vertical bars show S.E.M. The mild injury by heat produced significant oedema in the paws of depleted ($P < 0.001$) and of undepleted ($P < 0.001$) rats. The oedema produced by xylene was highly significant ($P < 0.001$). There was no significant difference between depleted and undepleted rats, either in the swelling of paws ($P > 0.1$) or of ears ($P > 0.5$).

sulphate 6.5 mg/kg i.v., although - as we have seen - plasma kininogen was reduced very drastically, and complement significantly. There was no correlation between complement titres and swelling, either in control rats or in depleted rats. Pretreatment with cellulose sulphate 6.5 mg/kg i.v. also failed to reduce the oedema produced by longer heating (45.5°C for 5, 10, 15 and 30 min).

Administration of cellulose sulphate 80 mg/kg i.p. over 3 days as used by Leme et al. (1967), did not deplete kininogen more completely than did 6.5 mg/kg i.v.; complement was only very slightly reduced. Inflammatory oedema was significantly depressed, but again there was no correlation between total complement titres and swelling of heated paws. This was also true of the swelling induced by xylene. In our experience, rats treated with cellulose sulphate 80 mg/kg i. p. were in a poor state of health. They had gained less weight than control rats. The blood clotting time was 20-60 min instead of the normal 1-2 min, and haemorrhagic lesions were found in skeletal muscles, lungs, kidneys, liver and intestine. Shock and toxic conditions may elicit reflex sympathetic vasoconstrictor activity which will reduce the blood flow in the skin and other tissues, and thus counteract the increased permeability of damaged blood vessels. Toxic conditions may also depress inflammation by other mechanisms, not yet fully understood (Seegmiller et al., 1962; Perillie et al., 1962).

Our results suggest that oedema in response to heat or chemical damage can fully develop inspite of depleted plasma kininogen and reduced complement.

The possible role of factor XII in the activation of the complement system was further examined by comparing the activation of plasma pre-kallikrein and of C'1s, and the acceleration of plasma clotting, which are produced by substances known to activate clotting factor XII (Fig. 2). Several related substances were also tested. Plasma kallikrein was assayed by its action on N-benzoyl-L-arginine ethyl ester (BAEe); to prevent the activation of the numerous other plasma proteases which hydrolyse BAEe, heparin (500 U/ml serum), trans-aminomethyl-cyclohexane carboxylic acid (AMCHA; 6 μmol/ml serum) and sodium edetate (3 μmol/ml serum) were added to the assay system. C'1 esterase is the only intrinsic plasma protease occurring in solution which hydrolyses N-acetyl-L-tyrosine ethyl ester (ATEe), and this substrate was used for assays. The hydrolysis of the esters was measured by continuous automatic titration (Eisen & Loveday, 1972). The 'kaolin clotting time' method (Margolis, 1958) was adapted for measuring the clot-promoting effect of the factor XII activated by the various materials tested. Experiments were carried out with normal sera and with sera from 2 patients with hereditary angio-oedema (HAO). These sera lack C'1 esterase inhibitor, an α_2 neuramino-glycoglobulin

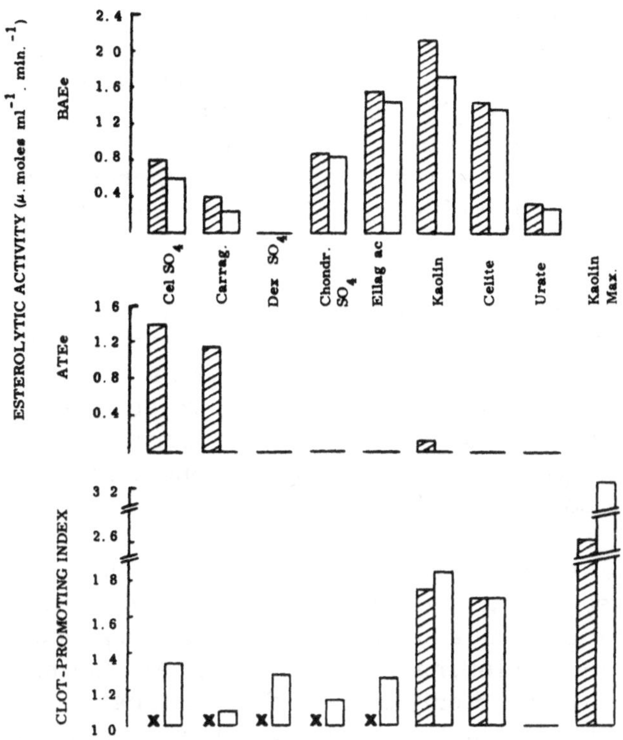

Fig. 2. Activation of esterolytic and clot-promoting activities. The basal hydrolysis of BAEe by plasma kallikrein and of ATEe by C'1 esterase was recorded by automatic titration for 2 min. One μmol of ellagic acid or 2 mg of the other activating materials, per ml of HAO serum (shaded columns) or normal serum (clear columns) were then added without interrupting the recording. The height of the columns shows induced minus spontaneous activity. Clotting tests were carried out with 400 times smaller concentrations of activating materials in plasma. Clot-promoting index = $\frac{\text{Clotting time without activating material}}{\text{Clotting time with activating material}}$. KM = maximum clot promotion by excess of kaolin (10 mg/ml). X= not tested.

which is essential for the inhibition of C'1 esterase and very important for the inhibition of permeability factor and plasma kallikrein (Levy & Lepow, 1959; Kagen & Becker, 1963). Clotting factors XII and XI and plasmin are also inhibited (Ratnoff et al., 1969; Forbes et al., 1970). The clot-promoting effect was measured mainly in normal plasma, because only limited amounts of HAO plasma were available. All assays were arranged so that the intrinsic plasma enzymes were not the limiting factors of the reactions. It was therefore possible to compare the relative efficacy of the various exogenous activating substances.

ATEe hydrolysis developed spontaneously, and could be enhanced by activating agents, only in HAO sera, and not in normal sera. BAEe was slowly hydrolysed both by normal and by HAO sera even before activation. The rates of this spontaneous BAEe splitting were usually higher in HAO sera. However, the enhancement of hydrolysis induced by any one activating substance was similar in both types of sera.

The most intense hydrolysis of ATEe was induced by the sulphated polysaccharides cellulose sulphate and carrageenin, which promoted clotting and BAEe hydrolysis only moderately. Potent activators of clotting factor XII, such as kaolin, celite or ellagic acid, induced only little activity against ATEe. This clearly differed from the hydrolysis of BAEe, which was most potently activated by those substances which also produced the greatest clot-promoting effect.

The time course of the induced esterolytic activities was also different. ATEe hydrolysis was enhanced within a few seconds of adding an activating agent, and remained constant after that. Increased BAEe hydrolysis developed after a lag period of 30 to 120 seconds, and the final rate was attained only after 3-4 minutes. Fig. 2 shows the rate of BAEe hydrolysis at 4 min after activation.

Further studies on the role of factor XII were carried out using the activation of C'1s by polysaccharides as well as the spontaneous generation of C'1 esterase (Donaldson, 1968b; Lundh et al., 1968). In the experimental conditions (37°C, pH 7.2, 50 μl of serum), detectable spontaneous hydrolysis of ATEe developed within 10-30 min, and attained maximum levels at 40-80 min. At room temperature the development was slower.

Previous removal of factor XII from HAO serum with celite (Nossel, 1964) slightly delayed the spontaneous development of C'1 esterase without reducing the final activity attained. Activation by cellulose sulphate or carrageenin was not affected.

The results suggest that clotting factor XII is not involved in the activation of C'1 esterase by polysaccharides, but may accelerate the spontaneous activation of this esterase.

Inhibition of C'1 esterase generation. The role of kinin-forming plasma enzymes in the generation of C'1 esterase was also examined with the aid of appropriate inhibitors. Hexadimethrine bromide, an inhibitor of the activation and activity of factor XII (Eisen, 1964; Ratnoff & Miles, 1964) inhibited the action of cellulose sulphate and carrageenin. The inhibitor may have acted by combining with the sulphate radicals of the two polysaccharides, rather than acting directly on factor XII. The spontaneous activation of C'1s was inhibited only by hexadimethrine concentrations higher than those effective against factor XII. Inhibitors of plasma kallikrein and plasmin (Soybean trypsin inhibitor 10mg/ml serum and Trasylol 5,000 U/ml serum) and of plasminogen activation (ε-amino caproic acid 10 μmol/ml serum or AMCHA 6 μmol/ml serum) did not suppress the activation of C'1s by the polysaccharides. Spontaneous activation was delayed, but the final rates of ATEe hydrolysis were not significantly reduced.

Our results suggest that factor XII, plasma kallikrein and plasmin may contribute to the spontaneous generation of C'1 esterase in HAO serum by activating small amounts of C'1s; the resulting C'1 esterase then autocatalytically activates the bulk of C1s. An autocatalytic phase was also suggested by the very pronounced S-shaped curve of the spontaneous C'1 esterase generation.

The present findings are compatible with the recent suggestion by Hadjiyannaki & Lachmann (1971) that in HAO plasma C'1 esterase may be generated autocatalytically, once all available C'1 esterase inhibitor is bound and digested by one or several of the serum proteases which it inhibits. The autocatalytic activation of C'1s is now being studied in more detail.

ACKNOWLEDGEMENTS

We are grateful to Professor C. A. Keele for valuable suggestions and comments, and to the Wellcome Trust for generous financial support.

REFERENCES

Austen, K. F. (1971). J. Hopkins Med. J., 128, 57

Donaldson, V. H. (1968a). J. exp. Med., 127, 411

Donaldson, V. H. (1968b). J. Appl. Physiol., 25, 416

Donaldson, V. H., Ratnoff, O. D., Dias Da Silva, W. & Rosen, F. S. (1969). J. Clin. Invest., 48, 642

Eisen, V. (1964). Brit. J. Pharmacol., 22, 87

Eisen, V. & Loveday, C. (1970). Brit. J. Pharmacol., 39 831

Eisen, V. & Loveday, C. (1971). Brit. J. Pharmacol., 42 383

Eisen, V. & Loveday, C. (1972). In press

Forbes, C. D., Pensky, J. & Ratnoff, O. D. (1970). J. Lab. Clin. Med., 76, 809

Hadjiyannaki, K. & Lachmann, P. J. (1971). Clinical Allergy, 1, 221

Kagen, L. J. & Becker, E. L. (1963). Fed. Proc., 22, 613

Kellermeyer, W. F. & Kellermeyer, R. W. (1969). Proc. Soc. exp. Biol. Med., 130, 1310

Leme, J. G., Schapoval, E. S. & Rocha E Silva, M. (1967). In: Int. Symp. on Vaso-active Polypeptides, ed. Rocha E Silva, M. & Rothschild, H. A., pp 213

Levy, L.R. & Lepow, I.H. (1959). Proc. Soc. exp. Biol., 101, 608

Lundh, B., Laurell, A., Wetterqvist, H., White, T. & Granerus, G. (1968). Clin. exp. Immunol., 3, 733

Margolis, J. (1958). J. Clin. Path., 11, 406

Nossel, H.L. (1964). The contact phase of blood coagulation. Blackwell, Oxford.

Perillie, P.E., Nolan, J.P. & Finch, S.C. (1962). J. Lab. clin. Med., 59, 1008

Ratnoff, O.D. & Miles, A.A. (1964). Brit. J. exp. Path., 45, 328

Ratnoff, O.D. & Naff, G.B. (1967). J. exp. Med., 125, 337

Ratnoff, O.D., Pensky, J., Ogston, D., Naff, G.B. (1969). J. exp. Med., 129, 315

Rocha E Silva, M. (1970). In: Bradykinin and Related Kinins, ed. Sicuteri, F., Rocha E Silva, M. & Back, N., pp 507, New York/London: Plenum Press.

Rothschild, M. (1968). Brit. J. Pharmacol., 33, 501

Seegmiller, J.E., Howell, R.R. & Malawista, S.E. (1962) J. Amer. Med.Assoc., 180, 469

THE ROLE OF PLASMIN IN THE ACTIVATION OF THE KININ SYSTEM*

Clement E. Burrowes, Henry Z. Movat and Mary J. Soltay

Division of Experimental Pathology and Institute of Immunology

University of Toronto, Toronto, Ontario, Canada

Some time ago plasmin was believed to be the main kinin-forming enzyme (1, 2). The studies of several investigators indicated that plasmin formed kinin from kininogen at a relatively slow rate (2, 3, 4, 5). However, when added to fresh plasma, plasmin induces rapid generation of kinin and this was interpreted to mean that plasmin activated prekallikrein and kallikrein induced the rapid formation of kinin (4, 6, 7, 8).

Our own studies were initiated to compare the kinin-forming activity of plasmin vs. kallikrein. In the course of these studies, when fresh normal and Factor XII-deficient plasmas were used as a source of kininogen, four times more kinin was released from normal plasma upon incubation for 1 minute. When incubated for 40 minutes, a two-fold increased was observed. This discrepancy disappeared when heated (61°C for 60 min.) plasmas were used for substrate (9). These observations lead to further studies with Factor XII (Hageman factor), since we had functionally pure preparations of this factor (10).

As described in detail elsewhere (9) plasminogen was first prepared by chromatography through Sephadex G-200 superfine, followed by chromatography on DEAE Sephadex, followed in turn by rechromatography on Sephadex G-200. More recently, it was prepared by affinity chromato-

---

* Supported by the Medical Research Council of Canada, the Canadian Arthritis and Rheumatism Society and the Ontario Heart Foundation.

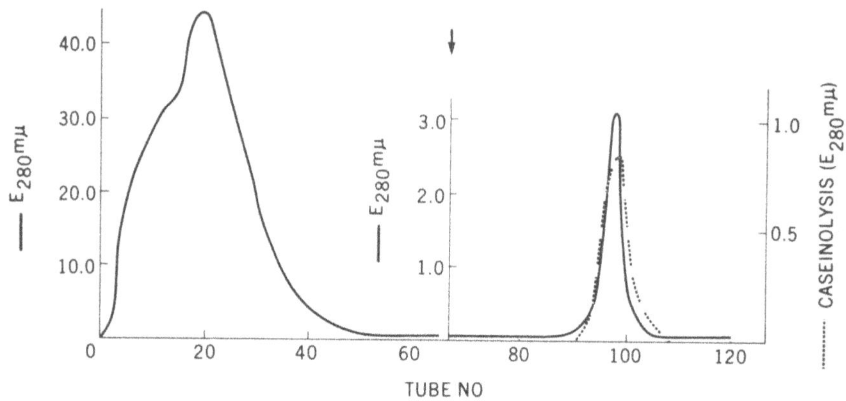

Fig. 1. A 2.5 x 30.0 cm column was packed with lysine-Sepharose (Miles Laboratories, Kankakee, Illinois) equilibrated with 0.1 M phosphate buffer, pH 7.4, containing 0.003 M EDTA. Two hundred ml of plasma containing 0.003 M EDTA was diluted to 375 ml with distilled water and passed through the column at a rate of 50.0 ml per hour. The column was then washed with 0.3 M phosphate buffer (pH 7.4) containing 0.003 M EDTA until the 280 mμ O.D. readings of the eluates were less than 0.02. The elution profile of the excluded protein is shown in the left half of the illustration. The plasminogen was then eluted with 0.3 M phosphate buffer containing 0.2 M epsilon aminocaproid acid and 0.003 M EDTA (right of the arrow). These procedures were done at room temperature (21-23°C). Subsequently, the plasminogen was freed of EACA by passing through Sephadex G-25 (0.005 M phosphate, pH 7.4) at 4°C and tested for caseinolysis, as shown in the right half of the illustration.

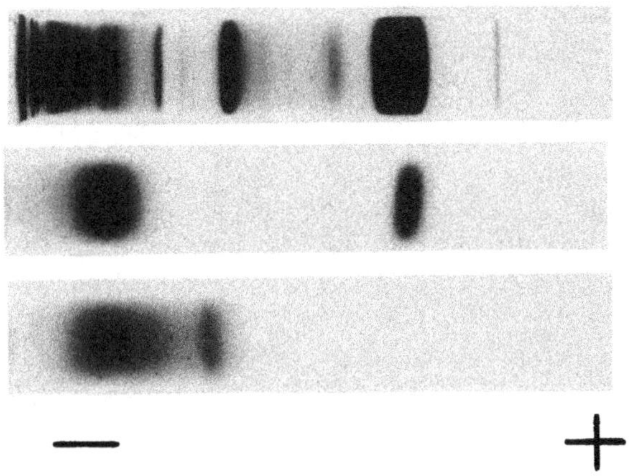

Fig. 2. Polyacrylamide electrophoresis of whole plasma (upper), of plasminogen obtained by affinity chromatography of whole plasma (middle) and of plasminogen obtained first by QAE Sephadex, followed by affinity chromatography (lower).

graphy (11) of whole plasma or of fractions eluting with 0.06 M NaCl in 0.1 M Tris HCl (pH 8.0) from a QAE Sephadex column (Figs. 1 and 2). Activated Hageman factor or Factor XIIa and prekallikrein were prepared as described in other publications (10, 12).

Plasmin (streptokinase activated plasminogen) added to prekallikrein caused no conversion to kallikrein, even with prolonged incubation. Factor XIIa by itself had likewise no effect on prekallikrein. However, a mixture of Factor XIIa and plasmin activated prekallikrein, which, as kallikrein, induced hydrolysis of BAEe and when added to kininogen liberated kinin. In these studies plasmin was first reacted with Factor XIIa and then the plasmin inhibited with lima bean trypsin inhibitor (Table 1). In our first experiments the Factor XIIa preparation had to be acidified to pH 3.0, followed by neutralization, to inhibit antiplasmin. Our more recent preparation of activated Hageman factor were free of this contaminant. In the reaction mixture of plasmin, Factor XIIa and prekallikrein, the conversion of the latter to kallikrein was a rapid process, reaching a peak within 2 minutes, as determined by BAEe hydrolysis. Plasmin liberated negligible amounts of kinin, but the reaction mixture of plasmin, Factor XIIa and prekallikrein liberated 240 ng of kinin from 1 ml kininogen (heated plasma) when incubated for 1 minute (9). Following 40 minutes incubation, some kinin was released from kininogen by plasmin, but in the presence of Factor XIIa and prekallikrein the amount of kinin release increased six-fold.

By polyacrylamide electrophoresis a prealbumin band was demonstrable in the reaction mixture of Factor XIIa and plasmin, identical to that obtained

TABLE 1. EFFECT OF PLASMIN ON ACTIVATED FACTOR XII

Reaction Mixture	Clotting of XII-Def. plasma (seconds)	Prekallikrein Activation (n-moles BAEe hydrol./min/ml)
Factor XIIa + plasmin + LBTI	262	19.5
Factor XIIa + LBTI	146	0
Plasmin + LBTI	-	7.5
Saline (veronal buffered, pH 7.35)	908	-

Table 1. One ml of a preparation of Factor XIIa was acidified (pH 3.0) with 0.5 ml 0.33 M HCl and reneutralized, after standing for 15 minutes, with 0.33 M NaOH, followed by dialysis against phosphate buffered (pH 7.4) or veronal buffered saline (pH 7.35). This treatment inhibited traces of antiplasmin in the Factor XIIa preparation. 0.3 ml of the treated Factor XIIa was mixed with plasmin, incubated for 10 minutes and the plasmin inactivated with 12 μg of lima bean trypsin inhibitor. In this process prekallikrein activator was released from Factor XII, which upon incubation with prekallikrein, converted the latter to kallikrein. Kallikrein induced hydrolysis of BAEe. For clotting assays the reaction mixtures were tested by the partial thromboplastin test (references 10 and 14).

from celite eluates (13, 14) and after trypsinization of Factor XIIa (10), presumably representing prekallikrein activator (15).

Concluding Remarks: The observations presented indicate that plasmin acts slowly on kininogen, but by acting on Factor XIIa, it releases prekallikrein activator. The latter rapidly converts prekallikrein to kallikrein and kallikrein in turn rapidly liberates kinin from kininogen. This mechanism seems to explain the observation that plasmin can liberate kinin rapidly when added to fresh plasma. Kallikrein seems to release more kinin from kininogen than does plasmin, but it still remains to be shown how much kinin can be released from a certain amount of kininogen by plasmin or kallikrein isolated from a given amount of plasma.

References

(1) Beraldo, W.T.: Formation of bradykinin in anaphylactic and peptone shock. Amer.J.Physiol., 171: 371, 1950.

(2) Lewis, G.P.: Formation of plasma kinins by plasmin. J.Physiol., 140: 285, 1958.

(3) Ratnoff, O.D.: Increased vascular permeability induced by human plasmin. J.Exp.Med., 122: 905, 1965.

(4) Back, N.: Fibrinolysin system and vasoactive kinins. Fed.Proc., 25: 77, 1966.

(5) Henriques, O.B., Lavras, A.A.C., Fichman, M., and Picarelli, Z.P.: Plasma enzymes that release kinins. Biochem.Pharmacol., 15: 31, 1966.

(6) Webster, M.E., and Pierce, J.V.: Studies on plasma kallikrein and its relationship to plasmin. J.Pharmacol.exp.Therap., 130: 484, 1960.

(7) Vogt, W.: Kinin formation by plasmin: An indirect process mediated by activation of kallikrein. J.Physiol., 170: 153, 1964.

(8) Seidel, G., Stücker, H.-U., and Vogt, W.: The role of plasmin for kinin-formation in human plasma. Naunyn-Schmiedebergs Arch. Pharmakol., 264: 305, 1969.

(9) Burrowes, C.E., Movat, H.Z., and Soltay, M.J.: The kinin system of human plasma. VI. The action of plasmin. Proc.Soc.Exp.Biol. Med., 139: Dec. 1971, in press.

(10) Soltay, M.J., Movat, H.Z., and Özge Anwar, A.H.: The kinin system of human plasma. V. The probable derivation of prekallikrein activator from activated Hageman factor (XIIa). Proc.Soc. Exp.Biol.Med., 139: Dec. 1971, in press.

(11) Deutsch, D.G., and Mertz, E.T.: Plasminogen: Purification from human plasma by affinity chromatography. Science, 170: 1095, 1970.

(12) Movat, H.Z., Soltay, M.J., Fuller, P.J., and Özge Anwar, A.H.: The relationship between the plasma kinin-system and the contact phase of blood-coagulation in man. In Vasoactive Polypeptides, Back, N., and Sicuteri, F., eds., Plenum Press, New York and London, in press.

(13) Movat, H.Z., and Özge Anwar, A.H.: The interrelationship between the kinin and blood-clotting systems. Fed. Proc., 30: 451, 1971.

(14) Özge Anwar, A.H., Movat, H.Z., and Scott, J.G.: The kinin system of human plasma. IV. The interrelationship between the contact phase of blood-coagulation and the plasma kinin system in man. Thromb. Diath. Haemorrh., in press.

(15) Kaplan, A.P., and Austen, K.F.: A prealbumin activator of prekallikrein. II. Derivation of activators of prekallikrein from active Hageman factor by digestion with plasmin. J. Exp. Med., 133: 696, 1971.

CLINICAL DETERMINATION OF THE KALLIKREIN IN HUMAN URINE

G. Porcelli, H. Croxatto, F. Porcelli

Istituto Chimica, Universita Cattolica, Roma, Italia
Fisiologia, Universidad Catolica, Santiago, Chile

Human urinary kallikrein, a potent vasoactive substance (Frey, 1925) has been estimated by comparying its hypotensive effect on dog's pressure (Frey et al., 1950), by its hydrolytic action on benzoyl-L-arginine ethyl ester (BAEE) (Trautschold and Werle, 1961), by its production of kinins after incubation with acid treated dog (Horton, 1959) or heated plasma (Miwa, 1965).

Of all these methods, the Trautschold and Werle technique appears to be destined for clinical application.

However, the Trautschold and Werle determination or urinary kallikrein is often impossible, as the urine extract has a high optical density for the spectrophotometrical measurements. In this case the AA. suggest use the alcohol dehydrogenase method or a fractionation of the extract.

In our laboratory a technique has been found to determine spectrophotometrically esterase activity of the human urinary kallikrein by a partial modification of the Trautschold and Werle method and by a standardized chromatographic procedure.

Methods

At toom temperature, add 125 ml acetone to 50 ml of a 24-hour specimen of urine, stored under a layer of toluene. Centrifuge off the supernatant. Take up the residue in 50 ml 5% HCl: Acetone= 1:4. Centrifuge off the soluble material. Take up the residue in 10 ml of water. Dialyse the clear supernatant over night in running water (centrigue off any precipitate).

The sample, introduced in a column (1x7 cm) of activated De-32, was eluted by 0,01 M tris-phosphate pH 4,1, containing 0,05 M NaCl (15 ml). The esterase activity finally was eluted by the same buffer containing 2MNaCl (3+3 ml).

The Hestrin method (1949), modified by Trautschold and Werle (1961) was adopted for the spectrophotometric measurements of the esterase activity, using N-Benzoyl-L-Arginine-ethyl-ester (BAEE). The method was checked with Padutin (Bayer, Leverkusen Germany) containing 0,1 K.U. per ml and a Zeiss spectrophotometer which enabled to measure hydrolysis at 253 mu in 15 minutes.

For a control, the kinin releasing activity was used, by incubation of aliquots of samples (0,01, 0,02, 0,05, 0,1 ml) mixed with 0.1 ml of Tyrode solution and 0,05 ml of dog serum (previously incubated for 88 hours at 37 C aseptically) for 2 minutes. Finally, the mixture was introduced in a bath where isolated rat uterus were placed. Tyrode solution, modified as described in G. Porcelli and H. Croxatto (1971), as bathing solution was employed. The oxytocic effect obtained with the samples were compared with that produced by a standard solution of pure bradykinin.

Results

The protein extract of 2 litres of human urine, prepared as described in "Methods", was separated on DE-32 column (1x7 cm) in parabolic NaCl gradient given by four cell solution (as indicated in note of fig. 1) communicating in series. The esterase and kininogenase activity was eluted in a single defined peak (fig.1).

Ovomucoid, even in doses of 1-5 mg per ml of the pooled fractions (66-72 fig. 1), does not inhibit the BAEE effect, whilst under similar condition 0,002 mg of Trasylol fully hindered the reaction (fig.2).

To demonstrate the esterase and kininogenase correspondence in each extract of human urine, prepared according to this method, few of these extracts were checked with the biochemical and biological test already described (G. Porcelli-H. Croxatto, 1971). These results are shown in Table 1.

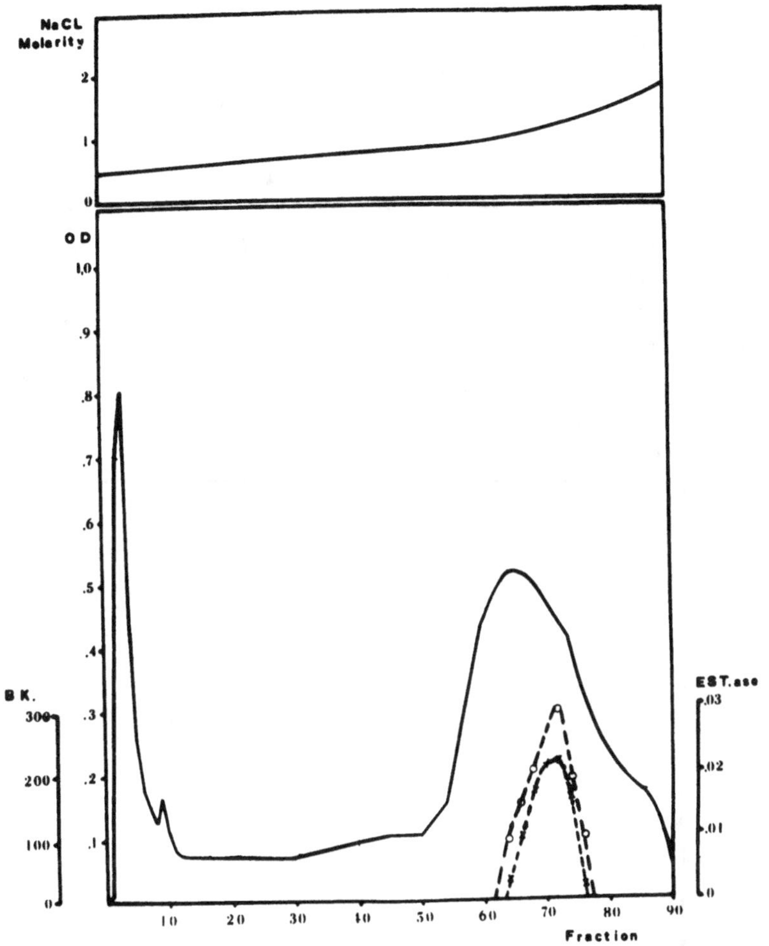

Fig. 1

DE-32 cellulose chromatography of 2 lt of human urine. Elution (20 ml/h; 5 fractions/h) was obtained using four-cell unit containing: cell no. 1 and cell no. 2 5 mM tris-phosphate, pH 4 and 0,05 M NaCl (100 ml); cell no. 3 same buffer and 0,1 M Nacl (100 ml); cell no. 4 same buffer and 2 M NaCl (100 ml).

Ordinate: (——) Optical density at 550 mu (Lowry, H. D., 1951); (x--x--x) BAEE esterase activity ($\triangle$ O.D.$_{253}$ mu/15') with 1 ml of each fraction; (o--o--o) Bk. ase activity in ng of Bradykinin effect of 1 ml of each fraction when incubated for 2 min. with (0,05 ml) dog serum.

TABLE I

Name	Volume of urine in 24 hour	K.U. in 50 ml of urine	Bk.ase activity in 50 ml of urine (ng of Bradykinin)	K.U. in total volume
Frontera	420 ml	0,115	930	0,963
Pantusa	900 ml	0,035	390	0,642
P.A.M.	780 ml	0,193	1770	3,050
D.S.G.II	960 ml	0,143	975	2,750
D.S.A.I	500 ml	0,150	930	1,500
Mariani V.	900 ml	0,043	360	0,772
Amatori L.	1200 ml	0,142	900	3,440

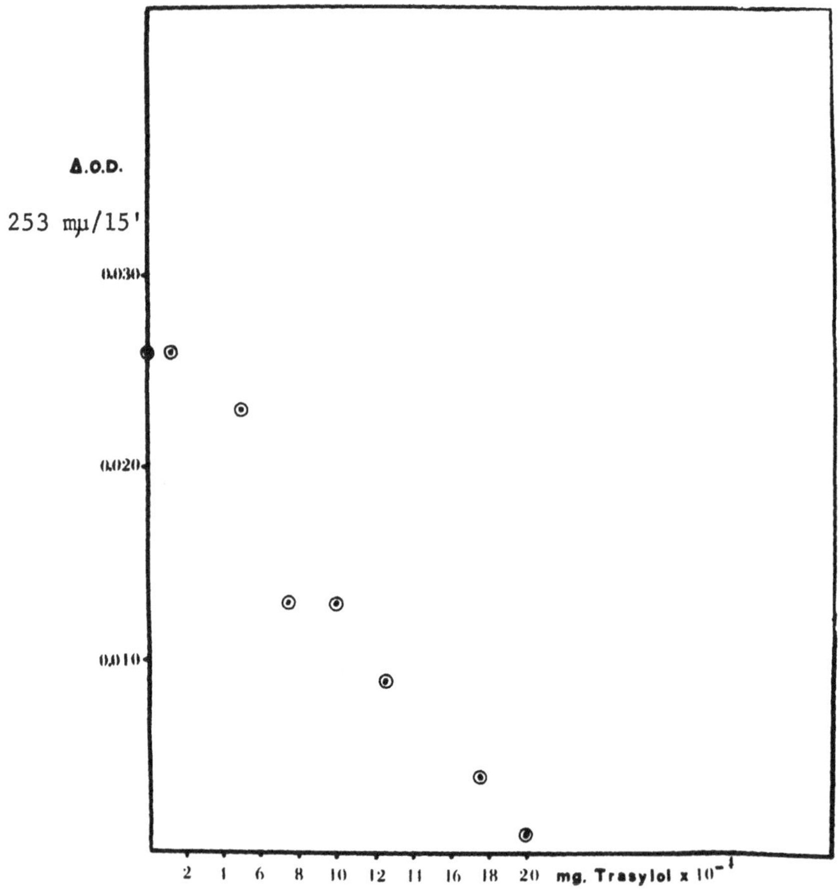

Figure 2 Inhibition of BAEE esterase activity with Trasylol per ml of the pooled fractions (66 - 72,fig. 1)

Discussion

Actually, the method here described appears to be feasible and reliable, and shows fair correlation with biological testing. It offers practical possibilities to study the variations of human urinary kallikrein under different physiological and physiopathological conditions.

However, we suppose that the results obtained with this method are subject to improvement in absolute values, controlling the influence of changes in incubation temperature and the kallikrein activators. Further, it would be convenient to use as standard a sample of pure human urinary kallikrein instead of PADUTIN. Efforts will be made in order to obtain pure material from human urine.

Acknowledgement: The AA. wishes to thank Prof. G. B. Marini-Bettolo for this helpful criticism and encouragement. The skillful assistance of Mario Di Jorio and Maria Ranieri is also acknowledged.

References

1) Frey E.E. and Kraut H. - J. Physiol. Chem., 32, 157 (1926)
2) Frey, E.K., Kraut H. and Werle E.-Kallikrein (Padutin) Ferdinand Enke, Stuttgart (1950)
3) Trautschold I. and Werle E. - Hoppe -Seylers Z. physiol. Chem. 325, 48 (1961)
4) Horton E.W. - J. Physiol., 267, 148 (1959)
5) Niea ,I. - Tohoku J. Exp. Med., 334, 86 (1965)
6) Hestrin ,S. - J. Biol. Chem., 180, 249 (1949)
7) Porcelli ,G., H. Croxatto, accepted for publication in J.I. of Biochemistry
8) Lowry ,H.D., Rosebrough ,N.J., Farr ,A.L. -J. Biol. Chem. 139, 265 (1951)

DISSOCIATION OF OEDEMA PROVOKING FACTOR OF AGKISTRODON PISCIVORUS VENOM FROM KININOGENASE

N. Bhargava, B.B. Vargaftig, C.J. de Vos
I.L. Bonta and T. Tijs
Research Laboratories of N.V. Organon, Oss, The Netherlands, Organon S.A. Eragny-sur-Epte, France and Department of Pharmacology, Medical Faculty Rotterdam, The Netherlands

The development of a method for measuring the hemorrhagic activity qualitatively as well as quantitatively by Bonta et al. (1) made it possible to characterise the hemorrhagic factors of some venoms. These factors were revealed to be basic polypeptides of less than 10,000 molecular weight. Since hemorrhages are known to accompany a variety of allergic inflammations and since it has also been found that oestriol succinate inhibits the hemorrhages (2) as well as the cobra venom induced rat paw oedema (3,9), it seemed worthwhile investigating whether some venom fractions which induce pulmonary hemorrhages are able to induce oedema. In addition some chemical and pharmacological properties of these fractions have been investigated.

MATERIALS AND METHODS

Venom of *Agkistrodon piscivorus* from Miami serpentarium; crystalline trypsin (Sigma Chemical Co); Sephadex G 50 (Pharmacia, Sweden;) Benzoyl-l-arginine ethyl ester (BAEE) from Serva Heidelberg; synthetic Bradykinin (Sandoz); kallikrein inhibitor (4-5) of bovine lung (Trasylol$^{(R)}$) from Bayer; Phenylbutazone (Geigy) and Cyproheptadine hydrochloride (Merck, Sharpe and Dohme (U.S.A.)).

Hemorrhagic activity was estimated on canine pulmonary surface by the method of Bonta et al. (1). Permeability effects were studied in groups of four male rats as described by Vargaftig et al. (6). Chromatography of snake venoms on Sephadex was performed as described in a previous paper by Bhargava et al. (7). The hydrolysis of BAEE was taken as a measure of esterase activity (8).

The quantitative estimation of kinins was performed on isolated uteri of oestrogen primed rats. The kinins were liberated by treatment with trypsin or snake venom fraction incubated with purified bovine plasma kininogen for 10 minutes at 37°C and the enzymatic reaction stopped by Trasylol. The extent of the oedema was measured and its inhibition was calculated by the procedure described by Bonta and de Vos (9).

RESULTS

The venom of A.piscivorus is rich in protease or esterase activity (10) which can be blocked by large amounts of Trasylol (see Fig. 1). The BAEE hydrolysing activity of one µg of crude venom was found to be equivalent to half a microgramme of crystalline trypsin or about 0.5 unit of kallikrein (11).

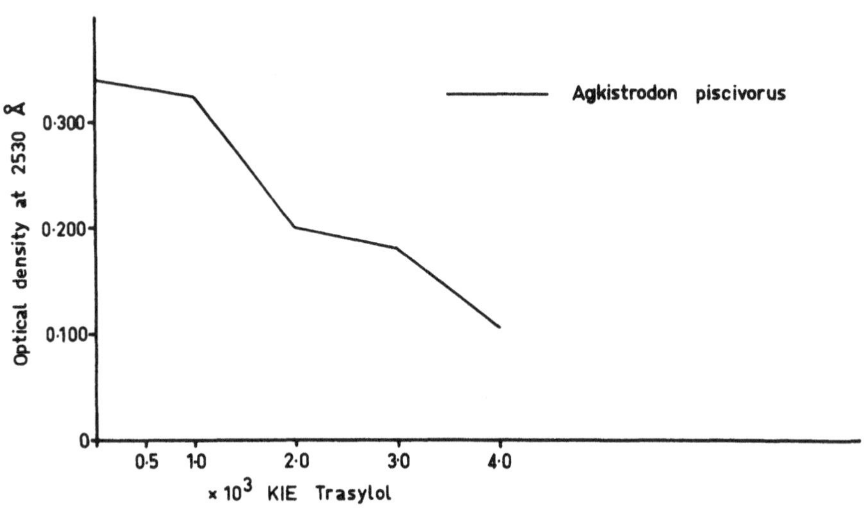

Fig. 1 The effect of Kallikrein Inhibitor Trasylol on esterase activity (BAEE) of 20 µg A. piscivorus.

If we consider that to inhibit one µg of crystalline trypsin or one unit of kallikrein, one kallikrein inhibitor unit (KIU) is needed (8) whereas to inhibit one µg of crude A.piscivorus venom at least 400 KIU are required (Fig. 1), then we may conclude that the kininogenase of A.piscivorus is not identical to pancreatic kallikrein or trypsin although the venom is definitely able to release kinins. Gel filtration of crude A.piscivorus venom on a Sephadex G 50 column yielded at least 4 main components which were designated G 50 I to G 50 IV in order of decreasing molecular weight. The major hemorrhagic activity as tested on canine pulmonary surface by the method of Bonta et al. (1) was found in fraction G 50 III, which contains polypeptides of molecular weight approximately 8,000. Our test showed that this most active hemorrhagic fraction was unable to hydrolyse either BAEE or purified kininogen. The fractions G 50 I and G 50 II contained most of the esterase activity (substrate: BAEE) and were also rich in kininogenase as shown by their ability to liberate kinins from kininogen of bovine plasma.

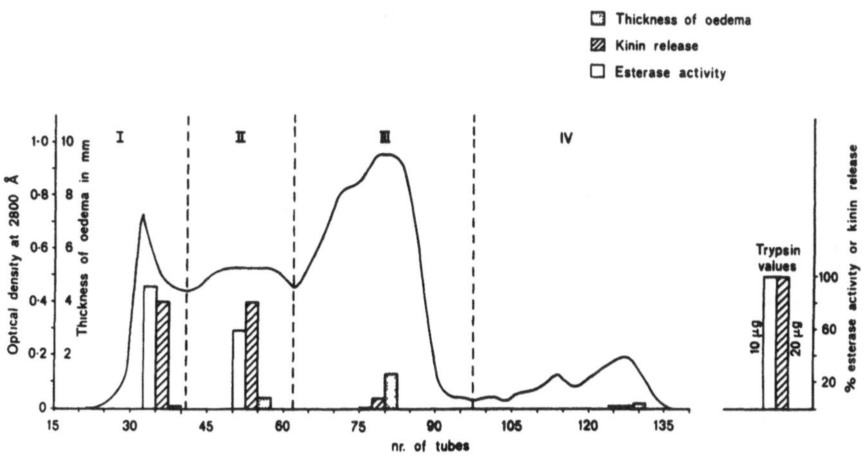

Fig. 2 Gel filtration of A. piscivorus venom on Sephadex G 50 column (2,5 x 200 cm) using 0.05 m acetic acid as eluant and determination of esterase, kininogenase and oedema provoking activity of the fractions.

Even when fraction G 50 III was free from kininogenase or esterase, it could still provoke oedema in the rat paw. Fraction G 50 II was also able to induce oedema, but was however not as active as fraction G 50 III in this respect (see Fig. 2).
Time course curves showed that the oedema produced by crude venom or the fraction G 50 III is of the rapid and transient type. It reaches a maximum after as little as half an hour and thereafter starts declining whereas the oedema provoked by kaolin reaches a maximum after 4 hours (see Fig. 3).
The oedema provoked by fraction G 50 III could be inhibited by Trasylol only after 4 hours. Phenylbutazone also inhibited the oedema after 4 hours whereas cyproheptadine required only half an hour (Table I).

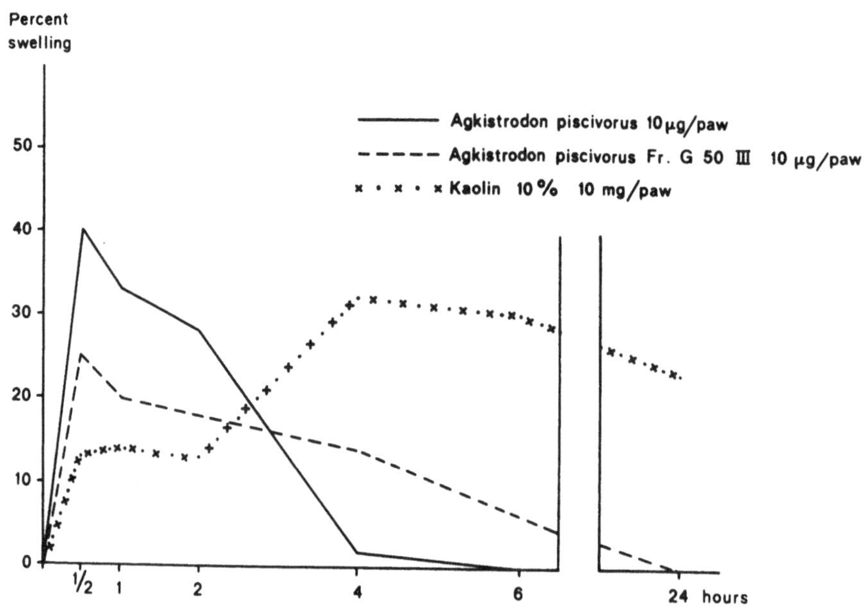

Fig. 3 Time course curves of rat paw oedema induced by crude A. piscivorus venom, fraction G 50 III and kaolin.

Table I

The effect of Trasylol, Cyproheptadine and Phenylbutazone on oedema provoked by G 50 III fraction of A.piscivorus venom.

Compound	Dose/paw	Inhibitory effect on rat paw oedema	
		After ½ hour	After 4 hours
Trasylol	1000 KIU	0	+
Cyproheptadine	0.1 mg	+	0
Phenylbutazone	5.0 mg	0	+

+ = Inhibition
0 = No Inhibition

Although electrophoresis of the fraction G 50 III showed more than one component, we have determined the amino acid composition of this fraction after 24 hours hydrolysis with 6 N hydrochloric acid. After dansylation with dansyl chloride and separation on polyamide layer chromatography, 16 different amino acids could be detected (Table II). Tryptophan could not be identified since it is destroyed by acid hydrolysis. No alkaline hydrolysis was performed for the determination. From its behaviour on Sephadex the fraction is expected to have a molecular weight of about 8,000.
The nitrogen content was 13.6%, a slightly higher than in the crude venom (12.7%) and the sugar (total reducing sugars) content of fraction G 50 III was 3.7%, also a little higher than in the crude venom (2.6%).

Table II

Chemical Analysis of G 50 III fraction of A. piscivorus venom

Qualitative Amino acid composition

Phe Arg Thr Ser Glu Asp Leu Ileu Val Pro Cys Gly
Met Ala Tyr

Molecular weight: approx. 8.000
Nitrogen content: 13.6% (Crude A.p. venom 12.7%)
Sugar content: 3.75% (Crude A.p. venom 2.6%)

DISCUSSION

According to Rocha e Silva et al. (12) the kininogenase of A.piscivorus venom is responsible for the oedema provoking activity. However the experiments described in this communication indicate that the kininogenase of this venom is not the only component responsible for inducing oedema in the rat paw.

We succeeded in separating a low molecular fraction, which proved more active in producing oedema, than the venom's kininogenase. Gel filtration revealed that the oedema provoking component has a molecular weight of 8,000, hence the absence of protease or esterase in this fraction was not unexpected.
Besides inducing oedema, the components of this fraction can increase the vascular permeability of rat skin and are also able to induce pulmonary hemorrhages. Since the fraction mentioned was not found to be homogeneous in an electric field, it is possible that each activity is due to a different factor.
According to Bonta and de Vos (13,14) the nature of the subplantar irritant employed determines the time course of the oedema and also the response to antagonistic drugs. For example serotonin, histamine and polyvinylpyrrolidone induce rapid transient oedema whereas a delayed long lasting oedema is produced by kaolin. Treatment of the rats with cyproheptadine inhibited serotonin induced oedema but did not significantly influence either polyvinylpyrrolidone or kaolin induced oedema. Phenylbutazone prevented the delayed oedema quite considerably but had no appreciable effect on rapid oedema. The fact that the early phase of oedema induced by the G 50 III fraction was inhibited by cyproheptadine is indicative of the release of serotonin or histamine.
Reduction of the oedema by Trasylol after 4 hours suggests that in the second phase kinins are involved. Possibly the administration of this fraction activates the endogenous zymogens of protease (i.e. prekallikrein, plasminogen etc.) and thus liberate kinins indirectly.
Inhibition of the oedema by phenylbutazone indicates some similarity with kaolin oedema, however, the time course curve in rat paw oedema induced by the venom was different. Thus our results correspond with those obtained by Willis (15) who showed that in carrageenin induced oedema in the first hour serotonin and histamine are liberated and that only after 2½ hours are kinins generated. After 6 hours prostaglandins are involved in this oedema (16).
Bonta and de Vos have also shown that cobra venom which is free of kininogenase is able to produce oedema in the rat paw (3,9). A corresponding G 50 III fraction of cobra venom

(which is also a peptide and possesses a molecular weight of 5 to 6000) is also capable of inducing oedema.
Thus we may conclude that the kininogenase of A.piscivorus is not the only factor in this venom responsible for producing oedema in the rat paw. This venom contains a peptide or most probably a glycopeptide which is more potent in inducing oedema than kininogenase.

SUMMARY

Protease or esterase of A.piscivorus venom is shown to be different from pancreatic kallikrein or trypsin. Since a higher dose of Trasylol is needed to inhibit the enzymatic activity of venom than that of kallikrein or trypsin. Kininogenase of the venom is not solely responsible for inducing oedema in the rat paw. A polypeptide of molecular weight of about 8,000 could be separated from this kininogenase and was found more potent in provoking oedema than kininogenase. The polypeptide provokes an oedema, which can be inhibited by Trasylol, cyproheptadine and phenylbutazone at different time intervals.

ACKNOWLEDGEMENTS

We thank Mr. Hamers and Mr. Omta for biochemical and Miss Rita From and Miss A.J.W. Boogaard for pharmacological technical assistance. Thanks are due to Mr. Bressers for photography work.

LITERATURE

(1) I.L. Bonta, B.B. Vargaftig, N. Bhargava and C.J. de Vos Toxicon 8, 3 (1970).

(2) I.L. Bonta and C.J. de Vos and A. Delver Acta Endocrinol. (Copenhagen) 48, 137 (1965).

(3) I.L. Bonta, Acta Physiol. Pharmacol. Neerl. 15, 188 (1969).

(4) F. Schultz, H. Kraut and N. Bhargava, Naturwissenschaften 50, 375 (1963).

(5) H. Kraut and N. Bhargava, Hoppe-Seyler's Z. physiol. Chem. 338, 231 (1964).

(6) B.B. Vargaftig (1971) to be published.

(7) N. Bhargava, P. Zirinis, I.L. Bonta and B.B. Vargaftig Biochem. Pharmac. 19, 2405 (1970).

(8) H. Kraut and N. Bhargava, Hoppe-Seyler's Z. physiol. Chem. 334, 236 (1963).

(9) I.L. Bonta and C.J. de Vos Acta Endocr., Copenh. 49, 403 (1965).

(10) I.L. Bonta, N. Bhargava and B.B. Vargaftig "Bradykinin and related kinins" Cardiovascular, Biochemical & Neural actions (Eds. F. Sicuteri and M. Rocha e Silva, Plenum Press 1970) pp. 191.

(11) I.L. Bonta, N. Bhargava and B.B. Vargaftig in: Proceedings of 2nd Intern. Symposium on Animal and Plant Toxins, Tel Aviv, eds. A. de Vries and E. Kochva (Gordon & Breach New York) in press.

(12) M. Rocha e Silva, R.Q. Cavalcanti and M.L. Reis Biochem. Pharmac. 18, 1285 (1969).

(13) I.L. Bonta and C.J. de Vos Proceedings of International Symposium on Vasoactive polypeptides: Bradykinin and Related Kinins, Sao Paulo 1966 pp. 205 (Eds. M. Rocha e Silva and H.A. Rothschild).

(14) I.L. Bonta and C.J. de Vos European J. Pharmacol. 1, 222 (1967).

(15) A.L. Willis J. Pharm. Pharmac. 21, 126 (1969).

(16) M.Di Rosa and D.A. Willoughby J. Pharm. Pharmac. 23, 297 (1971).

IDENTITY OF KININASE II WITH AN ANGIOTENSIN I CONVERTING ENZYME

R. Igic, K. Sorrells, T. Nakajima, and E. G. Erdös

Department of Pharmacology, University of Oklahoma

School of Medicine, Oklahoma City, Oklahoma

Blood and tissues contain various enzymes that hydrolyze bradykinin. Among the ones we characterized were a prolidase (imidopeptidase), a carboxypeptidase-type enzyme (carboxypeptidase N; kininase I) and a dipeptide hydrolase (kininase II; DH). Carboxypeptidase N inactivates kinins by removing the C-terminal arginine, while kininase II liberates the dipeptide Phe-Arg from the same end.

During these investigations the only other enzyme we knew that liberated C-terminal dipeptides, was the angiotensin I converting enzyme which released His^9-Leu^{10} from the decapeptide and converted it to angiotensin II.(Pertinent literature was summarized in ref. 1,2,3). When synthetic angiotensin I became available, we tested our plasma, lung and kidney kininase II preparations for angiotensin I converting activity. We found that the enzyme which inactivated the hypotensive peptide bradykinin also released the hypertensive angiotensin II from angiotensin I, thus it had a dual action.

The enzyme was found first in the plasma and in the kidney, but because the lung is very important in the metabolism of circulating bradykinin and angiotensin I, we purified DH from this organ as well.

Studies *In vitro*

Hog lungs were homogenized in 0.25 M sucrose and the particulate fractions were sedimented by differential centrifugations.

The final 100,000g supernatant was the source of DH. The enzyme was purified by the sequential use of gel filtration on a Sephadex G-200 column, Amicon membrane filtration, DEAE-Sephadex column chromatography and filtration on a Bio-Gel P 300 column. Column chromatography was followed by preparative disc gel electrophoresis. This yielded a single protein band indicating that the enzyme preparation was homogenous. The purified DH was tested by bio-assays and by chemical assays.

Even after purification DH retained both biological activities because it inactivated bradykinin and converted angiotensin I to angiotensin II. The former action was followed on the isolated rat uterus, the latter one on the isolated rat colon. Both actions of the enzyme were similarly inhibited by a variety of agents. The most active inhibitors were porcine insulin, the B-chain of insulin and two recently synthesized peptides, first found in snake venoms. These are the nonapeptide SQ 20881 (synthetized by M. Ondetti of Squibb Institute) and an undecapeptide, the potentiating Peptide C. They inhibited *in vitro* the hydrolysis of bradykinin and the conversion of angiotensin I in an about 10^{-6} M concentration. Glutathione, also inhibited the conversion of angiotensin I at a 10^{-4} M concentration.

The actions of SQ 20881 and Peptide C on DH seemed to be specific. When a human carboxypeptidase N (kininase I) preparation was used to inactivate bradykinin, neither Peptide C nor SQ 20881 blocked the cleavage of the Phe^8-Arg^9 bond in bradykinin at a 10 times higher concentration that inhibited DH (10^{-5} respectively 3×10^{-5} M).

DH cleaved also dipeptides of three optically active peptide substrates, as determined in the spectrophotometer. The hydrolysis of these short peptides was inhibited by compounds that in bio-assay blocked the cleavage of bradykinin and angiotensin I by DH. The inhibitors were also tested against concentrated guinea pig plasma DH that hydrolyzed the same three substrates as the lung enzyme. (Of all blood samples tested we found the highest DH activity in guinea pig plasma). The splitting of the substrates by guinea pig plasma or by swine lung DH was similarly inhibited by low concentrations of Peptide C and SQ 20881. The I_{50} values were between 3×10^{-7} M and 3×10^{-9} M, suggesting again that the lung and the plasma enzymes are quite similar or identical.

Because DH splits a variety of peptide dipeptide bonds, we prepared a radioactive, fluorescent substrate to be used to detect small quantities of the enzyme. The N-terminal end of triglycine was reacted with ^{14}C labeled dansyl (1-dimethylamino-naphthalene-5-sulfonyl; DNS) chloride. The resulting ^{14}C-DNS-Gly-Gly-Gly (0.6 nmole) was cleaved by DH to ^{14}C-DNS-Gly and to Gly-Gly.

The products were separated in thin-layer chromatography on silica coated glass microfiber sheets and were localized under the UV light. The Rf values of dansylated glycine derivatives on the chromatography sheets used were: 0.17 for DNS-Gly-Gly-Gly, 0.47 for DNS-Gly-Gly and 0.83 for DNS-Gly. The fluorescent spots were cut out from the chromatography sheets and the decrease in substrate concentration and the increase in labeled product (^{14}C-DNS-Gly) after hydrolysis by DH were measured in a liquid scintillation counter. ^{14}C-DNS-Gly-Gly-Gly was not hydrolyzed by DH in the presence of the inhibitor SQ 20881 (10^{-4} M).

DH in the Perfused Lung

The action of DH was also studied in the blood-free rat lung perfused *in situ* with Krebs solution. Radioactive (Asp^1-Ile^5-^{14}C-Leu^{10})-angiotensin I was added to the perfusion fluid (0.9 ug/ml) and the peptide and the products of its cleavage were collected after a single passage through the pulmonary circulation. The breakdown products and the labeled angiotensin I were separated in high voltage electrophoresis. The conversion of angiotensin I to angiotensin II by the lung was shown by the appearance of His-Leu-^{14}C and by the decrease in the concentration of labeled angiotensin I. After a single passage through the lung 42% of the recovered angiotensin I was hydrolyzed to angiotensin II and to His-Leu-^{14}C. Only 3% of the total count in the effluent was due to free Leu-C^{14} and 55 % to unhydrolyzed angiotensin I.

These experiments indicated that angiotensin I was converted in the lung by the removal of a dipeptide and not by the sequential release of leucine and histidine. Thus the conversion can be attributed to DH.

The liberation of angiotensin II from angiotensin I was inhibited by various compounds added to the perfusion fluid prior to the administration of the labeled peptide. Bradykinin was a remarkably potent inhibitor, probably because it is a competitive substrate of DH. Insulin and glutathione in the presence of dithioerythritol were also effective. The nonapeptide SQ 20881 inhibited DH, but the synthetic pentapeptide SQ 20475 had to be used in much higher concentration. Perfusion of the lung with EDTA partially blocked the conversion of angiotensin I.

^{14}C-DNS-Gly-Gly-Gly the radioactive and fluorescent substrate of DH was also perfused through the lung. As in the experiments *in vitro*, during the perfusion ^{14}DNS-Gly was cleaved from ^{14}C-DNS-Gly-Gly-Gly and no ^{14}C-DNS-Gly-Gly was liberated. Ten to 20% of the added dansyl peptide was hydrolyzed during a single passage through the rat lung, apparently by DH.

TABLE I.

Peptide Bonds Cleaved by DH

$-R_1-R_2-R_3-OH$	Source of enzyme: Lung	Source of enzyme: Plasma	Remark
$-Pro^7-Phe^8-Arg^9$	+	+	Bradykinin
$-Phe^8-His^9-Leu^{10}$	+	+	Angiotensin I
$-Pro^{28}-Lys^{29}-Ala^{30}$	+		B chain of insulin
Hip-Gly-Gly	+	+	
Hip-His-Leu	+	+	
Cbo-Pro-Leu-Gly		+	
_t_-BOC-Phe(NO_2)-Phe-Gly	+	+	
^{14}C-DNS-Gly-Gly-Gly	+		

Abbreviations: Hip = hippuryl
Cbo = carbobenzoxy group
_t_-BOC = _tert_.-butyloxycarbonyl

To establish the mode of action of DH on bradykinin *in situ* the protected tetrapeptide derivative of bradykinin was perfused through the rat lung. Ac-Ser-Pro-Phe-Arg was given at a rate of 0.4 micromoles per min. The advantage of using this peptide is that it has no biological action on the lung, thus it does not affect the perfusion pressure. Ten to 20% of the perfused peptide was cleaved during a single passage. DH released Phe-Arg dipeptide from this protected C-terminal tetrapeptide and no other major metabolite was detected either in thin-layer chromatography or in the amino acid analyzer. The liberation of the dipeptide was abolished when 100 ug/ml SQ 20881 was added to the perfusion fluid.

We conclude that the properties of DH coming from plasma, kidney or lung are quite similar. The enzyme purified from hog lung cleaved dipeptides *in vitro* from the C-terminal end of a variety of peptide substrates including angiotensin I, bradykinin, the B chain of insulin and ^{14}C-DNS-Gly-Gly-Gly (Table I).

In lung perfusion experiments Ang I was converted to Ang II by the release of His-Leu-^{14}C, Gly-Gly was liberated from ^{14}C-DNS-Gly-Gly-Gly and Phe-Arg was cleaved from a bradykinin derivative. Specific inhibitors blocked the inactivation of the hypotensive peptide and the release of the hypertensive one. The results indicate that DH has an important role in the lung in metabolizing active peptides by the hydrolysis of C-terminal peptidyl dipeptide bonds.

REFERENCES

1. Erdös, E.G., and Yang, H.Y.T. Kininases. In Bradykinin, Kallidin and Kallikrein. Handbook of Experim. Pharmacol., Vol. 25, ed. by Erdös, E.G., Heidelberg, Springer-Verlag, Pg. 289.

2. Yang, H.Y.T., Erdös, E.G., and Levin, Y. Characterization of a dipeptide hydrolase (kininase II; angiotensin I converting enzyme). J. Pharmacol. Exper. Therap. 117: 291, 1971.

3. Igic, R., Erdös, E.G., Yeh, H.S.Y., and Sorrells, K. The angiotensin I converting enzyme of the lung. Circul. Res. In press.

ACKNOWLEDGEMENTS

These studies were supported in part by grants HE 08764 and 5T01 HE 05859 from N.I.H., U.S.P.H.S. and by the O.N.R. N00014-68-A-0496 and N00014-69-A-0385. Dr. Igic is a Fulbright Fellow, supported by the Oklahoma Heart Association.

INHIBITION BY SULFHYDRYL REAGENTS OF THE EFFECTS OF BRADYKININ, ARACHIDONIC ACID AND "SLOW REACTING SUBSTANCE C"

B. B. Vargaftig and N. Dao Hai

Organon R & D, S. A. R. L., 60-Eragny-sur-Epte, France

Acidic antiinflammatory agents inhibit bronchoconstriction due to bradykinin (1), arachidonic acid (AA) (2), and Slow Reacting Substance C (SRS-C) (3), and suppress completely the hypotension due to SRS-C in the rabbit and in the dog (3, 4) as well as the secondary hypotension bradykinin evokes in the rabbit (5) and in the guinea-pig (6). It has been proposed that a new mediator, called "Rabbit Aorta Contracting Substance" (RCS), which is released by anaphylaxis, bradykinin, Slow Reacting Substance A (SRS-A) (7), SRS-C and AA (8), is involved in those effects of the mentioned agonists which are blocked by nonsteroidal antiinflammatory agents (7, 8); RCS release from the guinea-pig lung is blocked by phenylbutazone - like drugs.

We have investigated the interference of various thiol and antioxidant drugs with the effects of bradykinin, SRS-C and AA in the dog, the rabbit and the guinea-pig, and with the release of RCS from isolated guinea-pig lungs and in the circulation of dogs. It is demonstrated that the same agents block the effects of SRS-C and of AA in the three animal species and release of RCS from isolated guinea-pig lungs; moreover, inhibition by phenylbutazone of the hypotension induced by crude or partially purified SRS-C is accompanied by complete suppression of the formation of RCS in the circulation of the dog, giving further suport for the role of RCS in inducing those effects of SRS-C that are blocked by antiinflammatory drugs. The same antioxidants and -SH containing drugs inhibit those effects of bradykinin which have been adscribed to the formation of another mediator. Those results are paradoxical, because the sulfhydryl agents have been reported to potentiate the cardiovascular effects of bradykinin (9, 10).

MATERIAL AND METHODS

Conventional methods were used for recording dog and rabbit blood pressure on pentobarbitone anaesthetized animals. Bronchoconstriction was recorded by the Konzett and Rössler method (11), in guinea-pigs pretreated with the adrenergic _beta_ blocker propranolol, to avoid autoantagonism to bronchoconstriction by released catecholamines (12). After control responses were obtained to the various agonists, the putative antagonists were administered, and followed after five minutes, and thereafter at 15-30 minutes intervals, by the agonists. All injections were performed by the intravenous route. In certain experiments the putative antagonists were mixed with bradykinin, SRS-C or arachidonic acid, incubated at room temperature for 2 minutes and then injected to the animals. Concentrations of the antagonists were such that the final dose, in mg/kg, was at least tenfold lower than those doses which are necessary to block the _in vivo_ effects of the concerned agonists.

The effects of SRS-C upon the peripheral resistance were investigated in dogs, in which the distal end of the femoral artery was cannulated and perfused at constant flow by a Watson-Marlow pump with blood supplied by its proximal end. Paw perfusion pressure was measured from a T-fitting in the tubing between the pump and the artery. Drugs were injected into the paw by another T-fitting between the proximal end of the artery and the pump. Arterial blood pressure was measured from the carotid artery.

Release of RCS in anaesthetized dogs was detected the superfusion technique, (7, 8), in which blood is pumped from the animal onto a cascade of isolated organs, contractions being recorded via appropriate transducers on a Grass n° 7 polygraph. The isolated organs were : rat colon and stomach strip, hamster colon and rabbit aorta strip. The first three respond to prostaglandins E2 and F2α, whereas the last contracts to RCS and, as we report, to PGF2α. Blood was withdrawn from three sites : left and right heart, respectively through the carotid artery and the jugular vein, and from the paw, by collecting the outflow from the femoral vein and passing it upon the isolated organs. In this case arterial femoral flow was kept constant by use of the same procedure that allowed the observation of the peripheral resistance of the paw, as indicated above. Blood returned by gravity to the proximal end of the femoral vein, and arterial flow was fixed at 20-40 ml/minute. Before starting an experiment, and meanwhile the dog was prepared, the organs were mounted and tested with prostaglandins, noradrenaline and SRS-C under superfusion with Krebs solution. Crude SRS-C was an incubate of egg yolk emulsion with _Naja nigricollis_ venom, (Hans Hargte, Hambourg) as described (13). After hypotensive activity was released from the egg yolk lecithin by phospholipase A contained in the venom, the latter was neutralized with anti-venom serum (Institut Pasteur, Paris) to avoid interference of the venom with the test. Partially purified SRS-C was obtained as described by Vogt (14) through the courtesy of Dr P. Bruneau (Organon, France). Arachidonic acid (Nutritional Biochemicals Corp.); mepacrine dihydrochloride and mepyramine maleate (Rhone Poulenc); propranolol hydrochloride (Avlocardyl, Labs. Avlon, Paris); phenylbutazone (Geigy); 2-3 dimercaptopropanol (dimercaprol, Labs. Delalande, Paris); 2-mercaptoethanol and thioglycerol (Fluka); sodium diethyldithiocarbamate and dithiothreitol (Sigma); sodium dimethyldithiocarbamate (Pfaltz Bauer); sodium cromoglycate (Intal, courtesy of Fisons, United Kingdom), were the other reagents employed.

RESULTS

The hypotensive effects of crude SRS-C were blocked in dogs and rabbits, by various sulfhydryl and antioxidant agents. Effective antagonists in the dogs were the following, in decreasing order of potencies : pyrogallol, 2-mercaptoethanol, diethyldithiocarbamate and thioglycerol. Effective dose ranges for antagonism were between 0.5 and 12.5 mg/kg. A typical result obtained with 2-mercaptoethanol is in figure 1, from were it can be seen that the antihistaminic agent mepyramine did not inhibit the effect of SRS-C, whereas both 2-mercaptoethanol and phenylbutazone were active. At this dose level of the former responses to bradykinin were unaffected.

Dimercaprol had a varying effect, which depended upon dose and animal species. No intravenous injections were performed to dogs, due to its high toxicity, but a fixed injection of 40 mg/kg intraperitoneally evoked a partial to total blockade of the effect of SRS-C, meanwhile it potentiated the effects of bradykinin. In guinea-pigs, higher doses, by the intravenous route, depressed responses to SRS-C, to AA (figure 2) and also to bradykinin.

FIGURE 1 : BLOCKADE BY 2-MERCAPTOETHANOL AND BY PHENYLBUTAZONE OF THE HYPOTENSION DUE TO CRUDE SRS-C

Arterial blood pressure of a dog (female, 10 kg);
S = 0.6 ml/kg of crude SRS-C; M = 35 mg/kg of 2-mercaptoethanol;
Mp = 5 mg/kg of mepyramine; Ph = 10 mg/kg of phenylbutazone.
Unlabelled hypotensive responses are due to 2 ug/kg of bradykinin.
All injections were performed intravenously; Interval between injections of 15 minutes; Horizontal bar : time (four minutes); Vertical bar : blood pressure (cm Hg).

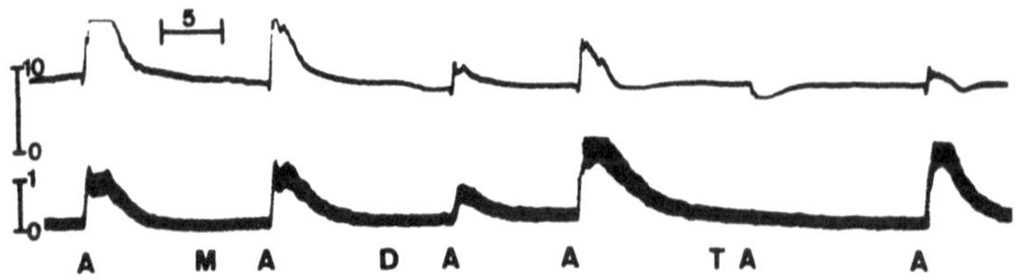

FIGURE 2 : INTERFERENCE OF DIMERCAPROL AND OF THIOGLYCEROL WITH BRONCHOCONSTRICTION AND HYPERTENSION DUE TO ARACHIDONIC ACID

Female guinea-pig (680 g);
A = 0.5 mg/kg of arachidonic acid; M = 5 mg/kg of morphine;
D = 5 mg/kg of dimercaprol; T = 100 mg/kg of thioglycerol;
All injections were performed intravenously; Upper : blood pressure (cm Hg); Lower : pulmonary resistance to inflation (cm Hg).

Mixing of sulfhydryl and antioxidant agents with SRS-C resulted in suppression of their hypotensive activity, when the concentrations of the antagonists were markedly lower than that would be effective *in vivo*, which confirms that antagonism is at least partly due to chemical inactivation. An example is provided in figure 3, in which diethyldithiocarbamate and thioglycerol were mixed with SRS-C, incubated at room temperature for 2 minutes, and then injected to a dog. Similar results were obtained with guinea-pigs, as *in vivo* administration or *in vitro* incubation of the antagonists with SRS-C and with AA resulted in suppression of the increase in pulmonary resistance to inflation. In figure 2 it may be seen that morphine did not interfere with the effects of arachidonic acid, whereas dimercaprol (5 mg/kg) partially blocked and thioglycerol (100 mg/kg) suppressed them completely.

The most active antagonist was pyrogallol, which suppressed the activity of SRS-C when mixed at concentrations down to 4 μg/ml; intravenous effective doses were in the 2-10 mg/kg range, for rabbits and guinea-pigs. A few ineffective agents are mentioned elsewhere (13), but it is noteworthy that the antiinflammatory antimalarial agent mepacrine, tested at 5 mg/kg in dogs, rabbits and guinea-pigs did not suppress the effects of bradykinin, SRS-C and AA.

Effects of bradykinin on the arterial pressure are known to be potentiated by sulfhydryl reagents (10, 11); nevertheless, it was

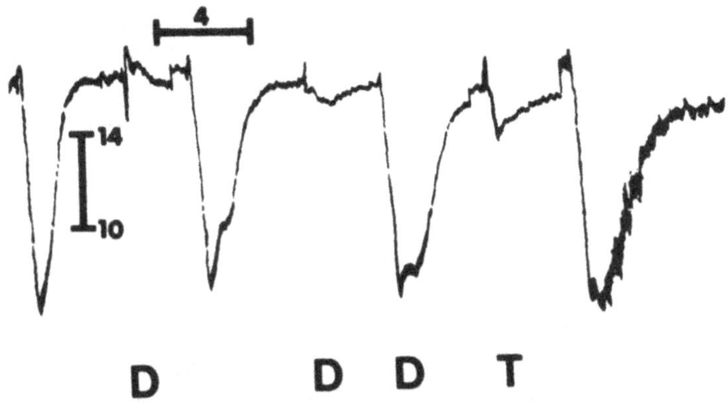

FIGURE 3 : SUPPRESSION OF THE HYPOTENSIVE ACTIVITY OF CRUDE SRS-C BY INCUBATION WITH DIETHYLDITHIOCARBAMATE AND THIOGLYCEROL

Arterial blood pressure of a male dog (17 kg). Unlabelled hypotensive responses are due to 0.2 ml/kg of crude SRS-C; the three responses labelled D are to the incubate of the same amount of crude SRS-C with respectively 34, 9.5 and 1.7 mg/ml of diethyldithiocarbamate, i.e. the equivalent to 10, 2.5 and 0.5 mg/kg, which is quite under the threshold effective intravenous dose. Response labelled T is to a similar incubate with 68 mg/ml of thioglycerol, i.e. 20 mg/kg. Horizontal bar : time (four minutes); Vertical bar : arterial blood pressure (cm Hg).

observed that the studied agents, at certain dose levels, curtailed markedly the effects of bradykinin on the arterial blood pressure of rabbits (figure 4); the same agents suppressed, apart from the bronchoconstrictor activity of SRS-C and of AA, that of bradykinin. This effect is seen in figure 5, where dithiothreitol, despite potentiating the hypotension induced by bradykinin in guinea-pigs, markedly diminushed the bronchoconstriction.

Injections of crude SRS-C into the femoral artery resulted in a marked and immediate decrease in the peripheral resistance, which was blocked by the intravenous administration of phenylbutazone (200-500 μg/kg) (figure 6). Intraarterial injections of PGE2 but not of PGF2α, elicited the same response as SRS-C, which was not blocked by phenylbutazone. Experiments with the superfusion technique indicated that crude and partially purified SRS-C, injected to the dogs at hypotensive doses (respectively 0.4 ml/kg and 5-10 mg/kg), elicited no noticeable change in the isolated organs, superfused with arterial or venous blood collected from the heart, except in one occasion when the injection of 10 mg/kg of partially purified SRS-C was followed by the contractions of the rat and rabbit organs.

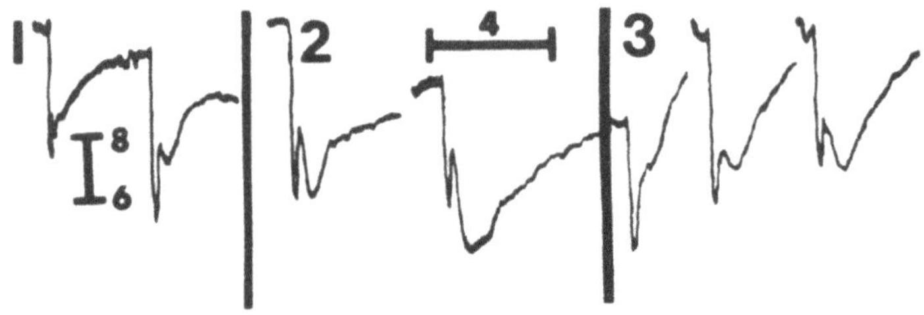

FIGURE 4 : SUPPRESSION BY THIOGLYCEROL OF THE SECONDARY HYPOTENSION DUE TO BRADYKININ ON THE RABBIT

Injection of 1 μg/kg of intravenous bradykinin to a 3.1 kg female rabbit. Panel 1 : before putative antagonists; Panel 2 : after 200 mg/kg of sodium dimethyldithiocarbamate; Panel 3 : after 200 mg/kg of thioglycerol. Interval between injection of fifteen minutes; observe potentiation of responses after dimethyldithiocarbamate and suppression of secondary hypotensive after thioglycerol, with return after thirty minutes.

FIGURE 5 : INHIBITION BY DITHIOTHREITOL OF BRONCHOCONSTRICTION DUE TO BRADYKININ

Upper to lower : recording of arterial blood pressure and of bronchial resistance to inflation; both in cm Hg. B = 5 μg/kg of bradykinin; C = 50 mg/kg of dithiothreitol. Observe marked reduction of bronchoconstriction, as compared to potentiation of effects upon the blood pressure. Intervals between injections of 15 minutes.

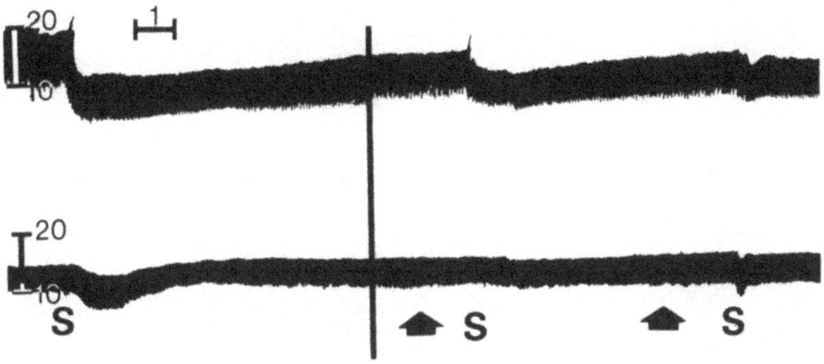

FIGURE 6 : EFFECTS OF CRUDE SRS-C UPON THE PERIPHERAL VASCULAR RESISTANCE

Upper to lower : resistance of a dog paw perfused at constant flow, and arterial blood pressure, both in cm Hg. S indicates intra arterial injection of 1 ml of crude SRS-C; the first and the second arrow indicate intra arterial injections of 100 and 500 μg of phenylbutazone, which reduced markedly the responses to SRS-C. Time in minutes.

In this case jugular blood was used. Very large contractions of the rabbit aorta, rat stomach and hamster colon were obtained when partially purified or crude SRS-C were injected directly into the tubing that transported venous blood from the distal end of the femoral vein to the organs (figure 7). Effective doses were of 1-5 mg of partially purified and of 0.2 ml of crude SRS-C. Injection of those doses into the paw, via the femoral artery were not followed by the appearance of contracting activity in the effluent blood, but a 2-10 fold increase in doses elicited similar responses. PGF2α at doses of 1-5 μg injected into the blood bathing the organs evoked marked contractions, including of the rabbit aorta, whereas PGE2 contracted the gastrointestinal strips, but never the vessels. Antiinflammatory agents (aspirin, indomethacin and phenylbutazone), administered at 2-5 mg/kg intravenously to the dogs, prevented the contractions of the organs after SRS-C administration, but not after PGF2α.

Release of RCS from isolated guinea-pig lungs was obtained after injections of AA, SRS-C and bradykinin; a ten minutes perfusion of the lungs with 2-mercaptoethanol, thioglycerol, dithiothreitol and pyrogallol (20-40 μg/ml), suppressed this release (figure 8). Mepacrine, although ineffective _in vivo_, blocked the release of RCS when evoked by bradykinin, but did not interfere with it when due to AA (20 μg/ml).

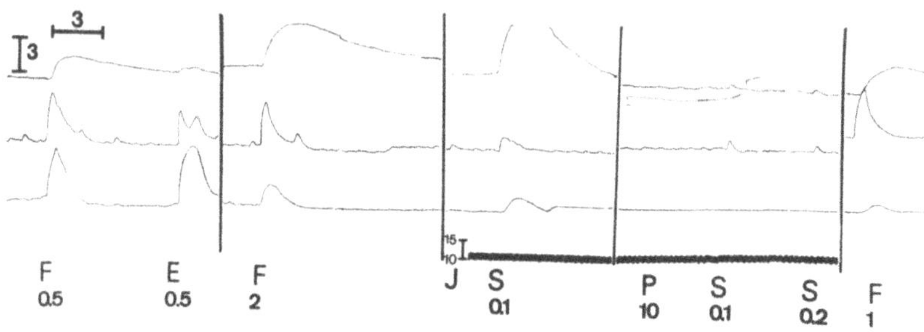

FIGURE 7 : RELEASE OF RCS BY CRUDE SRS-C AND ITS BLOCKADE BY PHENYLBUTAZONE

Upper to lower : contractions of spiral strip of rabbit aorta, of the rat colon and stomach strip and of the arterial blood pressure of a 23 kg dog. Injections were performed into the venous outflow superfusing the isolated organs, as follows : E and F, respectively prostaglandins E2 and F2 in µg; S, crude SRS-C in ml; phenylbutazone (P) was administered intravenously after control responses to the various agonists. Crude SRS-C released RCS from the blood, but this was blocked after phenylbutazone. J was control egg yolk superfused into the blood, which evoked no effect upon the isolated organs.

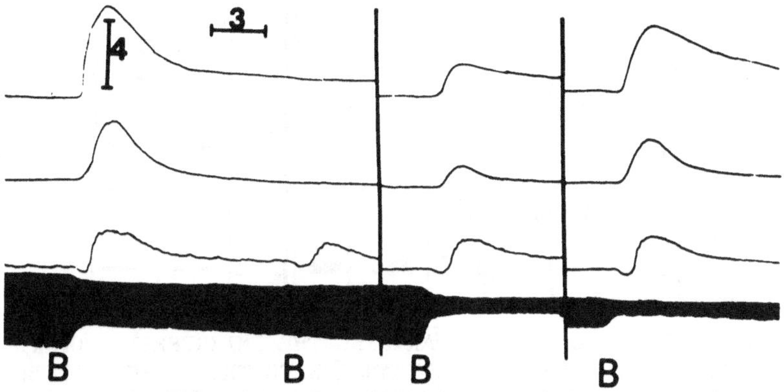

FIGURE 8 : PARTIAL INHIBITION OF THE RELEASE OF RCS FROM ISOLATED GUINEA-PIG LUNGS BY DITHIOTHREITOL

Upper to lower : contractions of the isolated rabbit pulmonary artery and aorta and of the rat stomach and intra-pulmonary pressure of an isolated guinea-pig lung. At B 10 μg of bradykinin were injected, as follows : first and fourth responses, injection into the lungs; second response, into the isolated organs, shortcuting the lungs; third response, into the lungs after a ten minutes perfusion with Krebs solution containing 40 μg/ml of dithiothreitol. Before test injections the isolated organs were perfused with dithiothreitol, at similar concentrations as those that were used for the lungs, to discard a direct effect of dithiothreitol upon the isolated organs. Horizontal scale : 3 minutes; Vertical scale : height of contraction (cm).

DISCUSSION

Figure 9 is a stylised diagram that summarizes our findings and from the literature, which demonstrates that phenylbutazone and mercaptoethanol, representative of the two series of antagonists we have investigated, suppress *in vivo* pulmonary effects due to bradykinin, to SRS-C and to arachidonic acid, curtail the hypotension due to bradykinin in the rabbit, and inhibit completely the hypotensive effects of SRS-C in dogs and rabbits. RCS release from guinea-pig lungs is also suppressed by the two antagonists, whether the releaser is bradykinin or arachidonic acid. Finally, formation in the dog's blood of a spasmogen identical to RCS, after SRS-C injections, is also suppressed by the acidic antiinflammatory drugs.

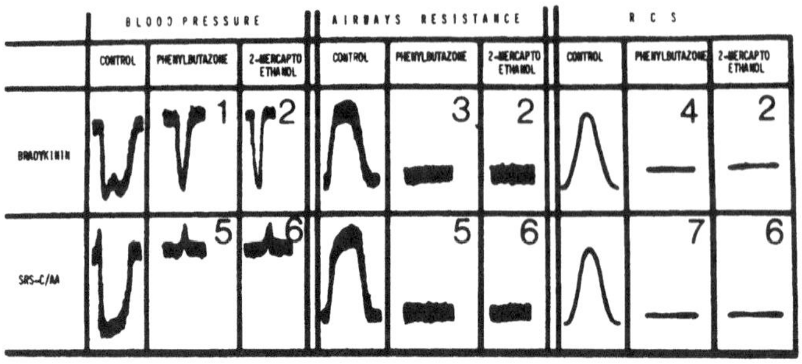

FIGURE 9 : STYLISED DIAGRAM SUMMARIZING THE INTERFERENCE OF PHENYLBUTAZONE AND MERCAPTOETHANOL WITH THE EFFECTS OF BRADYKININ, SRS-C AND ARACHIDONIC ACID UPON THE RABBIT BLOOD PRESSURE, THE GUINEA-PIG RESISTANCE TO INFLATION AND THE RELEASE OF RCS FROM THE ISOLATED GUINEA-PIG LUNGS.

1. Vargaftig, 1966; 2. Precise effect of thiol agents depends upon the dose; relatively low doses potentiate bradykinin; higher doses curtail the hypotension (Vargaftig and Dao Hai,ref.21); 3. Collier and Shorley, 1960; 4. Piper and Vane, 1969; 5. Vargaftig, Miranda and Lacoume, 1969; 6. Vargaftig and Dao Hai,ref.22; 7. Vargaftig and Dao Hai, 1971.

Those data suport the hypothesis that a common mediator is released by the various agonists which are similarly blocked by anti-inflammatory drugs (7, 8) and that this mediator might be a peroxide similar, if not identical, to the cyclic precursor involved in the synthesis of PGE2 and PGF2α (15). Interference of antiinflammatory agents with the synthesis of prostaglandins from arachidonic acid is now recognized (16).

The available biochemical data concerning the role of -SH agents in the synthesis of prostaglandins do not correlate directly with the inhibitory activity we now report for the formation of RCS, taken as a key intermediate in the biosynthetic process. Thiol agents may shift the in vitro synthesis from PGE2 to PGF2α, or alternatively, depending upon the precise conditions, but do not affect the final yield (17). These experiments may nevertheless not be directly pertinent to our results, the source of the enzyme, incubation procedures and other factors being different from our conditions.

Intraarterial injection of SRS-C into the dog paw was followed by a drop in resistance, demonstrating again that SRS-C acts peripherally, and not via the heart or the nervous system. This drop was blocked by phenylbutazone, which did not inhibit the drop due to

PGE2, or the contractions of the isolated aorta strip that followed injections of PGF2α into the blood bathing the isolated tissues. Effectiveness of PGF2α in contracting the aorta when injected into the blood might suggest that part of RCS activity is due to one of its metabolites, as the responses of gastrointestinal tissues to PGF2α decrease after contact with blood, whereas the responses of the aorta increase. Phenylbutazone does not block this activity of PGF2α showing that its point of impact is on an earlier stage, presumably by interfering with the cyclic peroxide precursor.

Recent yet unpublished results indicate that thiol agents may reduce anaphylactic shock evoked by ovoalbumin in guinea-pigs, if the effects due to histamine release are prevented by specific inhibitors. Similarly, RCS released during shock of isolated lungs is reduced, as are rat paw oedema due to carrageenin and U.V. erythema in guinea-pigs. Prostaglandins and lipid peroxides have been found in these experimental models of inflammation (18, 19), and activation of endogenous phospholipase A has been proposed as the common event leading to release of unsaturated fatty acids, taken up by the conversion system and transformed into prostaglandins, through the various unstable intermediates (4, 18, 20). Circunstancial evidence is accumulating and indicates that pharmacologically active lipids and peroxidized derivatives may be released in various physio-pathological conditions and play an important role in the final outcome of inflammation, shock, or anaphylactic phenomenae.

REFERENCES

1. COLLIER, H.O.J., and SHORLEY, F.G., Brit. J. Pharmacol., 15, 601, 1960.
2. BERRY, P.A., Ph. Thesis, London, 1966.
3. VARGAFTIG, B.B., MIRANDA, E.P., and LACOUME, B., Nature, 222, 883, 1969.
4. VARGAFTIG, B.B., and DAO HAI, N., J. Pharmacol. (Paris), 3,287, 1971.
5. VARGAFTIG, B.B., Experientia, 22, 182, 1966.
6. COLLIER, H.O.J., DINNEN, L.C., PERKINS, A.C., and PIPER, P.J., Arch. exp. Path. Pharmak., 269, 159, 1968.
7. PIPER, P.J., and VANE, J.R., Nature, 223, 29, 1969.
8. VARGAFTIG, B.B., and DAO HAI, N., Pharmacology, 6, 99, 1971.
9. FERREIRA, S.H., and ROCHA e SILVA, M., Biochem. Pharmac., 11, 1123, 1962.
10. ERDOS, E.G., and WOHLER, J.R., Biochem. Pharmac., 12, 1193, 1963.
11. KONZETT, H., and ROSSLER, R., Arch. exp. Path.,Pharmak., 195, 71, 1940.
12. COLLIER, H.O.J., in Hypotensive Peptides, Springer Verlag, 305, 1966.
13. VARGAFTIG, B.B., and COIRON, M., J. Pharmacol. (Paris), 2, 155, 1971.
14. VOGT, W., J. Physiol. (Lond.), 136, 131, 1957.
15. ANGGARD, E., and SAMUELSSON, B., J. Biol. Chem., 210, 3518, 1965.
16. VANE, J.R., Nature, 231, 235, 1971.

17. LANDS, W., LEE, R., and SMITH, W., in Prostaglandins, Ann. N. Y. Acad. Sci., 107, 1970.
18. WILLIS, A.L., Pharmacol. Res. Comm., 2, 297, 1970.
19. SØNDERGAARD, J., and GREAVES, M.W., J. Pathol., 101, 93, 1970.
20. KUNZE, J., and VOGT, W., in Prostaglandins, Ann. N.Y. Acad. Sci., 123, 1971.
21. VARGAFTIG, B.B., and DAO HAI, N., Experientia, 1972.
22. VARGAFTIG, B.B., and DAO HAI, N., Europ. J. Pharmacol., 1972.

ACKNOWLEDGEMENTS

We are grateful to Mrs. B. Charpentier and Misses C. Bellevrat and M. Douchin for an excellent technical assistance. This work and references 3, 4, 8, 13, 21 and 22 are part of a Thesis of Docteur-ès-Sciences Naturelles submitted by one of us (B.B.V.) to the University of Paris-Sud, Centre d'Orsay, in 1972.

STUDIES ON KININ-FORMING ENZYMES IN HUMAN PLASMA AND THEIR HETEROGENEITY

Hiroshi Moriya and Yoshio Hojima

The Laboratory of Physiological

Chemistry, Faculty of Pharmaceutical Sciences
Science University of Tokyo, Tokyo

Kallikrein-kinin system has been suggested widely related to pathological conditions. Some of its related pathological significance were reviewed by Kallermeyer and Graham (1), and by Webster (2). Although the activation process of kallikreinogen in human plasma has not been well established, Webster and Ratnoff (3) suggested that Hageman factor (HF) was necessary to convert human kallikreinogen into activated kallikrein. It was also suggested that HF would activate plasminogen (4) and permeability factor/dilute (PF/dil) (5) in the early stage followed by the successive activation of kallikreinogen by these enzymes and, on the contrary, human plasma kallikrein converted human plasminogen to plasmin (6). On the other hand, Nagasawa et al. observed that HF directly activated bovine kallikreinogen to kallikrein (7). Some reasons for these confused observations may be partly due to the heterogeneity of plasma kininogenases (2, 8-11) and/or the substrate specificities of these enzymes (12, 13). The clear illustration on plasma kininogenases and of intrinsic kinin system in plasma were not obtained yet and remain to be discussed including the problems of species difference and activation methods.

The authors described the partial purifications of human plasma kallikrein (HPK) activated with acetone and other plasma-kinin-forming enzymes (PKFE), and their relationships with activation methods in our previous papers (8, 14). The present paper deals with the further details of the heterogeneity of PKFE besides the purification of HPK.

Materials and Methods

Human Plasma — Human plasma was obtained from citrated blood (contained anti-coagulant solution in 20% v/v) (14), purchased from Tokyo Plasma Institute.

Animal Plasma — Bovine and dog plasma were obtained from 1% (w/v) citrated blood, and rat, guinea-pig and rabbit plasma from heparinized blood (final 50 units/ml).

Acetone Activation of Human Plasma — Acetone activation was done at pH 7.5 by standing for 3 hrs. at 25°C and 20% (v/v) acetone concentration (14). Acetone powder obtained at 75% (v/v) acetone was used for the HPK purification, and the powder at 50% was used for the preparations of various PKFE.

Salting-out Activation — This was referred to Jacobsen's report (15). Human or animal plasma was brought to 30% saturation with saturated ammonium sulfate solution, and after the separation of precipitate, the supernatant was raised to 45% saturation. The precipitate in this stage, collected by centrifugation, was dialysed against water for 10 hr. at less than 10°C. The dialysate was centrifuged to remove the precipitate formed during dialysis and then intactly used or lyophilized.

Casein Activation of Human Plasma — This activation was done in accordance with that of Habermann and Klett (16). Human plasma was mixed to 5 volumes of 0.5% (w/v) casein (Hammarsten casein, E. Merck AG.) solution dissolved in 0.005 N sodium hydroxide (pH 7.9, adjusted with acetin acid) at 20°C. The mixture was adjusted to pH 5.0 with 1 N acetic acid and stood for half an hour at 20°C. The precipitate formed was collected and dissolved in 0.01 M sodium bicarbonate (pH 8.0). To this solution solid ammonium sulfate was added and the precipitate between 27 - 45% saturation was dissolved in water. After that, the solution was adjusted to pH 4.7 and then the precipitate formed was removed. The supernatant was made to pH 8.0, dialysed against water and then lyophilized.

Purification of HPK — Early stages were carried out mainly according to the method previously described (14).

Step 1. Acetone powder, 43.4 g, prepared from 580 ml human plasma, was suspended in water at 5.2 mg/ml. The suspension was adjusted to pH 7.2, stirred for 1 hr., and centrifuged. The clear supernatant was obtained for the next step.

Step 2. Sixty three g of DEAE-cellulose (OH^- type, 0.94 mEq/g, Tohoku Pulp Co.) were added to the supernatant. The mixture was stirred for 40 min. at pH 6.0 adjusted with 5 N formic acid and 1 N ammonia, and afterward filtrated through a glass sintered funnel. The filtrate was used in step 3. Recovered DEAE-cellulose on the funnel was suspended into 1.2 liters of 0.1 M ammonium formate (pH 6.0), in order to check the activity adsorbed on DEAE-cellulose, and stirred for 20 min. at pH 6.4. After the separation of this suspension, DEAE-cellulose was successively treated with 0.4 M ammonium formate in the same way.

Step 3. The filtrate from Step 2 was adjusted to pH 7.3 and solid ammonium sulfate was gradually added to make 55% saturation. The precipitate collected was dissolved into 500 ml of water and then dialysed against water. The clear supernatant obtained from the dialysate was lyophilized and 4,880 mg of powder obtained.

Step 4. A part of the lyophilized powder, 1,010 mg, was dissolved in 20 ml 0.005 M phosphate buffer (pH 7.4) and dialysed against the buffer for 4 hr. After the clear supernatant of the dialysate was applied on a DEAE-Sephadex column (a - 50, OH^- type, 3.5 $\pm$ 0.5 mEq/g, 2.5 x 42 cm, Pharmacia) and pre-equilibrated with the same buffer, the column was washed with 30 ml of the buffer. The elution was subsequently carried out by the exponential gradient of 0.005 M to 0.2 M phosphate at pH 7.4 using 1,000 ml constant-volume mixing chamber. The flow rate was 20 ml/hr and each fraction of 5.1 ml was collected at 4°C.

Step 5. Lyophilized powder (contained salts), 360 mg out of 570 mg from the pooled active fractions in the above chromatography, was dissolved in 3 ml 0.005 M phosphate (pH 7.4) containing 0.15 M sodium chloride, and this sample was gel-filtrated through a column (1.5 x 75 cm) of Sephadex G - 150 (Pharmacia) with the same buffer. The flow rate was 15 ml/hr and each 4.3 ml was collected at 4°C.

Preparation of Various PKFE Activated with Acetone from Human Plasma — Acetone powder, 41.4 g prepared from 575 ml of human plasma, was suspended and extracted in 10 liters of water. Without the centrifugation in this case, 75 g DEAE-cellulose were added in this suspension, and this mixture was stirred for 40 min. at pH 6.0. Solid ammonium sulfate was added to the filtrate, which had been separated from DEAE-cellulose, to make up 60% saturation. The precipitate collected by centrifugation was dissolved in water, and after the dialysis, the clear supernatant was lyophilized. The obtained powder contained HPK (sample 1 in Table III).

Dr. C. A. Ryan, Washington State University) (24) and diisopropylfluorophosphate (DFP, Sigma) were employed as proteinase inhibitors. The concentration of enzyme (PKFE) in the pre-incubation mixture was 5 mg/ml and those of inhibitors were 1 mg/ml for LBTI and EWTI, 0.025 mg/ml for PCTI and 9.2×10^{-4} mg/ml (5×10^{-3} μmoles/ml) for DFP. For PKI, Trasylol and SBTI, the final concentrations were 0.001 - 0.05 mg/ml and their inhibitory potencies were calculated at less than 50 -60% inhibition. Both enzyme and inhibitor had been dissolved in 0.1 M Tris-HCl (pH 8.0). After pre-incubation (37°C, 30 min.), the sample was allowed to stand in an ice bath and the inhibition was determined by dog vasodilator assay, comparing the pre-incubated control in which the buffer was added instead of inhibitor.

Ampholine Isoelectric Focusing — The isoelectric focusing of enzymes was performed by a LKB 8101 Ampholine column (110 ml, LKB-Produkter AB), and according to the method of Vesterberg and Svensson (25). Ampholine-carrier ampholytes (LKB-Produkter AB) was used at 1% (w/v) final concentration. Enzyme sample was applied to the position in the column at which the enzyme would be expected to be focused from the preliminary experiment. Electrophoresis was carried out at constant 500 V for 30 - 40 hr under regulated temperature (10°C) by a Coolnics (CTR-LB) and its Circulator (CTE-1B) (Komatsu Solidate Co.) After electrophoresis, the column content was fractionated into 2 ml, pH of each fraction was measured at 10°C by a Hitachi-Horiba pH meter (F-5) and a Horiba 6028-10 T combined electrode, and then 1 ml or 2 ml M Tris-HCl (pH 8.0) was added into each fraction. Protein concentration at 280 mμ was determined at room temperature. The concentrations of protein and of enzyme activity in Figures have been expressed as those in original eluate.

Estimation of the Molecular Weights of Various PKFE — The molecular weight estimation of PKFE was carried out according to Andrew's method (26), using a Sephadex G - 200 column (Pharmacia, K 25/45). Solvent buffer was 0.05 M Tris-HCl (pH 7.5) containing 0.1 M sodium chloride and the column size was 2.5 x 43 cm. The filtration was carried out at 4°C and with the flow rate 15 ml/hr, and each 3.9 ml was fractionated. Each 1 ml of various PKFE (15 - 20 mg) and standard proteins (10 - 15 mg) dissolved in the buffer was separately applied on the column and gel-filtrated. As standard proteins, trypsin (2X cryst., Sigma), bovine serum albumin (BSA, Armour Pharm. Co.) and human γ-globulin (Fraction II, Tokyo Kasei Kogyo Co.) were used. The elution volume (Ve) of PKFE was determined by the dog vasodilator assay, and Ve of standard protein was estimated from the TAME esterolysis for trypsin (18) and the extinction at 280 mμ for BSA and human γ-globulin.

Results

Purification of HPK — The result of the purification up to Step 3 is summarized in Table 1.

Table I. Preparation of Crude HPK from Acetone Powder

Purification procedure	Protein recovery		Activity recovery		Specific activity	
	E_{280}/ml[a]	%	KU/ml[a]	%	KU/E_{280}	Purification
Step 1. Extract of acetone powder	46.1	100	7.72	100	0.17	1
Step 2. DEAE-cellulose filtrate	15.5	34	4.03	52	0.26	1.5
Eluates from DEAE-cellulose:						
0.1 M ammonium formate	14.3	31	0.52	7	0.04	0.2
0.4 M ammonium formate	10.3	22	0.46	6	0.05	0.3
Step 3. Salting-out, 0 - 55 % sat.	8.4[b]	18	3.36	44	0.40[c]	2.4

a) ml, equivalent to 1 ml of starting plasma;
b) 8.4 mg/ml;
c) 0.40 KU/mg.

The total activity in eluates with 0.1 M and 0.4 M ammonium formate (pH 6.0) from DEAE-cellulose was 13% against the starting extract (Table I). From this recovered activity, the authors judged that there were in the acetone powder of human plasma not only the so-called HPK but also the enzymes adsorbed on DEAE-cellulose (refer to Table II and III).

The gradient elution chromatography on a DEAE-Sephadex column of HPK (0.40 KU/mg, 1,010 mg, KU: Kallikrein unit) is shown in Fig. 1. The activity recovery after this chromatography was 78% of the starting samples. Fraction No. 62 to 83 were pooled and lyophilized, and 570 mg powder contained salts was obtained. Sephadex G - 150 gel-filtration used 360 mg (E_{280}-80) of this preparation is shown in Fig. 2.

On this gel-filtration (Fig. 2), the purity of starting sample was 1.2 KU/E_{280} and the activity recovery was 77%. The fraction No. 15 showed 12.7 KU/E_{280}, and this specific activity was the maximal through our investigations on HPK. But purified HPK was not stable in store for 1 month at -20°C and the trial to obtain further purified HPK was unsuccessful. However, the fraction with about 10 KU/E_{280} in Step 5, obtained in other experiment, showed homogeneous protein band, but slightly broad, in disc-electrophoresis (27). HPK seemed to migrate to γ-globulin position.

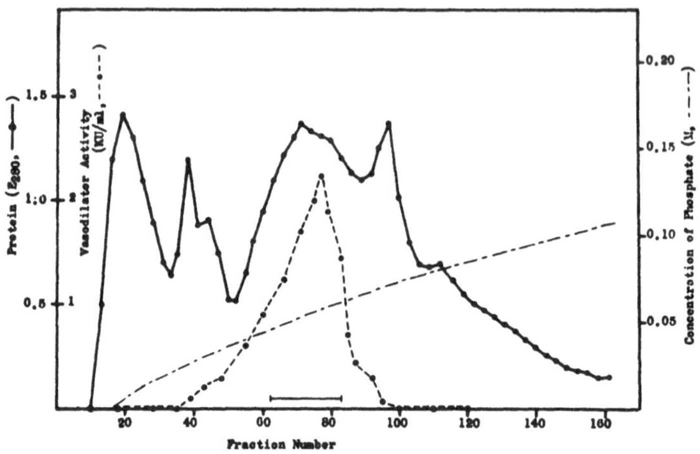

Fig. 1. Chromatography of HPK on DEAE-Sephadex (A - 50) Column (step 4).

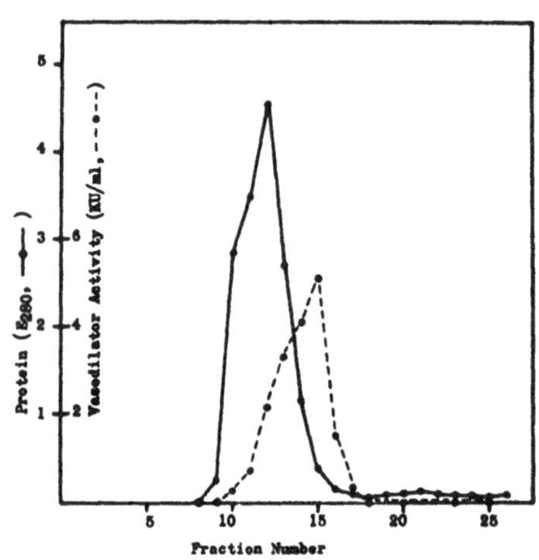

Fig. 2. Gel-Filtration of HPK on Sephadex G - 150 Column (step 5).

Preparations of Various PKFE — Recoveries of protein and of vasodilator activities and specific activities of various fractions from acetone powder are summarized in Table II.

More than 50% activity of the extract of acetone powder was found in eluates from DEAE-cellulose, while the activity of not-adsorbable enzyme (so-called HPK) to DEAE-cellulose was around 30%. Thus it appears that the total activity of acetone powder separated into three, roughly equal parts: HPK fraction, and two eluates with 0.1 M and 0.4 M ammonium formate. The salted-out (0 - 60% sat.), dialysed HPK solution in Table II was lyophilized and the powder of 2,640 mg was obtained as HPK. The eluate with 0.02 M ammonium formate produced 240 mg powder. Salted-out active fractions (0 - 45% sat. and 0 - 35% sat.) from 0.1 M eluate and 0.4 M eluate produced 250 mg and 1,950 mg, respectively. Table III shows the vasodilator activity, esterolytic activity, etc. of these samples (sample 1 - 4) and other PKFE preparations (sample 5 - 7).

The salting-out activated PKFE was found only in human (0.5 - 1 KU/ml original plasma) and rat (1 - 2 KU/ml) plasma pseudoglobulin fraction. This vasodilator activity was not detectable in bovine, dog, guinea-pig and rabbit plasma, in their euglobulin and albumin fractions as well. Back and Steger (28) and McConnel et al. (29) reported the inhibitory action of heparin to plasma kininogenase. Therefore, there would be the possibility of the presence of enzyme, although not detected in our investigation, which would be susceptible to heparin, existed in original plasma and/or used for dog assay.

Esterolysis and Fibrinolysis of Various PKFE — TAME and BAEE esterolytic activities of each PKFE were almost the same. The relation between the vasodilator activity and the esterolysis, namely, μmoles/min/KU, was not constant (Table LLL). Especially it appeared that so-called HPK was weak in esterolysis compared with vasodilator activity, and on the contrary, casein activated PKFE was much active in esterolysis. However, as some enzyme which is active to these synthetic substrates may exist in these crude samples, the conclusion about the relationship cannot be drawn.

On the fibrinolysis, sample 4 only showed this activity (Table III). This activity of this sample was eluated from a Sephadex G - 150 column (equilibrated and filtrated with 0.01 M phosphate buffer, pH 7.4, containing 0.15 M sodium chloride) ahead of the vasodilator activity. So, the fibrinolysis might be due to the contaminant (plasmin) in this sample.

Table II. Separation of HPK from Other PKFE

Procedure	Protein recovery E_{280}/ml [a]	%	Activity recovery KU/ml [a]	%	Specific activity KU/E_{280}	Purification
Extract of acetone powder	47.0	100	3.32	100	0.072	1
DEAE-cellulose filtrate	7.4	16	1.13	34	0.15	2.1
Salting-out, 0 - 60 % sat. (HPK fraction)[b]	5.6	12	0.87	26	0.27	3.8
Eluates from DEAE-cellulose:						
0.02 M ammonium formate	1.5	3	0.08	2	0.22	3.1
0.05 M ammonium formate	1.3	3	0.08	3	0.07	0.9
0.1 M ammonium formate	6.7	14	0.61	18	0.09	1.3
0.4 M ammonium formate	22.0	47	1.15	35	0.05	0.7
0.15 M sodium chloride	2.2	5	0.06	2	0.03	0.4

a) ml, equivalent to 1 ml of starting plasma;
b) dialyzed solution before lyophilization.

Table III. Some Enzymatic Activities and Molecular Weights of Human and Rat PKFE

Sample name	Sample	pI of PKFE in sample	KU/mg	TAME µmoles/min/mg	BAEE µmoles/min/mg	TAME/KU [a]	Fibrinolysis [b] cm^2/mg	Molecular weight
1	HPK, 0 - 60 % sat.	7.8	0.29	0.014	0.016	0.05	0	97,000
2	0.02 M eluate	7.8 > 6.9	0.19	0.041	0.030	0.22	0	-
	0.1 M eluate:							
3	0 - 45 % sat.	6.5, 6.9 > 7.8	0.30	0.031	0.028	0.10	0	74,000
	45 - 60 % sat.	-	0.02	0.002	0.002	0.10	0	-
	0.4 M eluate:							
4	0 - 35 % sat.	4.5	0.21	0.041	0.030	0.20	3.7	107,000
	35 - 45 % sat.	-	0.01	0.006	0.004	0.60	0	-
	45 - 60 % sat.	-	0.01	0.001	0.001	0.10	0	-
5	Salting-out activation	4.5	0.10	0.008	0.008	0.08	-	113,000
6	Casein activation	4.3, 4.6	0.28	0.128	0.087	0.46	-	111,000
7	Rat, salting-out act.	4.2, 5.3[c]	0.05	0.008	0.019	0.16	-	-

a) µmoles/min/KU;
b) digested area of heated fibrin plate;
c) described in our previous paper (8).

Effects of Inhibitors on PKFE — These PKFE were strongly inhibited by two PKI, Trasylol and SBTI (Table IV). Although Trasylol showed stronger inhibitory potency than PKI per mg, since molecular weights of two PKI are about 25,000 (22), PKI would be 2 - 3 times stronger than Trasylol per molarity. And also two PKI showed stronger activities than SBTI. In our previous investigations, PKI were weak inhibitors against glandular kallikreins (22, 23).

LBTI, EWTI and PCTI (a strong inhibitor of human salivary and human urinary kallikreins) (23) had hardly the inhibitory effect on these enzymes (Table IV). The salting-out activated enzyme of rat plasma (sample7) qas only slightly inhibited by LBTI and EWTI (both 0.4 KIU/mg, KIU: kallikrein inhibitor unit), but not by PCTI.

DFP inhibited these enzymes below 70% inhibition, but each value of KIU/mg DFP in Table IV would be lower than the value to be expected from the ratio of molecular weights of enzyme and DFP. This might be probably induced by the crudeness of sample. Various PKFE seemed to be the so-called serine enzyme because of DFP inhibition.

Table IV. Inhibitory Potencies of Proteinase Inhibitors on Various PKFE[a)]

Sample name	Sample	PKI pI 5.6	PKI pI 6.4	Trasylol[b)]	SBTI	LBTI EWTI PCTI	DFP
1	HPK, 0 - 60 % sat.	990	960	1490	420	0	490
3	(0.1 M eluate) 0 - 45 % sat.	550	360	840	250	0	740
4	(0.4 M eluate) 0 - 35 % sat.	620	370	1430	140	0	400
5	Salting-out activation	120	190	290	93	0	84
6	Casein activation	190	110	200	55	0	960
7	Rat, salting-out act.	84	72	86	42	SI[c)]	190

a) all values are shown with KIU/mg;
b) trasylol, commercially available, 1 KIU/0.145 µg;
c) slightly inhibited by LBTI and EWTI.

Ampholine Isoelectric Focusing of Various PKFE — The isoelectric fractionation of HPK is shown in Fig. 3. The active peak showed pH 7.80. In the previous paper (8), the isoelectric point (pI) of HPK was 7.77, 7.82 or 7.90. From these results, it was supposed that pI of HPK would be around pH 7.8. The authors, however, have not examined the small active peak in the fraction of pI 8.6 as yet.

The isoelectric fractionation of sample 2 with ampholytes in the pH range 5 - 10 was reported in the previous paper (8).

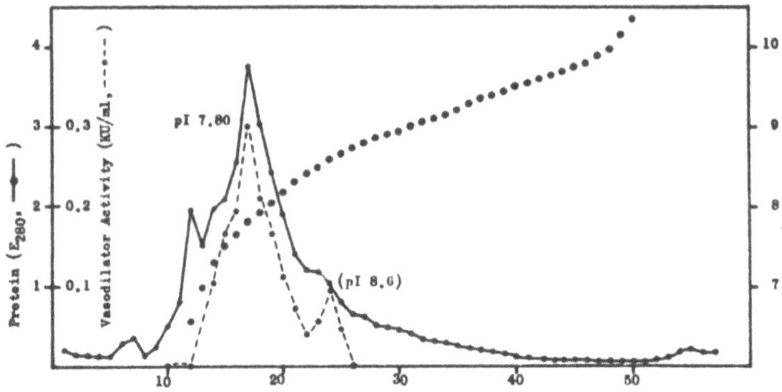

Fig. 3. Ampholine Isoelectric Focusing of HPK (PKFE pI 7.8). HPK (sample 1, 73 mg) was isoelectric-focused with ampholytes (pH 7 - 10) for 30 hr. The activity recovery after focusing was 39% against the starting activity.

Two peaks showing pI 7.8 and 6.9 were observed. Since the amount of the peak pI 7.8 was much more than that of pI 6.9, the manifestation of two PKFE in this sample has been expressed as pI 7.8 > pI 6.9 in Table III.

Fig. 4 shows the isoelectric fractionation of sample 3. In the previous paper (8), the activity was spread over fractions from pH 6 to 8. But, as shown in Fig. 4, two distinct peaks at pH 6.59 and pH 6.99 and a small peak at pH 7.77 were obtained. The activity of this small peak appeared to be about 10% of the

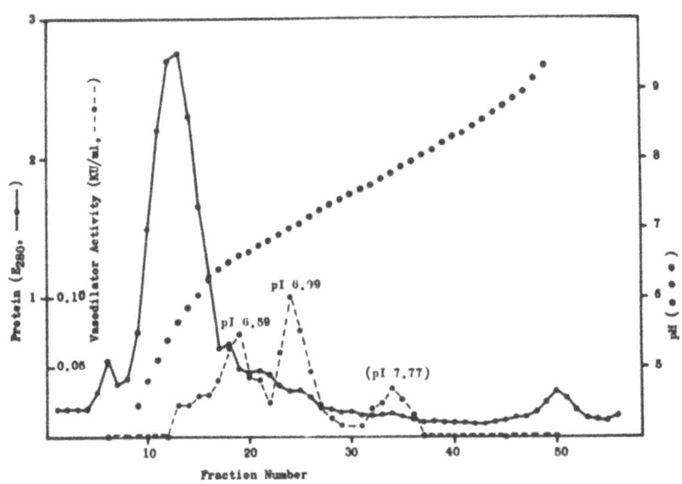

Fig. 4. Ampholine Isoelectric Focusing of Sample 3 (PKFE pI 6.9 and 6.5). Sample 3, 43 mg, was isoelectric-focused with ampholytes (pH 6 - 8) for 40 hr, and the activity recovery was 10%.

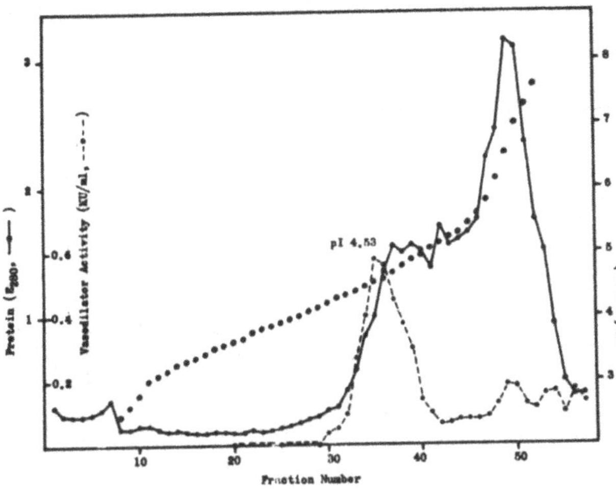

Fig. 5. Ampholine Isoelectric Focusing of Sample 4 (PKFE pI 4.5). Sample 4, 126 mg, was isoelectric-focused with ampholytes (pH 3 - 5) for 30 hr. and the activity recovery was 40%.

samples, so that the authors did not study this active component for further investigation with sample 3.

The isoelectric focusings of sample 4, 5 and 6 are shown in Fig. 5, 6 and 7. PKFE in these samples showed pI 4.53 (sample 4), 4.48 (sample 5) and 4.34 and 4.58 (sample 6), respectively. These PKFE seemed to be migrated like pre-albumin. In the previous paper (8) with ampholytes with pH gradient 3 to 10, active peaks of these samples had been located in pH 4.52 (sample 4), 4.52 (sample 5, with a small peak of pH 5.56), and 4.30 and 4.56 (sample 6), respectively. From these results, pI of main PKFE in these samples seemed to be around pH 4.5 (sample 4), 4.5 (sample 5), and 4.3 and 4.6 (sample 6) (Table III).

The pooled solution from fractions No. 32 - 42 (32 ml) in Fig. 6 was gel-filtrated through a Sephadex column (equilibrated with 0.02 M ammonium formate, pH 8.0, 2.5 x 43 cm). In this gel-filtration the separation of protein from ampholytes, sucrose and salts of Tris-HCl buffer was complete. The protein fraction was again isoelectrofocused with ampholytes of pH gradient 3 to 5, and only one active peak around pI 4.46 was obtained.

The authors cannot interpret why two PKFE have been detected only in sample 6 with casein activation, but it may happen that the precipitation of casein (?) which had been a little contaminated in the sample, gave two PKFE fractions during electrofocusing.

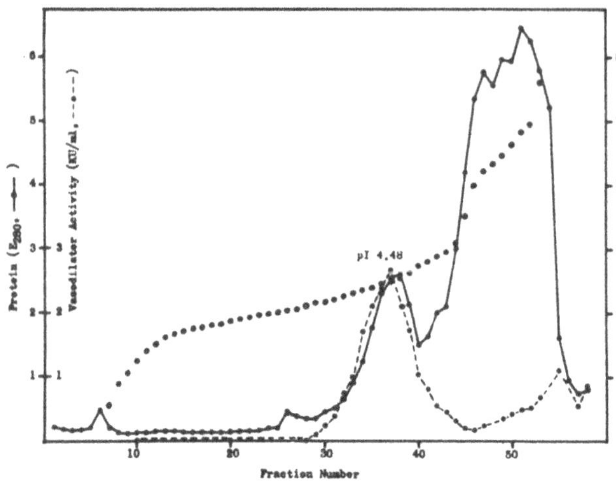

Fig. 6. Ampholine Isoelectric Focusing of Salting-out Activated PKFE (pI 4.5). Human pseudoglobulin fraction (8.2 ml, 9.0 KU/ml, 0.41 KU/E_{260}) was isoelectric-focused with ampholytes (pH 3 - 5) for 37 hr, and the activity recovery was 68%.

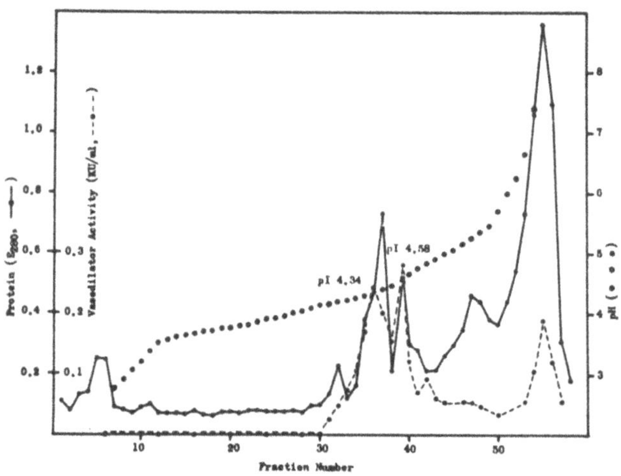

Fig. 7. Ampholine Isoelectric Focusing of Casein Activated PKFE (pI 4.3 and 4.6). Casein activated PKFE (43 mg, 0.13 KU/mg) was isoelectric-focused with ampholytes (pH 3 - 5) for 30 hr, and the activity recovery was 86%.

The active peaks in fractions No. 48 - 56 in Fig. 5, 6 and 7, which were not observed in previous experiments (8), were inexplicable, and these activities were inhibited in dog assay by some large doses of PKI, Trasylol and SBTI.

Molecular Weights of Various PKFE — The molecular weights of various PKFE by a Sephadex G-- 200 column were determined by the application of each elution volume (Ve) of PKFE on the standard slope obtained by the standard proteins (Fig. 8). The following molecular weights were estimated (summarized in Table III): 97,000 for HPK (pI 7.8), 74,000 for two PKFE (pI 6.9 and 6.5), 107,000 (PKFE pI 4.5), 113,000 for salting-out activated PKFE (pI 4.5) and 111,000 for casein activated PKFE (pI 4.3 and 4.6). In the determinations of Ve of PKFE contained in sample 3 and 6, only one active peak was obtained in each gel-filtration.

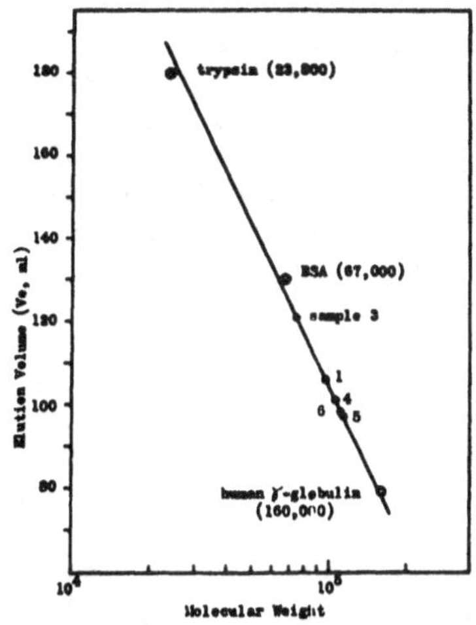

Fig. 8. Determination of Molecular Weights of Various PKFE (Sample 1 and 3 - 6) by Gel-Filtration on Sephadex G - 200 Column.

Discussion

The authors could obtain highly purified HPK (10 - 13 KU/E_{280}), but purified HPK seemed to be unstable even in the store at -20°C. Also, the desalted HPK was apt to be inactivated in lyophilization. For these reasons, the authors were unable to succeed in the preparation of lots of HPK with high KU/mg yet. Accordingly, the reliable KU or esterolytic activity per mg of HPK was still obscure. However, Pierce obtained HPK activated with acetone with 25 μmoles (TAME)/min/mg (personal communication) and Colman et al. reported human plasma kallikrein I with 5.57 μmoles (TAME)/min/mg from Cohn fraction IV (9). From animal serum, kallikreins activated by casein with 65 - 100 KU/mg (30) and 60 - 68 KU/mg (22.2 μmoles TAME/min/mg) (16) from hog serum and with 3.6 μmoles (TAME)/min/mg from bovine serum (10) were isolated. The authors have been investigating

HPK and congener enzymes, especially with regard to their arginineesterolysis. For example, the authors recently obtained purified HPK (10.6 KU/E_{280}) with 2.89 μmoles (TAME)/min/E_{280} (30°C) and 1.21 μmoles (BAEE)/min/E_{280} (25°C).

In the previous paper (8), the authors showed by Ampholine isoelectric focusing that the different PKFE had been activated from human plasma depending on the activation methods, namely, four PKFE (pI 7.8, 6.9, 6.5 and 4.5) from the acetone activation, one PKFE (pI 4.5) from the salting-out activation, and probably two PKFE (pI 4.3 and 4.6) from the casein activation. As shown in this paper, also the same corresponded pI of various PKFE was obtained. But these PKFE appeared to have some different properties, comparing with the results obtained by other investigators on plasma kallikrein, as discussed in the following.

Webster reported five kinin-liberating enzymes from human plasma activated with acetone from the result of DEAE-cellulose chromatography. She named these enzymes, enzyme I (so-called HPK), II (PF/dil), III, IV (HF) and V. In our previous paper, the authors showed four active components on DEAE-cellulose chromatography and also four in Ampholine isoelectrofocusing fractionation (8), and four pI of these PKFE have been distinctly detected in this paper. This discrepancy has been inexplicable.

The authors suppose that our purified HPK (PKFE pI 7.8) would be identical with Webster's enzyme I, kallikrein(ogen) isolated by McConnell and Mason (31) by DEAE-cellulose column, and/or Fraction A enzyme purified from antigen-antibody activated human serum reported by Movat et al. (γ-β globulin, M. W., 100,000) (11). Besides, Colman et al. (9) purified three human plasma kallikreins from Cohn fraction IV using kaolin activation. They called them from the eluting order by DEAE-cellulose chromatography, kallikrein I (slow γ-globulin, M. W., 99,800), II (fast γ-globulin, M. W., 163,000), and III (α-globulin, M. W., 124,000). Kallikrein I would seem to be an identical enzyme with our HPK, from its disc-electrophoretical migration and molecular weight.

Since the molecular weights (74,000) of PKFE pI 6.9 and 6.5 were generally smaller than other PKFE, there would be the possibility that either of these two PKFE was PF/dil or HF. According to Becker and Kagen (32), the sedimentation coefficients of human serum kallikrein (γ-globulin) and of PF/dil (β-globulin) were 10.1 - 12.2 and 5.2 - 6.6 respectively, and human HF showed 4.5 - 7 S in the inactive form (33), 20,000 - 100,000 (34) and 30,000 - 40,000 (35) as its molecular weight. Moreover, Movat et al. reported the existence of a low molecular activator (M. W., 45,000) of kinin system (11).

Jacobsen used human pseudoglobulin fraction (33 - 46% sat. by ammonium sulfate) as plasma kallikrein (15). But, the enzyme in this fraction showed pI 4.5 in the isoelectric focusing (Fig. 6). Therefore, human plasma kallikrein used by him was not identical with so-called HPK (pI 7.8). It would be proper to regard his plasma kallikrein as the similar one to acetone activated PKFE (pI 4.5) or casein activated PKFE (pI 4.3 and 4.6).

By the method similar to the salting-out activation, Armstrong and Miles found a kinin-forming substance (KFS) in human euglobulin (0 - 33% sat. by ammonium sulfate), which released the kinin not identical with bradykinin from human plasma (36). They observed an unaccountable behaviour of KFS in membrane electrophoresis. Namely, KFS in their last purification step changed its electrophoretical property like albumin into that like γ-globulin. The authors also observed the similar phenomena of acetone activated PKFE (pI 4.5) and of casein activated PKFE on their DEAE-cellulose chromatographies performed for purification. These enzymes did not adsorb on DEAE-cellulose equilibrated with buffer (0.05 M Tris-HCl, pH 8.0) which might enable these enzymes to adsorb to cellulose. And, also, casein activated PKFE in the crude preparation was eluted at the nearly equal salt concentration as in the case of so-called KPK by DEAE-Sephadex (A - 50) chromatography. So that it seemed that these enzymes had shifted the acidic charge to the basic. Would these observations be due to binding of enzyme to a carrier protein, for example, to macroglobulin? The authors have not been able to explain this phenomenon thoroughly.

Habermann and Klett reported hog serum kallikrein activated with casein migrated to between β - and γ-globulin in electrophoresis and had 97,000 as its molecular weight (16). Yano et al. purified this enzyme from bovine serum which migrated to the position of β_1 -globulin (the position guessed by the authors), and its molecular weight was 95,000 (10). Since our prepared human plasma kallikrein activated with casein showed pI 4.3 and 4.6, the electrophoretical properties of these enzymes have been quite different from them. It has been obscure that this difference might be occurred from the species difference from which plasma enzymes were prepared.

As above mentioned, there would be some differences, mainly in the electrophoretical migration, between our various PKFE and plasma kallikreins investigated by others. But, with the relation between these enzymes and the inhibitors, we obtained the similar results to other investigators. (Table IV). Our human PKFE were inhibited by two PKI, Trasylol, SBTI and DFP (9, 10, 16, 36), but not by LBTI, EWTI (10), and PCTI.

From these studies on human plasma kininogenases, the authors would especially like to take notice on the following points: (1) formations of different PKFE depending upon the activation methods, (2) the same PKFE (pI 4.3 - 4.6) found in three methods, (3) alteration of some behaviours of these PKFE (pI 4.3 - 4.6) against DEAE-ionexchangers, (4) strong inhibitions of two PKI on various PKFE. And, further, the authors suppose that heterogeneity of plasma kininogenases and their participation or specificity in the activity would be the important problems in the physiological and pathological meanings.

Summary

The purification of human plasma kallikrein (HPK) activated with acetone was studied and its maximal purity was 12.7 KU/E_{280}, its isoelectric point (pI) 7.8; however, highly purified HPK was unstable. Besides this HPK, three plasma-kinin-forming enzymes (PKFE) showing pI 6.9, 6.5 and 4.5 in Ampholine isoelectric focusing were found in the eluates of the adsorbed proteins on DEAE-cellulose. In the human plasma activated by the salting-out with ammonium sulfate and by casein, one PKFE (pI 4.5) and two PKFE (pI 4.3 and 4.6) were detected respectively. The molecular weights of these enzymes were estimated by Sephadex G - 200 column and showed 74,000 - 113,000. These enzymes were inactivated by two potato kallikrein inhibitors, Trasylol, soybean trypsin inhibitor and DEP, but not by lima bean-, egg white-trypsin inhibitors and potato chymotrypsin inhibitor I. Other properties of these PKFE were compared and discussed with plasma kallikreins reported by other investigators.

References

1. R. W. Kellermeyer and R. C. Graham, New Engl. J. Med., 279, 754,802, 859 (1968).

2. M. E. Webster, Fed. Proc., 27, 84 (1968).

3. M. E. Webster and O. D. Ratnoff, Nature, 192, 180 (1961).

4. W. Vogt, J. Physiol., 170, 153 (1964).

5. B. Mason and A. A. Miles, Nature, 196, 587 (1962).

6. R. W. Colman, Biochem. Biophys. Res. Commun., 35, 273 (1969).

7. S. Nagasawa, H. Takahashi, M. Koida, T. Suzuki and J. G. G. Schoenmakers, Biochem. Biophys. Res. Commun., 32, 644 (1968).

8. Y. Hojima, H. Moriya and C. Moriwaki, Allergy (in Japanese), 19, 824 (1970).

9. R. W. Colman, L. Mattler and S. Sherry, J. Clin. Invest., 48, 11, 23 (1969).

10. M. Yano, S. Nagasawa and T. Suzuki, J. Biochem., 67, 713 (1970).

11. H. Z. Movat, H. B. Sender, M. P. Treloar and Y. Takeuchi, "Advances in Experimental Medicine and Biology," Vol. 8, Ed. by F. Sicuteri, M. Rocha e Silva and N. Back, Plenum Press, New York-London, p. 123 (1970).

12. G. Seidel, Naunyn-Schmiedebergs Arch. Pharmakol., 264, 18 (1969).

13. R. Jahrreiss and E. Habermann, Naunyn-Schmiedebergs Arch. Pharmakol., 269, 85 (1971).

14. H. Moriya, K. Yamazaki and H. Fukushima, J. Biochem., 58, 315 (1965).

15. S. Jacobsen, Brit. J. Pharmacol., 26, 403 (1966).

16. E. Habermann and W. Klett, Biochem. Z., 346, 133 (1966).

17. H. Moriya, K. Yamazaki and H. Fukushima, J. Biochem., 58, 201 (1965).

18. B. C. W. Hummel, Can. J. Biochem. Physiol., 37, 1393 (1959).

19. G. W. Schwert and Y. Takenaka, Biochem. Biophys. Acta, 16, 570 (1955).

20. T. Astrup and S. Mullertz, Arch. Biochem. Biophys., 40, 346 (1952).

21. M. Lassen, Acta Physiol. Scand., 27, 371 (1952).

22. H. Moriya, Y. Hojima, C. Moriwaki and T. Tajima, Experientia, 26, 720 (1970).

23. Y. Hojima, H. Moriya and C. Moriwaki, J. Biochem., 69, 1019, 1027 (1971).

24. C. A. Ryan, Biochemistry, 5, 1952 (1966).

25. O. Vesterberg and H. Svensson, Acta Chem. Scand., 20, 820 (1966).

26. P. Andrews, Biochem. J., 96, 595 (1965).

27. B. J. Davis, Ann. N.Y. Acad. Sci., 121, 404 (1964).

28. N. Back and R. Steger, Proc. Soc. Exptl. Biol. Med., 133, 740 (1970).

29. D. J. McConnell, L. J. Kagen and E. L. Becker, Proc. Soc. Exptl. Biol. Med., 119, 652 (1965).

30. E. Werle and I. Trautschold, Ann. N.Y. Acad. Sci., 104, 117 (1963).

31. D. J. McConnell and B. Mason, Brit. J. Pharmacol., 38, 490 (1970).

32. E. L. Becker and L. Kagen, Ann. N.Y. Acad. Sci., 116, 866 (1964).

33. V. H. Donaldson and O.D. Ratnoff, Science, 150, 754 (1965).

34. R. S. Speer, H. Ridgway and J. M. Hill, Thromb. Diath. Haemorrhag., 14, 1 (1965).

35. A. P. Kaplan and K. F. Austen, J. Immun., 105, 802 (1970).

36. D. A. J. Armstrong and G. L. Milles, Biochem. Pharmacol., 13, 1393 (1964).

Acknowledgments

The authors wish to thank Miss Y. Taki, Mr. H. Nakamura and Mr. S. Kaneko, students in our laboratory, for their co-operations in the preparation of acetone activated PKFE and the fibrinolysis assay.

PHYSIOLOGY-PHARMACOLOGY

STRUCTURE-ACTIVITY RELATIONSHIPS OF NEUROHYPOPHYSEAL POLYPEPTIDES ON DIFFERENT TYPES OF ISOLATED MAMMALIAN BLOOD VESSELS*

Burton M. Altura[†]

Departments of Anesthesiology and Physiology, Albert Einstein College of Medicine, Bronx, N. Y. 10461

Up until the isolation and synthesis of vasopressin in 1953 by du Vigneaud vasopressin was given little serious thought as a useful vasoactive drug because of its well documented coronary constrictor action. Interestingly, however, there is now good evidence to indicate that at least two synthetic neurohypophyseal hormone (NHPH) analogues, namely 2-phenylalanine-8-lysine vasopressin (PLV-2) and 8-ornithine vasopressin, may have significant antiarrhythmic properties in man (see references in 9). Additional experimental studies indicate that at least one of these NHPH analogues, namely PLV-2 (the only one so far investigated in animal coronary circulation), increases the relative myocardial blood flow in rats (16) and relaxes a variety of isolated bovine coronary arteries (1). Furthermore, it has been demonstrated that when PLV-2 was used as a local vasoconstrictor in skin flaps it did not produce any decrease in tissue oxygen tensions, while epinephrine did (20). Possibly more importantly, our laboratory found that PLV-2 improved survival in animals subjected to various forms of shock (9, 10, 11, 13, 19) and suggested this might be due to its unusual microcirculatory effects as observed by direct in-vivo microscopy (4, 7-12, 19). In view of such findings we initiated systematic pharmacologic and structure-activity studies of the NHPH, and particularly their synthetic analogues, not only at the microcirculatory level but on various types of isolated mammalian blood vessels since the latter would be divorced from any in-vivo effects of metabolism and blood flow.

It has been generally thought that the degree of basicity of the amino acid residue in position 8 of the NHPH (see Figs. 1 and 2) is probably, per se, the

* This study was supported by research grants HE-12462 and HE-11391 from the National Institutes of Health, USPHS, Bethesda, Maryland.

† Recipient of Research Career Development Award 5-K3-GM-38, 603 from the N.I.H., USPHS.

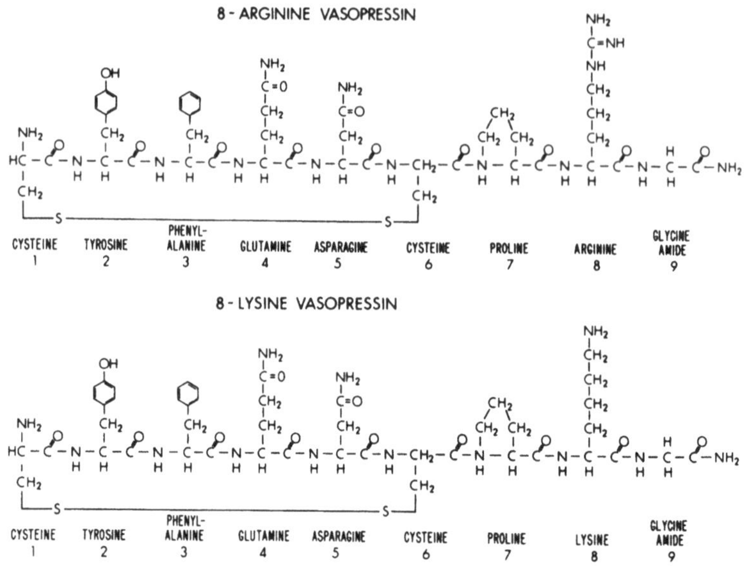

Fig. 1. Chemical structure of naturally occurring mammalian vasopressin hormones.

single critical structural factor for optimizing the pressor (vasoconstrictor) and antidiuretic properties of the molecules (18). However, recent work using some newly available synthetic NHPH analogues, and involving in-vivo and in-vitro experiments, raises some question as to the validity of such a concept (2, 4, 7, 8, 15). Quite unrelated to the basicity of the amino acid residue in position 8 of these new analogues, the length of the side chain in the same position appears also to be an important structure-activity determinant (2, 4, 7, 8). Delineation of these structure-activity determinants are, of course, strategic requirements for the design of new NHPH analogues with more precise vasomotor selectivity (17).

In view of the foregoing considerations of selective vasomotor actions, it is apparent that the conventional pharmacologic assays and activity indices of vasopressor drugs may have to be modified or entirely changed. Since many of these synthetic NHPH peptides show a predilection for the muscular venules (4, 7-10, 12), the rat blood pressure assays (see Stürmer, 1968 for review), used routinely to assess pressor (contractile) activity of a particular NHPH or analogue, may not accurately reflect its ability or inability to effect contraction, since contraction of muscular venules only would not increase blood pressure. The fact that threshold constrictor effects on venules are masked could explain most of the observations in animal assays which show no differences between systemic blood pressure patterns for 8-lysine

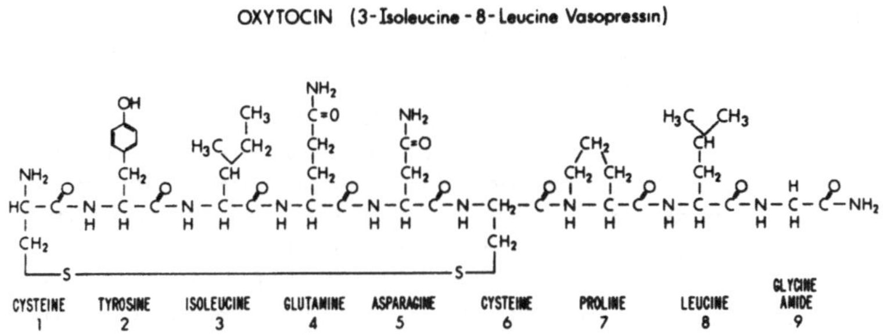

Fig. 2. Chemical structure of naturally occurring oxytocin (3-isoleucine-8-leucine vasopressin).

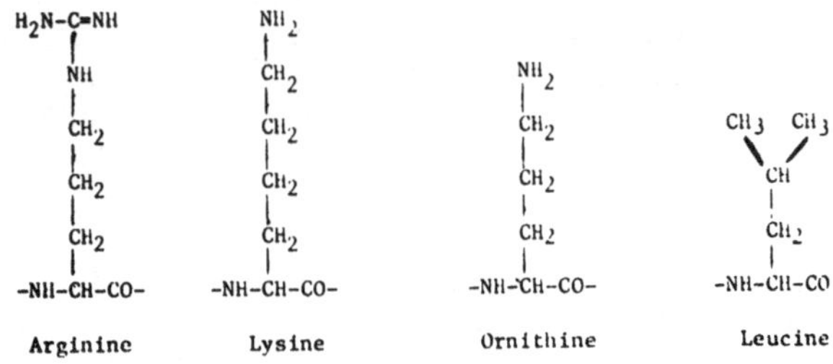

Fig. 3. Length of amino acid side chains in position 8 of NHPH analogues.

vasopressin, 8-arginine vasopressin, as well as PLV-2 (21). Therefore, interpretation of the functional importance of the amino acid residue in position 8 (Figs. 1 and 2) as well as in other positions of the NHPH molecules, to contractile activity, in vascular smooth muscle, on the basis of the essentially crude rat-pressor assays has obvious limitations (2).

Recent in-vitro studies, performed in our laboratory, strongly support the above opinion on the need for more sophisticated assay methods. For example, in-vitro studies on a variety of canine blood vessels show that 8-ornithine vasopressin (a molecule with a shorter side chain as well as less basicity in position 8 than 8-arginine vasopressin) exhibits greater intrinsic activities (i.e., greater developed tensions) and affinities for the vasopressin receptor when compared to 8-arginine vasopressin or 8-lysine vasopressin (2, 4). In addition, although oxytocin (3-isoleucine-8-leucine vasopressin) (Fig. 2) can induce contractile responses on isolated mammalian blood vessels (2, 4, 7, 8, 14, 22), it does not exhibit any significant basicity in position 8; its side chain in this position, like that of ornithine, also is shorter than that of arginine or lysine (Fig. 3). These factors suggested that an optimum interaction between length of side chain and basicity in position 8 may be necessary for maximal activation (contraction) of smooth muscle cells by the vasopressin receptor in mammalian somatic vascular smooth muscle; maximal basicity alone in position 8 does not in itself appear to promote optimum contractile activity (2, 4).

Modification of the vasopressin molecule in other than the 8-position can also affect its contractile activity, at least on canine blood vessels (2, 4). For example, loss of the OH group of the phenolic amino acid tyrosine in position 2 (as in 2-phenylalanine-8-ornithine vasopressin and 2-phenylalanine-8-lysine vasopressin) or replacement of the amino acid phenylalanine in position 3 (as in 3-isoleucine-8-ornithine vasopressin) not only reduces the affinity of the altered vasopressin molecules to their receptors but also results in losses in intrinsic activity. The phenolic and aromatic groups in positions 2 and 3 (Fig. 1), respectively, thus seem to be very important not only for affinity, but also for intrinsic activity (2, 4). They, thus, appear to somehow interact with the amino acid moiety in the eighth position, of the vasopressin molecule, to promote optimum contractility in mammalian vascular smooth muscle.

In order to more accurately assess the implications of the findings on canine blood vessels for mammals in general, I thought it would be advisable to determine whether these generalizations concerning positions 2, 3 and 8 hold for all types of mammalian blood vessels. I therefore turned my attention to an examination of isolated blood vessels in rats and man.

MATERIALS AND METHODS

Thoracic aortas were excised, after sacrifice, from male Wistar strain rats (275-375 grams) and prepared as helical vascular strip preparations 1.5 mm in width x 25mm in length, similar to methods described previously (5, 6). Helical and longitudinal vascular strip preparations were also prepared from

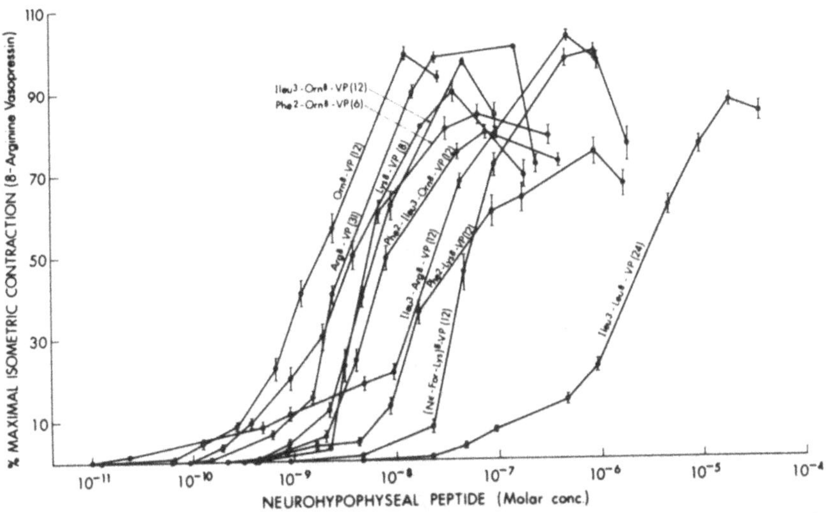

Fig. 4. Comparative contractile actions of NHPH analogues on isolated rat aorta. Numbers in parentheses denote the number of different rats utilized. Values are means ± S.E.M. Cumulative log dose-isometric response curves.

human umbilical arteries (1.3 - 1.5 in width by 25mm in length) and veins (2.5 - 3mm in width by 25mm in length) similar to previously described methods (15). All of the vascular strip preparations were mounted vertically, under isometric conditions in 20ml muscle baths, using procedures and instrumentation previously described elsewhere (5, 6), and allowed to equilibrate under their respective resting tensions (e.g., 1.5 grams for rat aorta and 2 grams for umbilical vessels) for 2 hrs. in Krebs-Ringer bicarbonate solution, the composition of which has been given previously (5, 6). The loading tensions were maintained and periodically adjusted throughout incubation and the experiments. The Krebs-Ringer solutions were oxygenated continuously with a 95% O_2 - 5% CO_2 mixture and kept at 37°C (pH 7.2 - 7.4).

NHPH-induced contractions. Complete cumulative dose-response curves were obrained on the vascular strip preparations for different pure, synthetic NHPH analogues (2, 5). Each rat aortic strip preparation and human umbilical vessel preparation was stimulated to contract by epinephrine (in supra-maximal concentrations), in the case of the rat tissues, or serotonin (in supra-maximal concentrations) in the case of the human blood vessels, periodically throughout the experiments since these vasoactive substances induce the greatest maximum responses in the respective vascular preparations (14, unpublished findings). Pure, synthetic 8-arginine vasopressin (AVP) was utilized as the standard for comparison on the rat aorta since it is the native vasopressin hormone in the rat's pituitary gland. NHPH analogues were not tested on any of the vascular smooth muscle preparations until a particular preparation yielded two consecutive maximum responses to epinephrine (in case of rat aorta) and serotonin (in case of umbilical vessels) that were

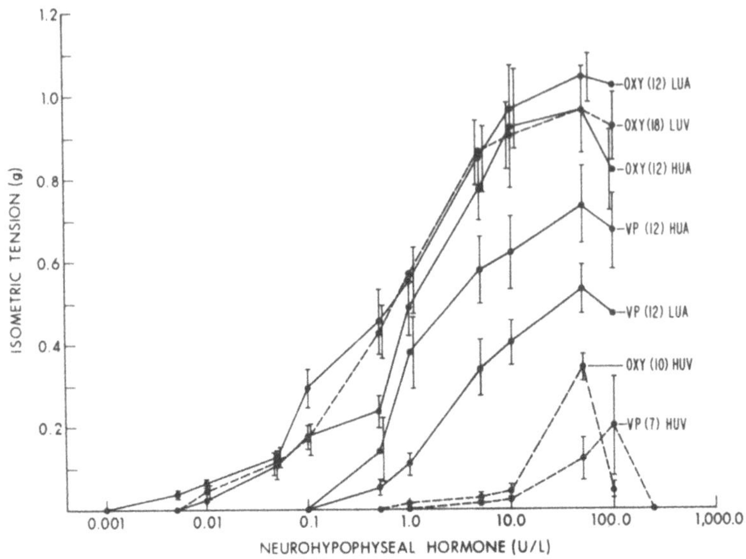

Fig. 5. NHPH action on isolated longitudinally and helically cut human umbilical arteries and veins. Numbers in parentheses denote the number of different cords utilized. Values are means ± S.E.M. Cumulative log dose-isometric response curves. L = longitudinally cut. H = helically cut.

within 5% of one another. Since all of these preparations, in preliminary experiments, exhibited tachyphylaxis to AVP, as well as cross-tachyphylaxis to other NHPH peptides, only one NHPH analogue was tested on any one vascular smooth muscle preparation. The results in the rat experiments were originally expressed in % maximum epinephrine and then normalized to % 8-arginine vasopressin.

COMPARATIVE ACTIONS OF NHPH ANALOGUES ON RAT AORTA

In these studies we examined 10 different NHPH analogues including three which we did not investigate on the canine blood vessels: namely 2-phenylalanine-3-isoleucine-8-ornithine vasopressin, 3-isoleucine-8-arginine vasopressin (arginine vasotocin) and an N-epsilon-formyl-8-lysine derivative of vasopressin, the latter being an analogue with a side chain approximately the same length as lysine but exhibiting no significant basicity (17). The results of these experiments are shown in Figure 4. This figure reveals several interesting phenomena: First, although 8-ornithine vasopressin (Orn^8-VP) does not induce a maximal contractile response greater than Arg^8-VP, it exhibits a much greater affinity for the vasopressin receptor as noted by a 2.5 fold shift to the left of Arg^{8}-VP. Second, as is the case with canine vessels (2, 4), the log dose-response curve for 8-lysine vasopressin (Lys^{8}-VP) is not only shifted to the right of Orn^{8}-VP but in addition the maximum response elicited by Lys^{8}-VP is reduced even though the basicity of lysine is not appreciably different from ornithine. Third, as is the case with most of the canine vessels, so far studied (2, 4), insertion of the amino acid phenylalanine (Phe) in place of the phenolic amino acid tyrosine in position 2 (e.g., 2-phenylalanine-8-ornithine vasopressin [Phe^{2}-Orn^{8}-VP] and 2-phenylalanine-8-lysine vasopressin [Phe^{2}-Lys^{8}-VP]) not only shifts the log dose-response curves to the right of Orn^{8}-VP and Lys^{8}-VP, respectively, but also decreases the maximum responses obtained with these two analogues. Fourth, insertion of the aliphatic amino acid isoleucine (Ileu) in place of the aromatic amino acid phenylalanine in position 3 (e.g., Ileu^{3}-Orn^{8}-VP and Ileu^{3}-Arg^{8}-VP) shifts the log dose-response curves four and seven fold to the right of 8-ornithine and 8-arginine vasopressin, respectively. Fifth, as is the case with canine vessels (2, 4) and rat microscopic blood vessels (4, 7, 8), oxytocin, that is 3-isoleucine-8-leucine vasopressin (Ileu^{3}-Leu^{8}-VP), an analogue lacking a free basic group in position 8 although shifting the log dose-response curve almost 1,000 fold to the right of Arg^{8}-VP is capable of eliciting a potent contractile response on the isolated rat aorta (i.e., 90% of Arg^{8}-VP). And sixth, although (N-epsilon-formyl)-8-lysine vasopressin has effectively no basicity in position 8 and its log dose-response curve is shifted 15 fold to the right of 8-lysine vasopressin, it can elicit a maximum contractile response unlike its parent molecule Lys^{8}-VP.

These structure-activity data on isolated rat aorta thus confirm and extend our hypothesis on isolated canine blood vessels and on rat microcirculatory blood vessels. That is, overall the data on a variety of mammalian somatic blood vessels, suggest that an optimum interaction between length of side chain and basicity in position 8 is, indeed, necessary for maximal activation (or contraction) of smooth muscle cells by the vasopressin receptor in mammalian somatic vascular smooth muscle (2, 4). Maximal basicity alone in position 8 does not in itself promote optimum contractile activity (2, 4). Furthermore, the present data on isolated rat aorta support the previous suggestion that the phenolic and aromatic groups in positions 2 and 3, respectively, are not only important for affinity, but also for intrinsic activity as well. In other terms, the phenolic hydroxyl of tyrosine and phenylalanine in positions 2 and 3 appears to enhance the activity of the amino acid residue in position 8 in promoting optimum contractile activity in mammalian somatic vascular smooth muscle (2, 4).

COMPARATIVE ACTIONS OF NHPH ON HUMAN UMBILICAL VESSELS

It is tempting to speculate that the different relative affinities and intrinsic activities of the various NHPH analogues on different canine (2, 4) and rat blood vessels (9, 10 and present work), as well as different magnesium requirements (3, unpublished findings), may indicate that the vasopressin receptor on different blood vessels between and even within a single mammalian species may not be identical (2-4). Although various explanations could be invoked to account for the relative differences in affinities and maximum responses obtained for the vasopressin analogues on different canine and rat blood vessels (2), the studies described below on human umbilical vessels could be construed as evidence to support my previous suggestion that the NHPH receptor may not be identical from one mammalian blood vessel to another.

Figure 5 represents some preliminary observations on isolated helically and longitudinally cut human umbilical arteries and veins. It should be noted that the log dose-response curves for oxytocin are not only shifted to the left of those for vasopressin on both helically and longitudinally cut human umbilical arteries suggesting a greater affinity of oxytocin for the neurohypophyseal receptor, but in addition oxytocin exhibits a greater intrinsic activity than vasopressin on these vessels as well as on human umbilical veins. These pharmacologic phenomena exhibited by oxytocin and vasopressin are thus very different from that seen for either other somatic mammalian blood vessels (2, 4 and present report) or small mammalian blood vessels (4, 7, 8) and might, indeed, be used to support the notion that the neurohypophyseal peptide receptor in human umbilical arteries and veins is probably different from that in other mammalian blood vessels.

SUMMARY AND CONCLUSIONS

In-vitro structure-activity relationship (SAR) studies were undertaken with rat aortic strips and human umbilical arteries and veins to determine: 1) what part or parts of the neurohypophyseal peptide molecules are essential for contractile activity; and 2) whether or not identical SAR's are obtained for NHPH analogues on different types of mammalian blood vwssels. Overall, the data on canine and rat blood vessels indicate: (a) the length of the amino acid side chain as well as basicity in position 8 are important not only for affinity but intrinsic activity (i.a.) on mammalian somatic blood vessels, but not on human umbilical vessels; (b) the phenolic and aromatic groups in positions 2 and 3, respectively, of the NHPH molecules are not only important for affinity but for i.a. as well; these moieties interacting with the amino acid in position 8 to promote optimum contractile activity; (c) preliminary observations on human umbilical arteries and veins suggest that the SAR's for NHPH analogues on these vessel types are quite different from those obtained on a variety of somatic vessels in other mammalian blood vessels. In conclusion, the present data 1) emphasize the importance of a comparative approach in the analysis of SAR's for neurohypophyseal peptides in the cardiovascular system, and 2) could be used to buttress the concept that drug molecules with selective regional vascular actions can be realized.

ACKNOWLEDGEMENTS

The author would like to thank Mr. C.F. Reich for his excellent technical assistance. The author is most grateful to Dr. B. Berde of Sandoz Inc. and Dr. Roderich Walter of Mt. Sinai School of Medicine for generously providing the pure, synthetic NHPH analogues used in these studies.

REFERENCES

1. B.M. Altura, "Differential actions of polypeptides and other drugs on coronary inflow vessels," Am. Heart J., 72: 709-711, 1966.
2. B.M. Altura, "Significance of amino acid residues in vasopressin on contraction in vascular muscle," Am. J. Physiol., 219: 222-229, 1970.
3. B.M. Altura, "Influence of magnesium and cysteine on vasopressin-induced contractions in various canine blood vessels, "Experientia, 26: 1089-1090, 1970.
4. B.M. Altura, "Pharmacology of neurohypophyseal hormones and analogs on isolated vascular muscle and in the terminal vascular bed," in: J.A. Bevan, R.F. Furchgott, R.A. Maxwell and A.P. Somlyo, Eds., Physiology and Pharmacology of Vascular Neuroeffector Systems. Basel, Karger, pp. 274-290, 1971.
5. B.M. Altura and B.T. Altura, "Heterogeneity of drug receptors in different segments of rabbit thoracic aorta," European J. Pharmacology 12: 44-52, 1970.
6. B.M. Altura and B.T. Altura, "Differential effects of substrate depletion on drug-induced contractions of rabbit aorta," Am. J. Physiol., 219: 1698-1705, 1970.
7. B.M. Altura and S.G. Hershey, "Pharmacology of neurohypophyseal hormones and their synthetic analogues in the terminal vascular bed. Structure-activity relationships," Angiology, 18: 428-439, 1967.
8. B.M. Altura and S.G. Hershey, "Structure-activity relationships of neurohypophyseal polypeptides in the microcirculation: A molecular basis for shock therapy," in: B. Schildt and L. Thorén, Eds., Intermedes Proceedings 1968: Combined Injuries and Shock. Stockholm, Almqvist and Wiksell, pp. 247-259, 1968.
9. B.M. Altura and S.G. Hershey, "A structure-activity basis for vasotropic peptide therapy in shock," in: These Proceedings.
10. B.M. Altura, S.G. Hershey and B.T. Altura, "Microcirculatory actions of polypeptides and their use in the treatment of experimental shock," Adv. Exptl. Med. Biol., 8: 239-247, 1970.
11. B.M. Altura, S.G. Hershey and V.D.B. Mazzia," Microcirculatory approach to vasopressor therapy in intestinal ischemic (SMA) shock," Am. J. Surgery, 111: 186-192, 1966.
12. B.M. Altura, S.G. Hershey and B.W. Zweifach, "Effects of a synthetic analogue of vasopressin on vascular smooth muscle," Proc. Soc. Exptl. Biol. Med., 119: 258-261, 1965.
13. B.M. Altura, R. Hsu, V.D.B. Mazzia and S.G. Hershey, "Influence of vasopressors on survival after traumatic, intestinal ischemia and endotoxin shock in rats," Proc. Soc. Exptl. Biol. Med., 119: 389-393, 1965.

14. B.M. Altura, D. Malaviya and L.R. Orkin, "Comparative contractile effects of vasoactive substances on isolated human umbilical arteries and veins," Federation Proc., 30: 383, 1971.
15. B.M. Altura, D. Malaviya, C.F. Reich and L.R. Orkin, "Effects of vasoactive agents on isolated human umbilical arteries and veins," Am. J. Physiol., 222, 1972 (in press).
16. B. Berde, "Some observations on the circulatory effects of oxytocin, vasopressin and similar polypeptides," in: J.H.M. Pinkerton, Ed., Advances in Oxytocin Research. London, Pergamon Press, pp. 11-35, 1965.
17. B. Berde and R.A. Boissonnas, "Basic pharmacological properties of synthetic analogues and homologues of the neurohypophysial hormones," in: B. Berde, Ed., Handbook of Experimental Pharmacology, vol. 23: Neurohypophysial Hormones and Similar Polypeptides. Berlin, Springer-Verlag, pp. 802-870, 1968.
18. R.A. Boissonnas, "The chemistry of oxytocin and vasopressin," in: M. Schachter, Ed., Polypeptides which Affect Smooth Muscles and Blood Vessels. London, Pergamon Press, pp. 7-19, 1960.
19. S.G. Hershey, V.D.B. Mazzia, B.M. Altura and L. Gyure, "Effects of vasopressors on the microcirculation and on survival in hemorrhagic shock," Anesthesiology, 26: 179-189, 1965.
20. P. Klingenström, B. Nylen and L. Westermark, "A clinical comparison between adrenaline and octapressin as vasoconstrictors in local anesthesia." Acta Anaesth. Scand., 11: 35-42, 1967.
21. K. Saameli, "The circulatory actions of the neurohypophysial hormones and similar polypeptides," in: B. Berde, Ed., Handbook of Experimental Pharmacology, vol. 23: Neurohypophysial Hormones and Similar Polypeptides. Berline, Springer-Verlag, pp. 545-612, 1968.
22. A.V. Somlyo, C. Woo and A.P. Somlyo, "Effects of magnesium on posterior pituitary hormone action on vascular smooth muscle," Am. J. Physiol., 210: 705-714, 1966.
23. E. Stürmer, "Bioassay precedures for neurohypophysial hormones and similar polypeptides," in: B. Berde, Ed., Handbook of Experimental Pharmacology, vol. 23: Neurohypophysial Hormones and Similar Polypeptides. Berlin, Springer-Verlag, pp. 130-189, 1968.

PARTICIPATION OF KININS IN THE REGULATION OF CEREBRAL VASOPERMEABILITY

G. Seidel* and U.Wendel

Max-Planck-Institut für experimentelle Medizin
Department of Biochemical Pharmacology
34 Göttingen, Germany

With numerous experiments the role of kinins has been studied in inflammation with increased vasopermeability, for instance after turpentine injection into pleural cavity, injection of kaolin, serotonin, carrageenin or polyvinylpyrolidon into rat paws and after thermal injury or exposure to ultra-violet irradiation (for review:Bonta, 1969; Kellermeyer and Graham, 1968). Obviously the significance of kinins is limited to local physiological and/or pathophysiological conditions (Creutzfeld, 1969; Ofstad, 1970).

The lymphostatic encephalopathy of rats studied by Földi and his colleagues (Csanda et al., 1968; Csillik and Földi, 1967; Földi et al., 1967) seems to be a model for a pathophysiological situation due to lymphogenic edema. Following surgical blockade of the cervical lymph vessels lymphostatic encephalopathy develops, which is characterized by extra- and intracellular edema, morphological disturbance and functional disorder of the brain. The cerebral blood vessels, too, are involved in this pathophysiological process (lymphostatic hemangiopathy). Within 5 days the blockade of the cervical lymph flow causes edema followed by hemangiopathy with increased permeability of the blood-brain barrier, which reinforces the edema and results in a circulus vitiosus (Joó et al., 1969; Várkonyi et al., 1969).

*These studies are part of a work for habilitation of G. Seidel.

We tried to find out whether kinins are involved in the increase of vasopermeability in the brain of encephalopathic rats. We studied the penetration of i.v. injected Evans blue into the brain of operated rats. Kininogen depleting substances, inhibitors of kallikrein or carboxypeptidase B were applied. Kininogen depletion by cellulose sulphate and pancreatic kallikrein should reduce or exclude the formation of kinins because of lack of substrate for kinin forming enzymes. Trasylol and SBTI were expected to inhibit kininogenases potentially activated in rats with lymphostatic encephalopathy. Kininolytic activity was increased by injection of carboxypeptidase B in order to destroy formed kinins faster than under normal circumstances (for review: Wilhelm, 1971).

In Nembutal anaesthetized female Wistar rats (220 - 260 g) all lymph nodes which could be discovered in the ventral cervical region were removed and the lymph vessels tied up as far as possible. Usually 6 lymph nodes were found on each side, 4 adjacent to the gl. submandibularis and 2 between the muscles near the plexus brachialis. The surgical blockade of the cervical lymph vessels caused an encephalopathy-hemangiopathy with deteriorated central nervous function and increased vasopermeability. Fig. 1 shows a typical graph of the spontaneous motility obtained from groups of 4 rats on the 5th day after cervical lymphatic blockade or sham operation. Put together into a cage, operated animals got less interest faster in their surroundings than sham operated controls. The permeability of the cerebral vessels was investigated on the 5th postoperative day. Operated urethane anaesthetized rats were injected i.v. with Evans blue (3 ml/kg of a 5% solution). 30 min later the animals were decapitated and the brains were removed from the skull. The total dye from the cerebral vessels and brain tissue was extracted by formamide and estimated photometrically (623 nm). During lymphostatic encephalopathy which lifts up cerebral vasopermeability the amount of extractable dye increased significantly. In the experiments with impaired kallikrein-kinin-system Evans blue was injected i.v. always 60 min after termination of specific acute or chronical treatment.

Depletion of plasma kininogen was performed by cellulose sulphate and/or pancreatic kallikrein i.v. (Rothschild, 1966, 1968; Starr, 1970). The effect of this treatment was controlled by estimating total plasma kininogen (modified method of Dawson et al.,1966,

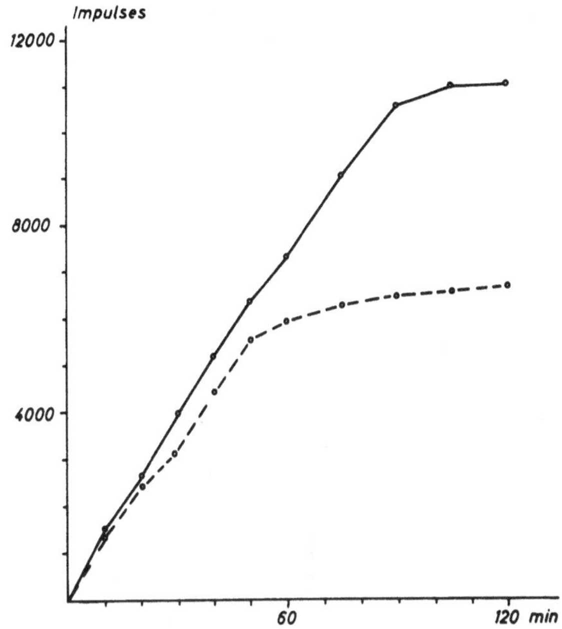

Fig. 1: Typical graph of spontaneous motility obtained from groups of four rats five days after cervical lymphatic blockade (--o--) or sham operation (—o—). Motility was recorded by an Animex Activity Meter.

and Urbanitz et al., 1969) and plasma protein before and after depletion. In some other cases the depleting effect was controlled by recording blood pressure changes in Nembutal(R) anaesthetized rats.

One group of rats suffering from lymphostatic encephalopathy was injected i.v. with 8 consecutive doses of 150 KU/kg of pancreatic kallikrein, one dose every 10 min. Rats of a second group received 3 mg/kg cellulose sulphate i.v. followed by 8 times of 150 KU/kg every 10 min. For control another group was injected with corresponding volumes (9 times 0.07 ml) of saline while a fourth group served as another control measuring the effects of anaesthesia itself. Before and 60 min after the last injection of each compound or in the fourth group at corresponding times venous blood samples were estimated for total plasma kininogen and protein concentration. Results are shown in fig 2.

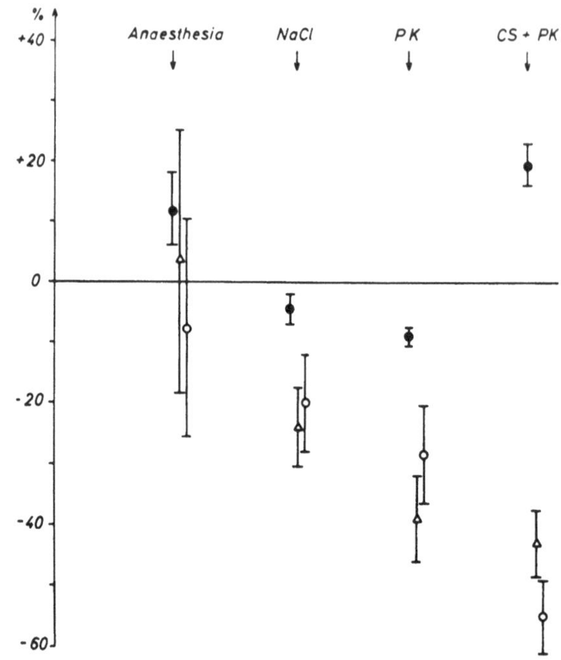

Fig. 2: Per cent change of plasma protein and kininogen by anaesthesia, by anaesthesia plus saline (NaCl), pancreatic kallikrein (PK) or the combination of cellulose sulphate and pancreatic kallikrein (CS + PK). • Protein, ▵ kininogen, o kininogen corrected for change in protein content. Means ± S.E.M.

Anaesthesia caused a slight but not a significant hemoconcentration, while the injection of saline in contrast causes a not significant hemodilution. Animals injected with pancreatic kallikrein showed a significant hemodilution while an additional application of cellulose sulphate increases the plasma protein concentration considerably. The estimations of plasma kininogen had to be corrected for the amount of hemoconcentration respectively hemodilution for each single animal. Only this method provides the actual amount of kininogen depletion.

Injection of cellulose sulphate results in tachyphylaxis. 3 mg/kg of cellulose sulphate i.v. lowered the blood pressure from about 120 to 60 mm Hg due to kinin release. Within 15 to 20 min blood pressure recovered. A second injection of cellulose sulphate 60 min later was without effect, but injection just 20 to 30 min after the first dose still lowered the blood pressure. The used doses of hog pancreatic kallikrein had been given previously 8 times (150 KU/kg every 10 min). 60 min after a dose of 3 mg/kg of cellulose sulphate or after the last of 8 doses of 150 KU/kg of pancreatic kallikrein blood pressure reached its starting level.

To investigate the pathogenetic role of the kallikrein-kinin-system in lymphostatic encephalopathy on the 5th postoperative day rats were injected i.v. with cellulose sulphate, hog pancreatic kallikrein and later Evans blue. Fig. 3 shows the effect of acute depletion of kininogens on the total amount of extractable dye in encephalopathic and non-operated rats. Operated and non-operated controls obtained corresponding volumes of saline. The third column in comparison with the first one shows that under conditions of lymphostatic encephalopathy the amount of extractable dye is significantly higher (23%, $p < 0.001$) than in non-operated rats. The depletion of both kininogens caused a significant decrease of extractable dye by 8% ($p < 0.02$). Kininogen depletion in non-operated rats had an effect opposite to operated animals. This effect did not reach a significant level.

In a second series of experiments we tried to inhibit the development of lymphostatic encephalopathy by repeated application of either pancreatic kallikrein or both cellulose sulphate plus pancreatic kallikrein from the 1st to the 4th postoperative day according to the scheme shown in figure 4. 60 min before i.v. injection of Evans blue on the 5th postoperative day high doses of both compounds were applied i.v. to lower that amount of kininogen which remained or had renewed. Chronical kininogen depletion with pancreatic kallikrein already reduced the extractable amount of dye from the brain significantly by 10% ($p < 0.005$) (Fig.4). This effect was increased to 12% when cellulose sulphate had been injected in addition $p < 0.005$).

Trasylol or SBTI were either injected from the 1st to the 5th postoperative day during the development of the disease or in acute experiments on the

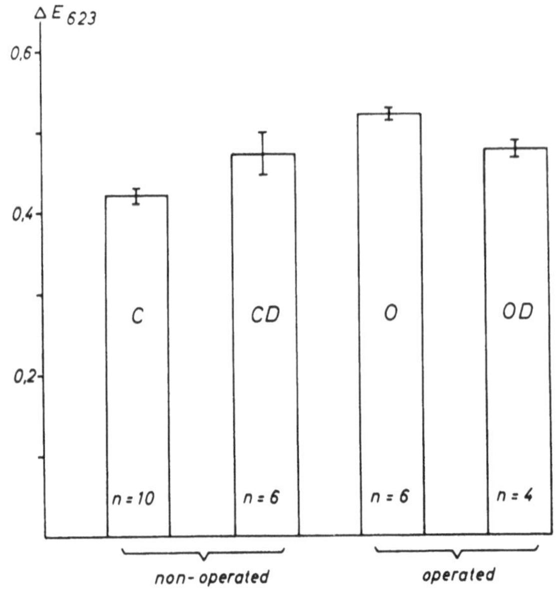

Fig. 3: Effect of acute plasma kininogen depletion on the amount of Evans blue (ΔE_{623}) extracted from brains of non-operated and encephalopathic rats five days after cervical lymphatic blockade. C: non-operated controls. CD: kininogen depletion in non-operated controls. O: cervical lymphatic blockade. OD: kininogen depletion in rats with cervical lymphatic blockade. Means $\pm$ S.E.M.

5th day 60 min before dye application. Doses are shown in table I. Both serum kallikrein inhibitors failed to lower the extractable dye. In contrast a slight not significant increase of Evans blue was found in the brains of rats with lymphostatic encephalopathy as well after chronical as after acute treatment.

The level of kinin destroying enzymes was elevated by i.v. injection of carboxypeptidase B. Simultaneously with an initial injection of 600 U of carboxypeptidase B an infusion of the enzyme was started (100 U/kg . min). During a corresponding treatment a nearly constant and highly increased activity of carboxypeptidase was achieved in the blood (Urbanitz et al., 1969). Controls were injected and infused with saline. 1 hour later Evans blue was injected the infusion going further on.

Group	1st - 4th postop day (i.p.)	5th postop day (i.v.)
C	saline	saline
PK	1000 KU/kg PK every 12h	600 KU/kg PK
CS+PK	3mg/kg CS every 24h 500 KU/kg PK every 12h	3mg/kg CS 300 KU/kg PK

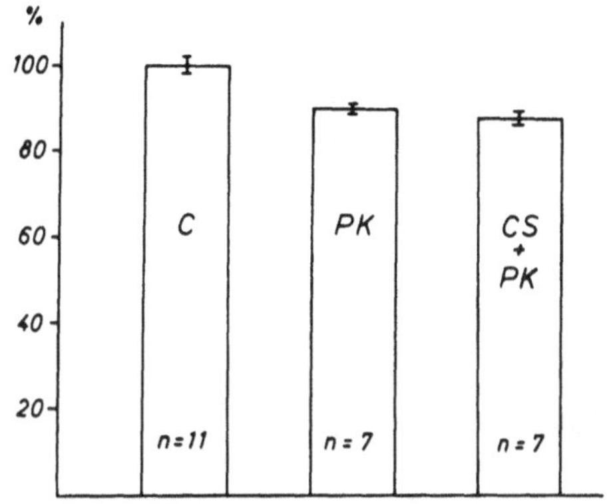

Fig. 4: Effect of chronical depletion of plasma kininogens on the amount of Evans blue in rat brains five days after cervical lymphatic blockade. C:controls (=100%). PK:depletion by pancreatic kallikrein. CS + PK: depletion by cellulose sulphate and pancreatic kallikrein. Means ± S.E.M. Scheme of treatment is shown in the table above.

	Acute treatment	Chronical treatment
Trasylol	5 th postop. day: 100000 KIU/kg i.v. 1h before Evans blue	1st - 4th postop. day: 100000 KIU/kg i.p. every 12 h 5 th day: 100000 KIU/kg i.p. 3h before Evans blue
SBTI	5 th postop. day: 20 mg/kg i.v. 1h before Evans blue	1st - 4th postop. day: 20 mg/kg i.p. every 12 h 5 th day: 20 mg/kg i.p. 3h before Evans blue

Table I: Scheme of acute and chronical treatment with kallikrein inhibitors in encephalopathic rats.

30 min later the rats were decapitated. In non-operated animals carboxypeptidase B did not change the amount of extractable dye (Fig. 5). From brains of encephalopathic

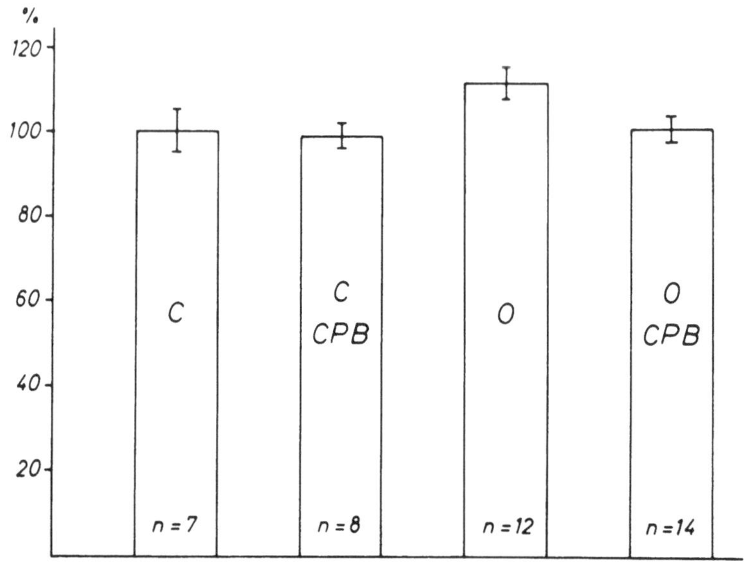

Fig. 5: Effect of carboxypeptidase B (CPB) treatment on the amount of Evans blue in brains of non-operated and encephalopathic rats. C: non-operated controls (=100%). C CPB: CPB application to non-operated controls. O: cervical lymphatic blockade. O CPB: CPB application to rats with cervical lymphatic blockade. Means ± S.E.M.

rats 12% more dye could be extracted ($p < 0.05$). This increased amount of dye was reduced significantly by 10% during the infusion ($p < 0.05$).

The reduction of spontaneous motility and the increased amount of extractable Evans blue from total brain of rats with blocked cervical lymph vessels indicate that in the CNS of these animals an experimental disease developed with functional disorder and increased vascular permeability for protein bound dye. Corresponding to our results the increased vasopermeability seems to be mediated at least partly by kinins because previous exhaustion of plasma kininogen lowers the permeability for protein bound dye into the tissue of the brain and the systemic amplification of the kininolytic potential results in the same effect. It seems that in brain of rats suffering from lymphostatic encephalopathy kininogenases are activated receiving their substrate from plasma. After pretreatment with pancreatic kallikrein and cellulose sulphate the kinin precursor is widely reduced.

We did not see any effect of Trasylol or SBTI. Both substances have strong inhibiting effect upon kinin forming enzymes activated in plasma (Vogel et al., 1966). In the brain of encephalopathic rats kinin forming enzymes are possibly activated which originate from the tissue and are not inhibited in vivo by Trasylol or SBTI. Werle and Vogel (1961) could activate a kallikrein-like enzyme in homogenized rat brains with corresponding qualities by trypsin.

Much evidence exists for the involvement of the kallikrein-kinin-system in tissue injury and inflammation. Lymphostatic encephalopathy is accompanied by edema partly originating from the increased vascular permeability. Thus this experimental disease might be compared with models for microvascular disturbance, possibly of inflammatory origin (for review: Bonta, 1969). Probably a multiplicity of mediators might be involved to induce increased vasopermeability (Arrigoni Martelli et al., 1969; Ratnoff, 1969). Even in lymphostatic encephalopathy activation of kinin forming enzymes might not initiate the development of symptoms, but certainly kinins play some role.

Our results correspond to other experimental data about formation of kinins in the CNS for instance by subarachnoid or cerebral hemorrhage (Sicuteri et al., 1970). Onabanjo et al. (1970) consider that in mallarial infection kallikrein and kinins amongst other things cause an increase of endothelial permeability with breakdown of the blood-brain barrier. Perhaps many pathophysiological situations in the brain are connected with decreased pH in brain tissue, thus kinin destroying enzymes are widely inactive (Iwata et al., 1969).

Some studies show cerebral vasodilatation after injection of bradykinin into the a.carotis (Athanasoulis et al., 1969; Watters, 1971) or intravenously (Carpi and Pinto Corrado, 1961). According to Carpi and Pinto Corrado cerebral vessels are particularly sensitive to the vasodilatatory action of bradykinin. They even assumed a possible role of bradykinin in the pathophysiological control of brain circulation. Vasodilatation and increased vasopermeability are symptoms which are provoked by nearly the same concentration of kinin (Lewis, 1970). Our method comprises the total dye in brain tissue and cerebral vessels. Therefore dye increase might partly be due to a possible vasodilatation. Further series estimating the dye content of the brain after rinsing the cerebral vessels free of blood, possibly will present more prominent effects, just as studies on different parts of the brain with already normally different vascular permeability (for review: Földi, 1963).

We point to differences in kininogen depleting effects of i.v. injected cellulose sulphate in rats comparing with Rothschild (1966, 1968). After 3 mg/kg i.v. of cellulose sulphate in rats he found a 70% depletion of plasma kininogen. Eisen and Loveday (1970) only found a 20-25% depletion as we did. In contrast to Rothschild we observed a strong hemoconcentration after injection of cellulose sulphate.

It seems interesting to study the role of kinins in other situations of lymphogenic edema, perhaps the role of kinins in regulation of cerebral blood flow which up to these days seems to be an unsolved problem.

References

Arrigoni Martelli, E., Corsico, N. and Fogagnolo, E.: Significance of the release of bradykinin in local inflammatory reactions and related effect of antiphlogistic drugs; in A.Bertelli and J.C.Houck (Eds.), Inflammation biochemistry and drug interaction, p. 185 - 196. Excerpta Medica, Amsterdam (1969).

Athanasoulis, C.A., Ferris, E.J. and Shapiro, J.H.: Bradykinin in cerebral angiography: Experimental observations in monkeys. Invest. Radiol. 4, 323 - 326 (1969).

Bonta, I.L.: Microvascular lesions as a target of anti-inflammatory and certain other drugs. Acta Physiol. Pharmacol. Neerl. 15, 188 - 222 (1969).

Carpi, A. and Pinto Corrado, A.: Cerebral vascular action of bradykinin in the dog. Experientia 17, 326 - 327 (1961).

Creutzfeld, W.: Kininfreisetzung bei Pankreatitis; in G.L.Haberland, P.Matis (Eds.), Neue Aspekte der Trasylol-Therapie 3, p. 89 - 101 FK. Schattauer Verlag, Stuttgart (1969).

Csanda, E., Földi, M., Obál, F. and Zoltán, Ö.T.: Cerebral oedema as a consequence of experimental cervical lymphatic blockage. Angiologica 5, 55-63 (1968).
Csillik, B. and Földi, M.: Severe alterations in myelin structure in experimental lymphogenous encephalopathy. Experientia 23, 835 - 836 (1967).
Dawson, W., Starr, M.S. and West, G.B.: Inhibition of anaphylactic shock in the rat by antihistamines and ascorbic acid. Br.J.Pharmac.Chemother. 27, 249 - 255 (1966).
Eisen, V. and Loveday, C.: Effect of cellulose sulphate on serum complement. Br.J.Pharmac. 39, 831 - 833 (1970).
Földi, M.: Die Rolle der Lymphzirkulation im Säftekreislauf des Auges und des Zentralnervensystems. Arch.Kreislaufforsch. 41, 186 - 212 (1963).
- , Csillik, B., Joó, F. and Zoltán, Ö.T.: Electron microscopic alterations in the central nervous system in experimental lymphogenic encephalopathy I. Angiologica 4, 50 - 56 (1967).
Iwata, H., Shikimi, T. and Oka, T.: Pharmacological significances of peptidase and proteinase in the brain (report I) - enzymatic inactivation of bradykinin in rat brain. Biochem. Pharmacol. 18, 119 - 128 (1969).
Joó, F. , Takács, Ö., Varga, L., Domján, G., Börcsök, Ethel, Rott, Susan, Rosta, Margit, Zoltán, Ö.T., Csillik, B. and Földi, M.: Collapse of the blood brain barrier in lymphostatic cerebral hemangiopathy and in triethyltin poisoning. Med. Exp. 19, 342 - 350 (1969).
Kellermeyer, R.W. and Graham, R.C.: Kinins - possible physiologic and pathologic roles in man. New England J.Med. 279, 754 - 759; 802 - 807; 859 - 866 (1968).
Lewis, G.P.: Physiologically active peptides and vascular permeability. Thrombos. Diathes.haemorrh. (Stuttg.) Suppl. 40, 225 - 229 (1970).
Ofstad, E.: Formation and destruction of plasma kinins during experimental acute hemorrhagic pancreatitis in dogs. Scand.J. Gastroent. 5, Suppl. 5 (1970).
Onabanjo, A.O., Bhabani, A.R. and Maegraith, B.G. : The significance of kinin-destroying enzymes activity in plasmodium knowlesi malarial infection. Br.J.exp.Path. 51, 534 - 540 (1970).
Ratnoff, O.D.: Some relationships among hemostasis, fibrinolytic phenomena, immunity, and the inflammatory response. Advan.Immunol. 10, 145 - 227 (1969).

Rothschild, A.M.: Pharmacodynamic properties of cellulose sulfate and related polysaccharides - a group of bradykinin-releasing compounds; in M.Rocha e Silva, H.Rothschild (Eds.), International Symposium on Vasoactive Polypeptides: Bradykinin and related kinins, p. 197 - 203. Sao Paulo (1966).

- , :Some pharmacodynamic properties of cellulose sulphate, a kininogen-depleting agent in the rat. Br.J.Pharmac. Chemother. 33, 501 - 512 (1968).

Sicuteri, F., Fanciullacci, M., Bavazzano, A., Franchi, G. and Del Bianco, P.L.: Kinins and intracranial hemorrhages. Angiology 21, 193 - 210 (1970).

Starr, M.S.: Inability of dextran to release kinin in rats. J.Pharmac. Pharmacol. 22, 139 - 140 (1970).

Urbanitz, U., Wiegand, H. and Habermann, E.: Zur Frage der Bedeutung des Kininsystems beim thermischen Ödem der Rattenpfote. Naunyn-Schmiedebergs Arch. Pharmak. 246, 476 - 493 (1969).

Várkonyi, T., Joó, F., Csillik, B., Zoltán, Ö.T. and Földi, M.: Increased permeability of the blood-brain barrier in lymphostatic encephalopathy. Angiologica 6, 275 - 279 (1969).

Vogel, R., Trautschold, I. and Werle, E.: Natürliche Proteinasen-Inhibitoren. Georg Thieme Verlag, Stuttgart (1966).

Watters, J.W.: The effects of bradykinin and histamine on cerebral arteries of monkeys. Radiology 98, 299 - 304 (1971).

Werle, E. and Vogel, R.: Über die Freisetzung einer kallikreinartigen Substanz aus Extrakten verschiedener Organe (1). Arch. int. pharmacodyn. 131, 257 - 261 (1961).

Wilhelm, D.L.: Kinins in human disease. Ann.Rev.Med. 22, 63 - 84 (1971).

INFLUENCE OF THE AUTONOMIC NERVOUS SYSTEM ON FIBRINOLYTIC ACTIVITY CAUSED BY BRADYKININ

M. Tesi, L. Caramelli

Department of Angiology, Main Regional Hospital S.M.Nuova of Florence

During the last few years, some data have posed the problem of the connection between the fibrinolytic and plasmakinin system. The relationships between the two systems have been investigated in previous studies at various levels, and even though up to this moment many questions remain unresolved, it is possible however to affirm that reciprocal states of reactive conditioning have been surely ascertained, as you see in Fig. 1.

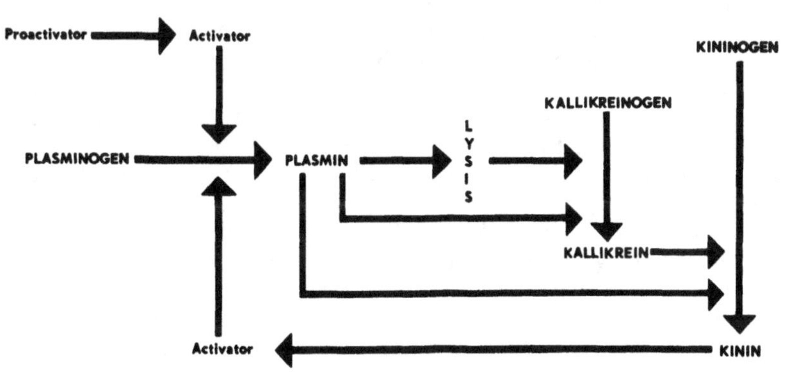

Fig. 1 - Drawing of the relations between plasminogen-plasmin system and kininogen-kinin system.

It has been observed that some activations of the fibrinolytic system are followed by the activation of the kinin one, that is to say plasmin and some products of lysis would behave as kallikreinogenase and as kininogenase. And vice versa some activations of the plasmakinin system are followed by a lysis activation, that is to say bradykinin would act as plasminogenase.

Furthermore other enzymatic systems seem to partake of possible reactive chains connected with lysis and kinins. It is enough to remind of the coagulative system whose bonds are the most ascertained, of the complement and of others.

The observation that plasmin is capable of activating the system of plasmakinins was first reported by Beraldo (1950), and then by Lewis et al. (1958). Successively it was ascertained that activation is not brought about directly, but by means of a process set up by the kallikreinogen-kallikrein system (Webster and Pierce, 1960; Eisen, 1963; Vogt, 1964). In some experimental or pathological conditions with particularly high concentration of plasmin one may find kininogen activation directly plasmin-induced (Back, 1966; Seidel et al., 1969). Other products of lysis too, and in particular, some fibrinopeptides may activate plasmatic kinins (Gladner, 1966).

Concerning bradykinin we know that in vitro it does not activate lysis (Neri-Serneri et al., 1965; Tesi and Brunetti, 1967). It activates lysis on the contrary in vivo, if injected intra-arterially (Neri-Serneri et al., 1962; Deidda et al., 1964; Nuti et al., 1964; Tesi and Brunetti, 1967; Brunetti and Tesi, 1967).

At last is to be considered the influence of the autonomic nervous system on the fibrinolytic activity, as regards both the sympathetic and the parasympathetic system.

Activation of lysis following administration of catecholamines were referred by Biggs et al. (1947), and confirmed by Fearnley et al. (1952), Sherry et al. (1958), Genton et al. (1961), Ambanelli et al. (1963), Ponari et al. (1964). We ourselves with intra-arterial administration of noradrenaline have reported activation of lysis with increases of 33% (Brunetti et al., 1967).

For the parasympathetic system Soulier and Koupernik (1948) described increases in lysis after acetylcholine-induced shock. Cervini and Ficola (1950) also noted increases in lysis due to acetylcholine, Allori et al. (1961), Andreotti et al. (1961) pointed out similar activations when the drug were injected intra-arterially. In repeated experiments carried out with acetylcholine intra-arterially injected, our group has noted high increases in lysis, with average of 42% (Tesi and Brunetti, 1967 a/b; Brunetti and Tesi, 1967; Tesi and Caramelli, 1970). Consequently we wish to recall that Kwaan et al. (1957) suggested the possibility that activation of the fibrinolytic system is always connected with the coming in action of cholinergic fibres which would seem to act leading to liberation of activator.

At last some findings pointed out that variations produced on fibrinolytic activity by sympathetic and parasympathetic blockers are referred. Ishioka et al. (1967) reported reduction in lysis due to adrenaline, when previously a beta-adrenergic blocker, nethalide, had been injected; a second alpha-adrenergic blocker, phenoxibenzamine, produced no variations whatsoever. Ponari et al. (1964) using dihydroergotamine reported a slight tendency to increase in fibrinolysis. Some of our previous papers concerning acetylcholine–induced lysis, have enabled us to establish that atropine abolishes the increases due to acetylcholine; that regitin, an alpha-blocker, and that butidrine, a beta-blocker, lowers them remarkably (Tesi and

Brunetti, 1967; Brunetti and Tesi, 1967).

MATERIALS AND METHODS

The research was carried out in 40 patients hospitalised in the Department of Angiology in the Main Regional Hospital of S.M.Nuova.

1) - In a first group of 10 subjects the fibrinolytic activity due to bradykinin was estimated by means of the following method.

The basal fibrinolysis was determined by taking a sample of venous blood from the antecubital vein.

Synthetic bradykinin, 15 gamma in 10 ml saline, was injected into the brachial artery of the same arm. After 5' of arterial administration of the drug, lysis was estimated by taking a sample of venous blood from the antecubital vein of the treated arm.

Increases in lysis were calculated, by estimating the difference between the fibrinolytic activity after bradykinin and the basal one.

2) - In a second group of 10 subjects we have determined the fibrinolytic activity produced by bradykinin when preceded by atropine.

The method employed was exactly analogous to the foregoing, with the single difference that 1' before arterial injection of bradykinin, atropine, 10 mg in 10 ml saline, was injected also intra-arterially. Increases in lysis were calculated.

3) - In a third group of 10 subjects the fibrinolytic activity due to bradykinin preceded by an alpha-blocker was determined.

The method was the same, except that 1' before the arterial injection of bradykinin an alpha-blocker, regitin, 20 mg in 10 ml saline, was administered intra-arterially too. Increases in lysis were then estimated.

4) - In a fourth group of 10 subjects we have determined the fibrinolytic activity due to bradykinin preceded by a beta-adrenergic blocker.

Here too 1' before the arterial injection of bradykinin, a beta-blocker, butidrine, 100 mg in 10 ml saline, was administered in the same artery. Increases in lysis were then estimated.

The fibrinolytic activity was determined following the method of Tilled et al. (1955), modified by Andreotti et al. (1964), The method is based on measurement of the amount of erythrocytes liberated by a clot of blood incubated at 37°C for a standard time. The samples of blood are carried out with siliconated and refrigerated material.

As the fibrinolytic activity due to intra-arterial injection of bradykinin is very high, we kept the samples in a warm-bath for a period of 2 hours, as indicated by method in hyperlytic situations.

In accordance with the quoted method the fibrinolytic activity is expressed in percentage of lysis.

RESULTS

1) - In the first group of 10 patients in whom bradykinin alone was intra-arterially administered, the value of basal lysis resulted 9,1%, and bradykinin induced lysis resulted 61,6%. The increases in lysis were shown to be 52.5%, as you see on table n. 1.

2) - In the second group of 10 subjects in whom the injection of bradykinin was preceded by atropine, the basal lysis was 13.5%, lysis after administration of

NAME AND AGE	BASAL FIBRINOLYSIS IN %	FIBRINOLYSIS IN % at 5'	INCREASES IN % (3-2)
1	2	3	4
B. R. a. 76	10	78	68
D. M. a. 51	6	66	60
M. G. a. 63	29	67	45
P. E. a. 60	3	52	49
R. O. a. 46	12	64	52
C. A. a. 59	6	56	50
L. G. a. 66	4	43	39
S. F a. 61	22	66	44
P. V. a. 48	3	61	58
C. B. a. 60	3	63	60
A. G. a. 62			
	$\overline{X}=9{,}1$	$\overline{X}=61{,}6$	$\overline{X}\pm s=52{,}5\pm9{,}03$

Table 1 - Fibrinolytic activity after intra-arterial administration of bradykinin (15 γ).

the two drugs was 36,5%. The increases in lysis were shown to be 23%, as you see in table n. 2.

The test of Student showed significance with values of $P < 0.001$ in respect of bradykinin.

3)- In the third group of 10 patients in whom bradykinin had been preceded by an alpha-blocker, regitin, intra-arterially too, the basal lysis was 17.6%, lysis after administration of the two drugs was 40.4%. The increases in lysis were shown to be 22.8%, as you see in table n. 3.

The test of Student showed significance with values of $P < 0.01$ in respect of bradykinin.

4) - In the fourth group of 10 subjects in whom bradykinin had been preceded by a beta-adrenergic blocker, butidrine intra-arterially too, the basal lysis was 13.4%, lysis after administration of the two drugs was 43%. The increases in lysis were shown to be 29.6%, as you see in table n. 4.

The test of Student showed significance with values of $P < 0.001$ in respect of bradykinin.

COMMENT

From the present research the fibrinolytic action of bradykinin in man when the drug is injected intra-arterially appears to be confirmed. Such action had been described in our previous studies (Tesi and Brunetti, 1967 a/b).

It is also shown that the various blockers administered intra-arterially reduced the described action. This is true for atropine, a parasympathetic blocker, and either for regitin and butidrine, sympathetic blockers.

NAME AND AGE	BASAL FIBRINOLYSIS IN %	FIBRINOLYSIS IN % at 5'	INCREASES IN % (3-2)
1	2	3	4
G. L. a. 46	12	37	25
L. A. a. 49	18	49	31
R. B. a. 60	4	34	30
C. I. a. 68	17	35	18
F. C. a. 74	8	8	0
B. P. a. 63	13	36	23
F. L. a. 67	18	52	34
S. B. a. 78	18	38	20
P. R. a. 50	15	50	35
C. Q. a. 71	12	26	14
	$\overline{X}=13,5$	$\overline{X}=36,5$	$\overline{X}\pm s=23\pm10,24$

Table 2 - Fibrinolytic activity after-intra-arterial administration of atropine (10 mg) and bradykinin (15 γ). Atropine injected carried out 1' before bradykinin.

NAME AND AGE	BASAL FIBRINOLYSIS IN %	FIBRINOLYSIS IN % at 5'	INCREASES IN % (3-2)
1	2	3	4
P. A. a. 58	35	73	38
V. E. a. 69	26	26	0
T. G. a. 63	29	80	51
M. F. a. 47	4	4	0
G. L a. 69	26	26	0
U A. a. 57	15	36	20
C. A. a. 60	15	66	51
R. A. a. 48	21	21	0
C. M. a. 51	3	28	25
S. L a. 71	2	45	43
	$\overline{X}=17,6$	$\overline{X}=40,4$	$\overline{X}\pm s=22,8\pm21,90$

Table 3 - Fibrinolytic activity after intra-arterial administration of adrenergic alpha-receptor blocking agent regitin (20 mg) and Bradykinin (15 γ). Regitin injection carried out 1' before bradykinin.

NAME AND AGE	BASAL FIBRINOLYSIS IN %	FIBRINOLYSIS IN % at 5′	INCREASES IN % (3-2)
1	2	3	4
B. U. a. 69	21	66	45
D. S. a. 56	14	50	36
C. E. a. 51	30	76	46
P. A. a. 62	1	1	0
C. A. a. 55	9	9	0
C. E. a. 46	22	42	20
V. P. a. 70	2	27	25
F. A. a. 64	10	45	35
S. O. a. 56	15	63	48
D. A. a. 61	10	51	41
	$\overline{X}$ = 13,4	$\overline{X}$ = 43,0	$\overline{X} \pm s$ = 29,6 ± 17,98

Table 4 - Fibrinolytic activity after intra-arterial administration of adrenergic beta-receptor blocking agent butidrine (100 mg) and bradykinin (15 γ). Butidrine injection carried out 1' before bradykinin.

1) - From the first group of experiments, as you see in Fig. n. 2, it is shown that bradykinin intra-arterially administered activates lysis at a high degree. Really the increases in lysis (activity after bradykinin, diminished of basal lysis) are in value of 52.5%.

These increases are greater, as you see in Fig. n. 3, than those produced by acetylcholine: 42%; and by nicotinic acid: 45%; but are inferior to those produced by eledoisin: 70%. All obtained by the same method (Pratesi et al., 1967 a/b; Tesi, 1967; Tesi and Brunetti, 1967; Tesi et al., 1967; Tesi and Caramelli, 1970 a/b).

2) - From the second group of experiments, as you see in Fig. n.2, it is shown that atropine injected intra-arterially 1' before bradykinin, diminishes activation of the fibrinolytic system produced by the latter. The increases in lysis fall from 52.5% (bradykinin alone) to 23% (atropine-bradykinin).

It is conceivable that atropine provokes a partial block of the fibrinolytic activity of the injected territory, probably at tissutal level.

3) - From the third group of experiments it is shown that the alpha-adrenergic blocker, regitin, introduced intra-arterially 1' before the administration of bradykinin, as you see in Fig. n. 2, diminishes the activation of the fibrinolytic system produced by the kinin. The increases in lysis fall from 52.5% (bradykinin alone) to 22.8% (regitin-bradykinin).

In this case too, one may think of a probable peripheric action of regitin at tissutal level. The drug might cause the regional conditions of response to the bradykininic stimulus to vary.

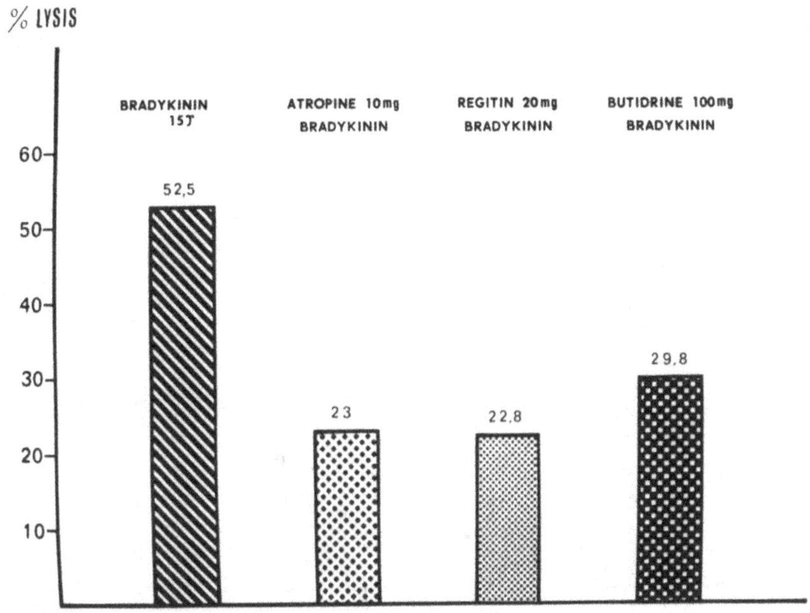

Fig. 2 Lysis increases after intra-arterial administration of:
bradykinin
atropine-bradykinin (atropine carried out 1' before bradykinin)
regitin-bradykinin (regitin carried out 1' before bradykinin)
butidrine-bradykinin (butidrine carried out 1' before bradykinin)

4) - From the fourth group of experiments, as you see in Fig. n. 2, it is shown that the beta-adrenergic blocker, butidrine, introduced intra-arterially 1' before administration of bradykinin, diminishes the activation of the fibrinolytic system produced by the kinin. The increases in lysis fall from 52,5% (bradykinin alone) to 29.6% (butidrine-bradykinin).

In a such manner the substances blocking the autonomic nervous system all reduced the bradykinin produced lysis, in rates not widely differing among themselves. The percentages of reduction for the various drugs are the following: atropine induced a reduction of 56%; regitin a reduction of an equal amount, i.e. 56%; butidrine a slightly lower reduction that is 42%; as you see in Fig. n. 2.

We wish to recall that, in an experiment carried out with a similar method to study the influence of the autonomic nervous system on fibrinolysis produced by acetylcholine intra-arterially injected, as you see in Fig. n.4 , we found for acetylcholine increases in lysis of 42%. On this percentage of lysis, atropine brought about a reduction of 90%, regitin a reduction of 80%, butidrine a reduction of 60% (Tesi and Brunetti, 1967 a/b).

The results obtained allow us to assume the existence of a regional regulation of the autonomic nervous system on determined tissutal vascular functions. At least one may speak of a kind of modulation of certain reactive conditions necessary for the fibrinolytic response to some stimuli, in this case to intra-arterially introduced stimuli with the technique described.

This affirmation concerning an automatic nervous control of the fibrinolytic

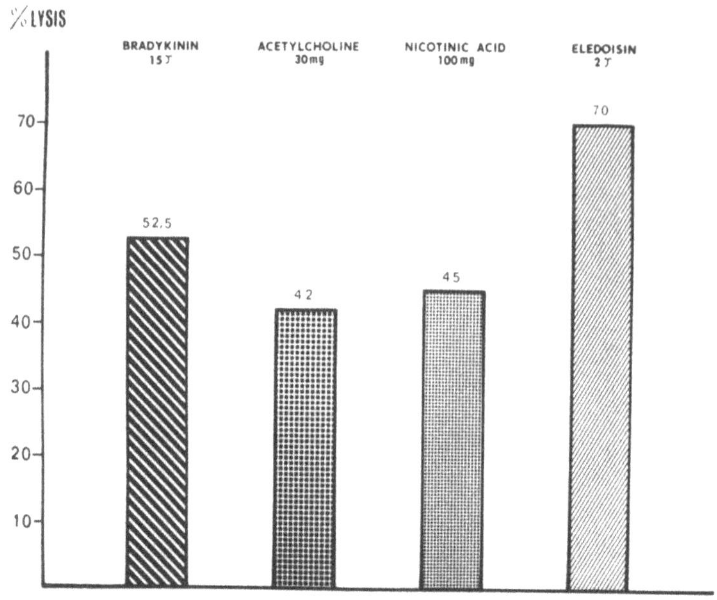

Fig. 3 - Lysis increases after brachial artery administration of:
- bradykinin; - acetylcholine; - nicotinic acid; - eledoisin
(lysis determined on venous blood of antecubital vein 5', after injection).

system, is in accordance with the experiences of Neri-Serneri et al. (1967) who with the xylocainic block of the stellate ganglion obtained reduction of the lysis activated with venous stasis in the corresponding arm; and with painful stimulation of the biceps brachii muscle observed increase in the lysis produced by the same method.

We also find reduction of bradykinin-induced lysis using pharmacological block of the sympathetic and parasympathetic system carried out in different sequences, by injecting intra-arterially now atropine, now an alpha-adrenergic blocker, and finally a beta-adrenergic blocker. In the same way, in our previous research the lysis activated by acetylcholine was proved to lower.

The variations described seem to be most probably bound up with tissutal vascular changes in injected district. In a particular way they seem to be referable to varied conditions of response of micro-vessels and tissues surrounding them, on which both the lytic stimuli (bradykinin, acetylcholine) and the blocking ones may act.

By abolishing or depressing the autonomic nervous system control over the injected territory, some enzymatic and metabolic conditions necessary for the functioning of the coagulative - fibrinolytic system reduce their potentiality. In this way the regional responses to various stimuli are changed, in respect to those present before the block quoted.

SUMMARY

The relationships between the fibrinolytic and plasmakinin systems are known

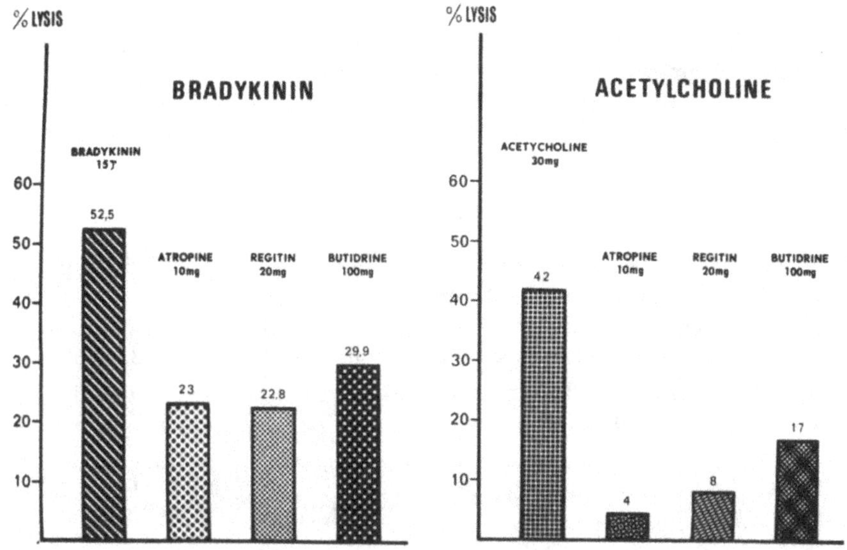

Fig. 4 - Action of Autonomic nervous system on fibrinolysis induced by bradykinin and acetylcholine intrarterially administrated. Changes of lysis' increases produced by atropine, regitin, and butidrine

enough: some lysis products and plasmine act as kallikreinogenase and as kininogenase, bradykinin on the contrary acts as plasminogenase.

Also some data about the influence that autonomic nervous system exerts on fibrinolysis are known, even though put in question.

On this basis the AA. have liked to study the modifications that some substances, whose effects develop at level of autonomic nervous system, may have on fibrinolysis by bradykinin.

The research has been carried out using an original method in man.

The AA. have injected intra-arterially the substance under examination, and they have estimated lysis variations on the vein flowing back from injected area. The research has been carried out to determine the following data:

1) - The fibrinolytic activity of bradykinin: injecting the drug into the brachial artery and estimating lysis on the antecubital vein.

2) - Influence of cholinergic blocking drugs: injecting atropine intra-arterially 1' before bradykinin.

3) - Influence of alpha-adrenergic blockers: injecting regitin intra-arterially 1' before bradykinin.

4) - Influence of beta-adrenergic blockers: injecting butidrine intra-arterially 1' before bradykinin.

From the experiments the fibrinolytic action of bradykinin in man, when administered intra-arterially, appears to be confirmed with increases in lysis in mean value of 52.5%.

It is also shown that the various blockers if injected into the same artery, reduced the quoted action.

Atropine induced a reduction of 56%. Regitin a reduction of equal amount i.e. 56%. Butidrine a slighty lower reduction that is 42%.

The results are then discussed, and the variations provoked by depressing autonomic nervous system control are referred to changes of response of micro-vessels and of surrounding tissues to lytic stimuli.

REFERENCES

Allori L., Isidori A., Petrignani S., Perego M.A.: Arterioclisi acetilcolinica nelle arteriopatie aterosclerotiche. Min. Cardioang. 9, 694, 1961.

Ambanelli U., Cattabiani G., Dettori A.G., Ponari O.: Studio sugli effetti dell'adrenalina sulla emocoagulazione e sulla fibrinolisi. Haemat. 48, 483, 1963.

Andreotti L., Nuzzaci G., Panerai A.: Effetto fibrinolitico dell'acetilcolina somministrata per via endoarteriosa. Bol. Soc. Ital. Biologia Sper. 37, 1273, 1961.

Andreotti L., Nuzzaci G.: Valutazione dell'attività fibrinolitica nel sangue umano. Caratteristiche e risultati di un semplice metodo di misura dell'attività fibrinolitica atto all'impiego clinico. Riv. Crit. Clin. Med. 64, 167, 1964.

Back N., Guth P.S., Munson A.E.: On the relationship between plasmin and kinin. Ann. N.Y. Acad. Sci. 104, 53, 1963.

Back N.: Fibrinolysin system and vasoactive kinin. Fed. Proc. 25, 77, 1966.

Beraldo W.T.: Bradykinin formation by the action of peptone and fibrinolysin on plasma globulins in vitro. Ciencia Cult. 2, 300, 1950.

Biggs R., Macfarlane R.G., Pilling J.: Observations on fibrinolysis. Experimental activity produced by exercise or adrenaline. Lancet 1, 402, 1947.

Brunetti S., Tesi M.: Azione di alcuni farmaci sulla dinamica della fibrinolisi nell'uomo. Riv. Crit. Clin. Med. 67, 233, 1967.

Brunetti S., Tesi M., Caramelli L.: Variazioni dell'attivabilità del sistema fibrinolitico in rapporto all'età. Riv. Crit. Clin. Med. 67, 456, 1967.

Brunetti S., Tesi M., Biddau A., Caramelli L.: Effetti della stimolazione selettiva dei recettori alfa beta adrenergici sulla fibrinolisi dinamica nell'uomo. Riv. Crit. Clin. Med. 67, 386, 1967.

Cervini C., Ficola G.: La fibrinolisi ematica. Clin. Nuova 11, 389, 1950.

Deidda C., Nuti A., Tesi M., Brunetti S.: Considerazioni comparative sull'azione fibrinolitica della bradichinina somministrata per via endoarteriosa nell'uomo. Rass. Neurol. Veg. 18, 769, 1964.

Eisen V.: Kinin formation and fibrinolysis in human plasma. Journ. Physiol. 166, 514, 1963.

Fearnley G.R.: Fibrinolysis Ed. Arnold London 1965.

Genton E., Kern F., Kaulla v. K.N.: Fibrinolysis induced by pressor amines. Am. J. Med. 31, 564, 1961.

Gladner J.A.: Potentiation of the effect of bradykinin. Hypotensive Peptides. Springer Verlag, pag. 344, 1966.

Ishioka T., Matsumare T., Shimamoto T.: Effect of adrenergic blocking agents on increased activity induced by adrenaline. Int. Symp. Coag. Fig. System. Ed.

STP Torino, pag. 271, 1967.

Kwaan H.C., Lo R., Mac Fadzean A.J.S.: On the production of plasma fibrinolysis activity within veins. Clin. Sci. 16, 241, 1957.

Lewis G.P.: Formations of plasma kinins by plasmin. J. Phys. 140, 285, 1958.

Neri-Serneri G.G., Rossi Ferrini P.L., Masotti G., Paoletti P., Silvestrini E.: Researches on the homeostasis of the coagulation fibrinolysis system; release of active substances from tissues and clearence phenomena. Int. Symp. Coag. Fib. System. Ed. STP Torino, pag. 395, 1967.

Neri-Serneri G.G., Procacci P., Rossi Ferrini P.L., Paoletti P.: The role of the sympathetic system in the local homeostasis of the coagulation fibrinolysis system. Int. Symp. Coag. Fib. System. Ed. STP Torino, pag. 383, 1967.

Neri-Serneri G.G., Rossi Ferrini P.L., Masotti G., Paoletti P., Silvestrini E., Panti A., D'Ayala Valva G.: Effects of bradykinin on coagulation and fibrinolysis study in vitro and in vivo. Bradykinin Symp., Florence Italy, July 16th 1962.

Nuti A., Deidda C., Brunetti S., Tesi M.: Azione fibrinolitica della bradichinina somministrata per via endoarteriosa nell'uomo. Rass. Neurol. Veg. 18, 769, 1964.

Ponari O., Cattabiani G., Mastroandrea R., Dettori A.G.: Sui rapporti fra sistema nervoso e processi della emocoagulazione e fibrinolisi. Rivista Critica e contributo personale Hemostase 1964.

Pratesi F., Tesi M., Brunetti S., Dabizzi R.P., Caramelli L.: Action angiodinamique et fibrinolytique de l'eledoisine. Actualités Cardiol. et Angéiol. Int. 4, 319, 1967.

Pratesi F., Tesi M., Brunetti S., Spinelli P., Corsi C., Caramelli L., Dabizzi R.P.: Quelques nouveautés dans le traitment medical des angéiopathies. Rapport au Congrés De la Soc. Franç. Angéiol. XX Anniversaire Paris Octobre 1967.

Seidel G., Stuker H.U., Vogt W.: The role of plasmin for kinin formation in human plasma. Arch. Pharmac. 264, 305, 1962.

Sherry S., Lindemayer R.I., Fletcher A.P., Alkjaersig N.: The apparence of a fibrinolysin activator in the blood of patients with enhanced fibrinolytic activity. J.Clin. Invest. 37, 931, 1958.

Soulier J.P., Koupernik G.: Constance de la fibrinolyse au cours du choc acétylcolinique. Sang. 19, 362, 1948.

Tesi M.: Acido Nicotinico e fibrinolisi nell'uomo. Nota I - Fibrinolisi distrettuale (cerebrale, dell'arto superiore, dell'arto inferiore) da somministrazione di acido nicotinico per via arteriosa. Riv. Crit. Clin. Med. 67, 465, 1967.

Tesi M., Brunetti S., Caramelli L.: Acido Nicotinico e fibrinolisi nell'uomo. Nota II - Tachifilassi e riattivazione fibrinolitica dinamica (generale e distrettuale) da somministrazione di acido nicotinico per via venosa e per via arteriosa. Riv. Crit. Clin. Med. 67, 484, 1967.

Tesi M., Brunetti S.: Etude sur le rapport entre système nerveux autonome et fibrinolyse chez l'homme. Proceedings of the VI international Congres of Angiology, Ed. Cientifico Medico Barcelona pag. 94, 1967.

Tesi M., Brunetti S.: Azione bloccante dell'atropina sulla fibrinolisi da acetilcolina e da bradichinina. Riv. Crit. Clin. Med. 67, 257, 1967.

Tesi M., Brunetti S., Caramelli L.: Variations in the cerebral fibrinolytic potential following carotid administration of nicotinic acid (nicotinic acid fibrinolytic test) in subjects with previous cerebral infarct. Proceedings of the Int. Cong. on clinical evaluation of vasoactive drugs effects. Rome May 1968, Ed. CEPI, pag. 129.

Tesi M., Caramelli L.: Fibrinolytic Potential and Haemorragic Tissue Angiophilia in chronic venous insufficiency due to idhiopatic varices. III Congrés Int. de Phlèbologie. Amsterdam 1968, Ed. Stenvert e Zoon, pag. 729.

Tesi M., Caramelli L.: Terapia fibrinolitica e distretto cerebrale. Studi effettuati mediante attivazione farmacologica del territorio encefalico. II Simp. Int. di Angiol. Encef., Firenze aprile 1970 (in press).

Tesi M., Caramelli L.: Contributo all'interpretazione patogenetica dell'infarto cerebrale mediante lo studio del potenziale fibrinolitico del distretto cerebrale. II Simp. Int. di Angiol. Encef. Firenze aprile 1970 (in press).

Tesi M., Caramelli L.: Aspetti della trombofilia nella malattia aterosclerotica: la fibrinolisi spontanea e la fibrinolisi attivata. Congresso Int. sulla Diagnosi Prognosi e Profilassi dell'Arteriosclerosi, Firenze Dicembre 1970 (in press).

Tillet W.S., Johnson A.J., Mac Carty W.C.: The intravenous infusion of the streptococcal fibrinolytic principle (streptokinase) into patients. J. Clin. Invest. 34, 169, 1955.

Vogt W.: Kinin formation by plasmin an indirect process mediated by activation of kallikrein. J. Physiol. 170, 153, 1965.

Webster M.E., Pierce J.V.: Studies on plasma kallikreins and its relationship to plasmin. J. Pharmac. Exp. Ther. 130, 484, 1960.

FINE STRUCTURE OF SECRETORY GRANULES IN MAMMALIAN SUBMAXILLARY GLANDS

Gundula Dorey and K.D. Bhoola

Department of Pharmacology, University of Bristol

Medical School, University Walk, Bristol BS8 1TD

INTRODUCTION

In exocrine glands macromolecules synthesized for secretion are mainly segregated and sequestered in specific organelles within cells. It is now established that the serine protease kallikrein, and the digestive enzyme amylase, are stored intracellularly in membrane-bounded organelles. Trypsin-like enzymes isolated from the rat submaxillary gland are probably similarly stored in granules. It is evident from this study and those of Shackleford and co-workers (see Shackleford & Wilborn, 1968) that mammalian submaxillary glands show marked differences in the morphology of the various acinar and duct cells and the granules which contain secretory enzymes and other macromolecules. The question therefore arises as to which granules within which cells contain kallikrein, trypsin-like enzymes and amylase. The precise cellular location of the granules which contain the secretory enzymes may provide some insight into the possible physiological role of submaxillary kallikrein and the trypsin-like proteases.

In the present report the fine structural organisation of the granules in the acinar and the various duct cells is described and specific differences between their morphological features are discussed. The six species examined are cat, dog, rabbit, rat, hamster and guinea-pig.

METHODS

The removal and preparation of the submaxillary gland for electron microscopy was the same for all the six species. The glands were fixed in a mixture of 4% paraformaldehyde, 2.5% glutaraldehyde and 1% acrolein in cacodylate buffer (pH 7.3) and subsequently in 1% osmium in cacodylate buffer (pH 7.3). After dehydration through the usual alcohol series and propylene oxide the tissue was embedded in Araldite. Sections were cut on an LKB Ultrotome and examined using a Hitachi HU IIB electron microscope.

OBSERVATIONS

Mammalian submaxillary glands consist typically of acini, intercalated ducts and striated ducts. Modifications of this arrangement may, however, be present in different species. In cat and dog the acini are capped by demilunar cells. In the rabbit cells with characteristic morphological features and granule population are seen between the acini and intercalated ducts. The intercalary region in the guinea-pig consists of two portions, of which the more distal contains granules which are similar to those in the non-specialised striated duct cells of the rabbit, cat and dog. In the rat and hamster, the proximal striated ducts are modified into granular tubules.

Acinar Cell Granules

Two distinct granule populations are present in the acinar cells of the cat. The populations are confined to separate cells but occur within the same acinus. The first population consists of moderately electron-dense granules of relatively uniform size and shape (see Table 1). Granules of the second population are larger and more variable in shape and may coalesce to form larger units. They contain a loosely arranged intragranular material which sometimes includes a small dense body. Cells containing these granules occupy the extensive demilunar region of the acinus.

In the dog the acinar granule population shows a wider variation in the organisation of intragranular material, although all variations are found within the same cell. The paler granules tend to form large amorphous sacs. The demilunar region consists of small cells with discrete granules of a more compactly arranged secretory material which includes circular profiles.

Rabbit acinar cells contain a single population of very pale organelles which regularly fuse to form larger units. Of particular interest in the rabbit are cells at the junction of the acinus

Table 1 Comparison of the size of acinar cell granules

Electron-density of granules	Species						
	Cat	Dog	Rabbit	Rat	Hamster M	Hamster F	Guinea-pig
Pale	2.5	2.6	2.7	1.4	2.2	-	1.8
Intermediate	1.8	2.1	-	-	1.7	2.0	1.5
Dense	-	-	2.7*	-	1.5	-	1.1

Each granule size is indicated in micrometres (μm) and represents the mean of a minimum of 25 measurements on discrete, non-confluent organelles. *The size of the granules in the junctional cells of the rabbit is included for comparative purposes.

M = male; R = female.

and the intercalated duct (Fig. 1). The granules contained in these cells are periodic acid-Schiff (PAS) positive and their affinity for PAS is unaffected by diastase. Ultrastructurally the granules appear very similar to those in some acinar cells of both guinea-pig (Fig. 2) and male hamster.

Acinar cells of the rat also contain a single population of electron-pale granules (Fig. 3). The granules are somewhat smaller than those in the rabbit. Intragranular figures which have a characteristic crystal-like pattern are a distinctive feature.

Acinar cell granules in the female hamster are uniform in size and show a characteristic dichotomy in electron density; the granules contain a central pale area of loosely arranged material and a peripheral, more compact dense area, usually crescent-shaped. In contrast, the acinar cell granules in the male hamster possess an electron-dense core and a crescent-shaped outer paler region. In addition some cells contain granules showing three different densities, similar to rabbit junctional and guinea-pig acinar cells. Acinar cell granules in the guinea-pig submaxillary gland show a marked variation in electron-density and size (see Fig. 2, see Table 1). Larger pale and smaller moderately dense granules are most frequent but small dense compact organelles are also present. The pale granules are discrete; few of them fuse to form larger units.

Intercalated Duct Cell Granules

Most species possess somewhat sparsely distributed very dense, small granules in the apical region of intercalated duct cells. In the guinea-pig, however, although the proximal intercalary region closely resembles that in the other species (see Fig. 2), the distal part consists of larger columnar type cells containing very many smaller, electron-dense granules in the apical portion of the cells (Table 2).

Organelles in Granular Tubules

Specialised striated duct cells seen in the rat and hamster are called granular tubules. In the rat granules of two electron-densities are seen in these duct cells (Fig. 4). One population consists of large dense granules; the other consists of granules similarly pale in electron-density but divisible into two groups on the basis of size (see Table 2). In the hamster the organelles can also be separated into two different size populations, but both groups possess the same density.

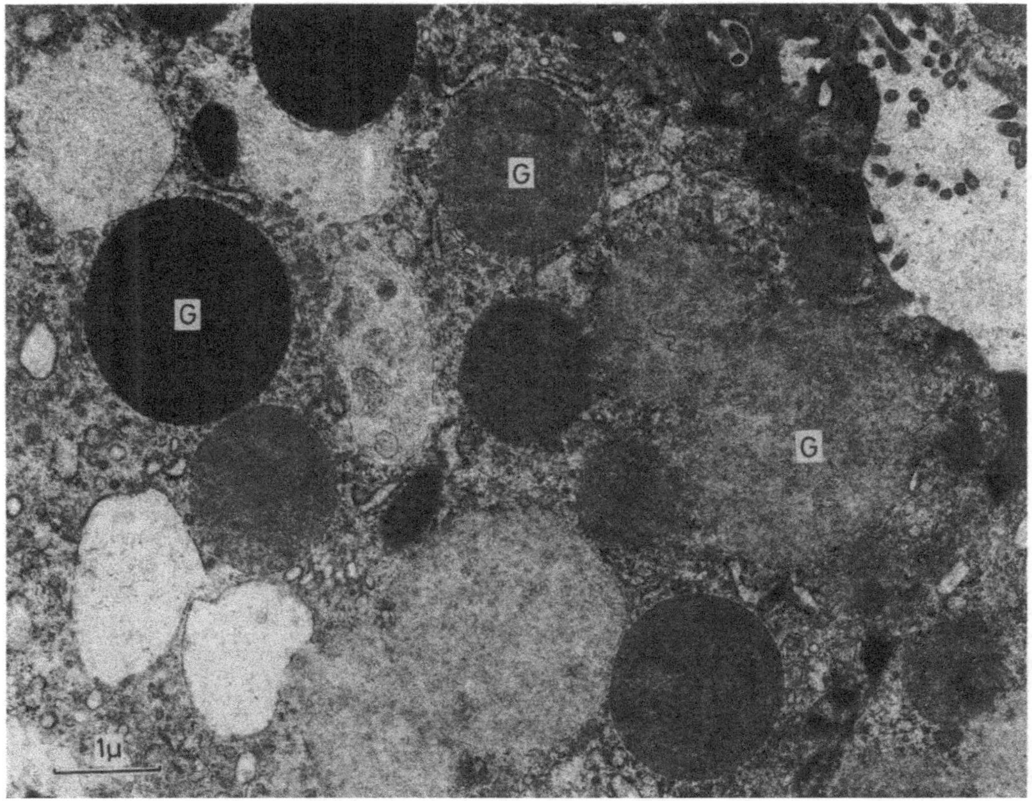

Fig. 1. Rabbit: Junctional cells between acini and intercalated ducts containing discrete granules (G) which show a marked variation in density.

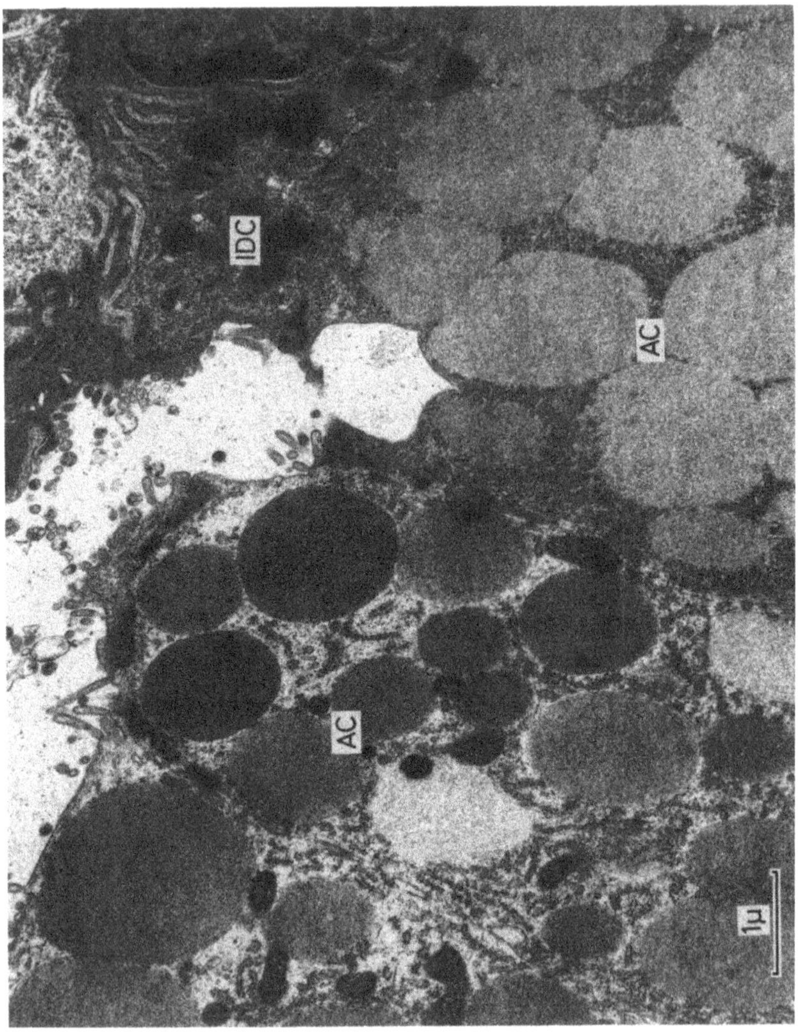

Fig. 2. Guinea-pig: Junction between acinar (AC) and intercalated duct cells (IDC). Acinar granules are larger and very variable in density compared with the small uniformly dense granules of the intercalated duct cells.

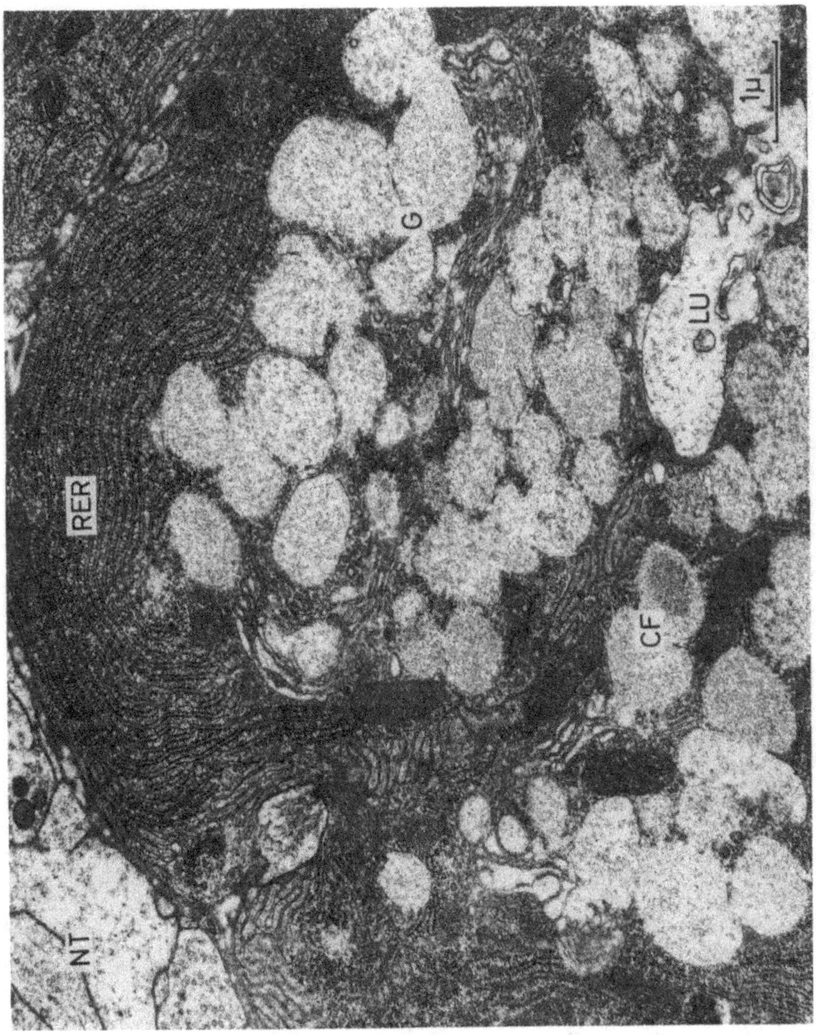

Fig. 3. Rat: Acinar cells containing pale, amorphous granules (G) which tend to fuse. Some granules show a characteristiccrystal-like intragranular figure (CF). Granules are always arranged apically near the acinar lumen (LU); the basal part of the cell is filled with rough endoplasmic reticulum (RER). Nerve terminal (NT).

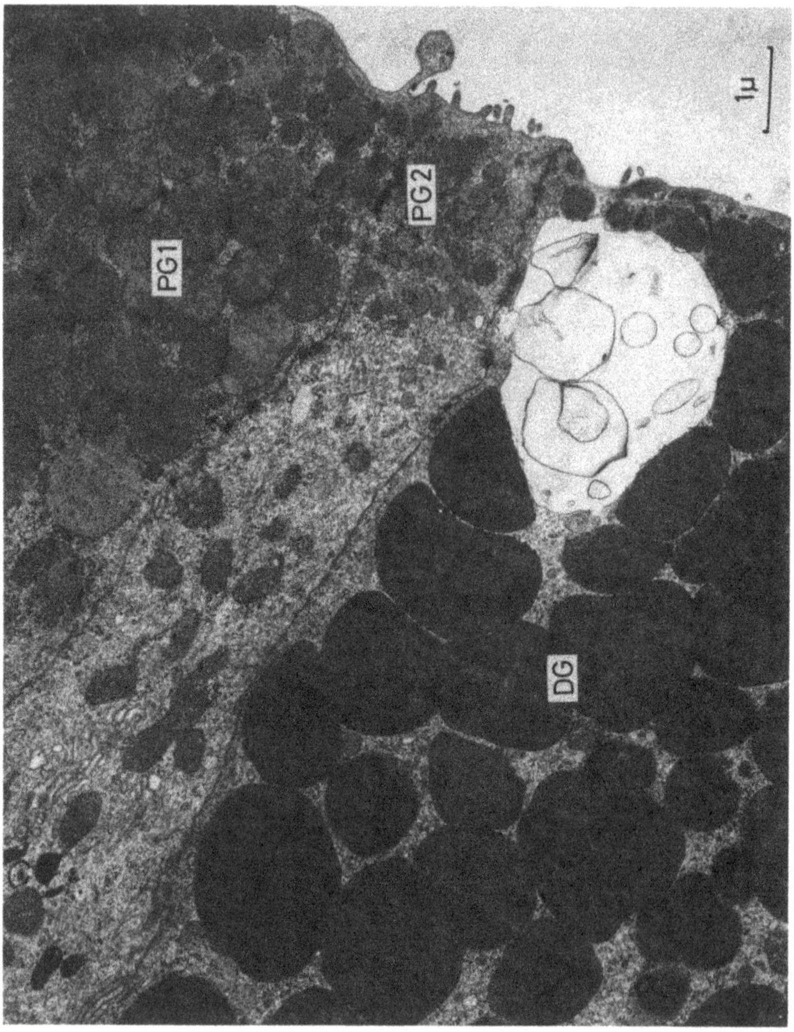

Fig. 4. Rat: The cells of the granular tubules contain granules of three types, large pale (PG1), and small pale (PG2) and large dense (DG). The three types are generally found in different cells.

Table 2 Comparison of the size of duct cell granules

	Cell types	Electron-density of granules	Cat	Dog	Rabbit	Rat	Hamster	Guinea-pig
1.	Modified acinar/ intercalated duct	Dense	-	-	2.7	-	-	-
2.	Intercalated duct	Dense	*	*	0.76	0.65	0.99	0.76
3.	Modified intercalated/ striated duct	Dense	-	-	-	-	-	0.57
4.	Striated granular tubules	Intermediate	-	-	-	(a) 2.0 (b) 0.4	-	-
		Dense	-	-	-	1.4	(a) 2.6 (b) 0.7	-
5.	Striated duct	Intermediate	0.42	0.3	0.15	-	-	-

Each granule size is indicated in micrometres (μm) and represents the mean of a minimum of 25 measurements. (a) and (b): these organelles are readily separable into two populations according to size, and measurements were therefore made on both the large and the small populations. * Indicates no measurements were made.

Granules in Striated Duct Cells

The non-specialised striated ducts of the cat, dog and rabbit contain very small, discrete granules of variable density in the apical regions of the cells.

DISCUSSION

The enzyme distribution study in submaxillary glands reported in the preceding paper indicates that, whereas all six species contain kallikrein (even though the gland content shows a wide variation) only the rat, hamster and cat contain trypsin-like proteases, and significant amounts of amylase are detected only in the guinea-pig.

Although it has been known for some time that amylase is stored intracellularly in granules, the storage of kallikrein in membrane-bounded organelles has only recently been established (Bhoola & Ogle, 1966; Bhoola, 1968; Erdös, Tague & Miwa, 1968; Chang, Erdös, Miwa, Tague & Coalson, 1968; Bhoola, 1969; Bhoola & Heap, 1970).

Light microscopy histochemical studies seem to indicate that amylase in the mouse is localised in the granular tubules (Chrètien & Zajdela, 1961; Smith, Frommer & Schiff, 1971). In contrast, Bhoola and Heap (1970) have shown that kallikrein- and amylase-containing organelles in the guinea-pig appear to be located in acinar cells. The subcellular studies of Erdös and co-workers (1968) in the rat and mouse, although confirming the localisation of kallikrein in granules, give little indication of the purity of the granule-containing fractions or the cellular origin of the organelles reported to contain kallikrein. From the limited species studies of the isolated organelles there seems, therefore, to be no uniformity in the cellular location of kallikrein.

Although previous studies in the mouse and the rat show an association between proteolytic activity and morphological changes in the granular tubules, no attempt has been made to determine whether or not a parallelism exists between the first appearance of the secretory organelles in these cells and the synthesis by the gland of kallikrein, trypsin-like protease and amylase.

At the time of sexual maturity, steroid hormones are believed to produce biochemical and morphological changes in the granular tubules. Both renin (Oliver & Gross, 1967) and one of the trypsin-like proteases, glandulain (Riekkinen & Niemi, 1968) show a remarkable dependence on androgens. Two questions therefore arise. First, which of the trypsin-like enzymes in the rat and hamster, and mouse (unpublished results) are stored in the granular tubules, and second, does kallikrein like renin in the mouse also show a sex hormone dependence.

Since the cellular origin and the possible sex hormone dependence of kallikrein and the trypsin-like proteases in submaxillary glands may relate to function, further studies are clearly required to determine the precise cellular origin of the organelles containing these secretory enzymes in the various mammalian species. This study forms the basis of future reports in which it is hoped to examine correlations between the gland content of secretory enzyme and changes in granule populations in the acinar and the various duct cells during post-natal development and after castration and ovariectomy.

We wish to thank the Wellcome Trust for a travel grant.

REFERENCES

Bhoola, K.D. (1968). J. Physiol., Lond. 196, 431-445.
Bhoola, K.D. (1969). Biochem. Pharmac. 18, 1252-1254.
Bhoola, K.D. & Heap, P.F. (1970). J. Physiol., Lond. 210, 421-432.
Bhoola, K.D. & Ogle, C.W. (1966). J. Physiol., Lond. 184, 663-672.
Chang, T.S., Erdös, E.G., Miwa, I., Tague, L.L. & Coalson, Jacqueline (1968). Circulation Res. 23, 507-517.
Chrètien, M. & Zakdela, F. (1961). C.r. hebd. Sèanc Acad. Sci., Paris. 260, 4263-4266.
Erdös, E.G., Tague, L.L. & Miwa, I. (1968). Biochem. Pharmac. 17, 667-674.
Oliver, W.J. & Gross, F. (1967). Am. J. Physiol. 213, 341-346.
Smith, R.J., Frommer, J. & Schiff, R. (1971). J. Histochem. Cytochem. 19, 310-319.
Shackleford, J.M. & Wilborn, W.H. (1968). Ala J. med. Sci. 5, 180-203.

DISTRIBUTION OF KALLIKREIN, TRYPSIN-LIKE PROTEASES AND AMYLASE IN SUBMAXILLARY GLANDS

K.D. Bhoola and Gundula Dorey

Department of Pharmacology, University of Bristol

Medical School, University Walk, Bristol BS8 1TD

INTRODUCTION

In recent years, Schachter has questioned the evidence for a humoral, transmitter-like role for submaxillary kallikrein. Subcellular experiments were designed to bring a new approach to this problem which originally arose with the finding of Heidenhain (1872) that chorda nerve mediated vasodilatation in the cat submaxillary gland was not blocked by atropine. In the autumn of 1962, experiments were commenced to determine the subcellular localization of kallikrein in homogenates of guinea-pig submaxillary gland. The early results appeared to indicate that kallikrein was stored intracellularly in granules (Bhoola & Ogle, 1966). In subsequent studies this finding was confirmed and the kallikrein-containing organelles characterised in greater detail (Bhoola, 1968; Bhoola & Heap, 1970). Essentially similar results have been reported by Erdös and his colleagues in the rat and mouse (Erdös, Tague & Miwa, 1968; Chang, Erdös, Miwa, Tague & Coulson, 1968). Furthermore, a similar subcellular distribution pattern has been obtained for kallikrein in homogenates of the submaxillary glands of the cat, dog and rabbit (Bhoola, 1969).

In order to identify the kallikrein-containing granules and to ascertain their precise cellular location, the *in situ* histochemistry and ultrastructure of the submaxillary gland was examined (Heap & Bhoola, 1969; Heap & Bhoola, 1970). The histochemical reactivity and morphological organisation of the various populations of granules showed significant variation. Consequently a detailed morphological study of the secretory organelles in the submaxillary gland of several species was carried out and our findings are reported in the succeeding paper.

The combined subcellular and morphological studies have established that kallikrein, like the digestive enzyme amylase, is stored in membrane-bounded organelles. Recently, similar proteases but more closely resembling trypsin in substrate specificity have been isolated from the rat submaxillary gland (Riekkinen, Ekfors & Hopsu, 1966; Riekkinen, Ekfors, Hollmèn, Hopsu-Havu, 1967; Riekkinen, Ekfors & Hopsu-Havu, 1967). The question therefore arises as to whether these trypsin-like enzymes also possess kininogenase activity and what, if any, is their physiological role. The species distribution of the proteases may also relate to function. The tissue content of serine proteases has been determined in a few species only. In the present study, therefore, the submaxillary gland content of kallikrein, amylase and trypsin-like enzymes has been determined in the cat, dog, rabbit, rat, hamster and guinea-pig.

METHODS

Submaxillary glands from each of six species were removed under pentobarbitone anaesthesia, rinsed several times in 0.9% NaCl, freeze-dried and pooled. Tyrode solution extracts were used to determine the kininogenase activity of kallikrein. Heated, dialysed dog plasma was used as substrate. The extracts were assayed on the isolated guinea-pig ileum, cat jejenum, rat uterus and rat duodenum in the presence of l-cysteine, atropine and mepyramine. The esterase activity of kallikrein was determined on benzoyl arginine ethyl ester (BAEe) using the alcohol dehydrogenase method described by Trautschold and Werle (1961). Enzymes with trypsin-like activity were measured on benzoyl arginine p-nitroanilide (BAPA) using Tris-HCl extracts (0.067M, pH8.0). Tissue extracts in Sörensen's buffer (pH 6.9) were used to measure amylase (see Bhoola & Heap, 1970). Protein was also measured in the tissue extracts (Lowry, Rosebrough, Farr & Randall, 1951). Enzyme activity was expressed as bradykinin equivalents on milliunits/mg protein tissue extract (see Bhoola & Dorey, 1971). Inhibition studies were carried out with trasylol and soya bean, lima bean and ovomucoid trypsin inhibitors.

RESULTS

Submaxillary glands of the rat possessed the highest kinin-releasing activity (see Table 1). Comparatively high concentrations of kallikrein were also identified in the guinea-pig, hamster and the cat (and the mouse, unpublished results). No previous values have been reported for the rabbit, hamster and the guinea-pig.

In the cat, dog and rat the kallikrein content was qualitatively similar to that reported by Werle and his colleagues (see Frey, Kraut & Werle, 1951). The action of tissue enzymes on plasma substrates could result in the release of smooth muscle-contracting substances other than kinins. Since the guinea-pig ileum is not specific for kinins, the presumed kininogenase activity was confirmed by assays on the cat jejunum, rat duodenum and rat uterus.

Extracts of each of the six species possessed BAEe hydrolysing activity (see Table 1). The esterase activity, however, did not correlate with the kininogenase activity, nor was there any quantitative correlation between the ratios of the two activities.

The kallikrein activity in the tissue extracts was examined after incubation with four inhibitors of trypsin. Although the inhibitory properties appeared similar there was no qualitative correlation between the inhibition of the kininogenase and esterase activities. In the rat, both activities were almost completely inhibited by trasylol. The kininogenase activity of kallikrein in the dog, hamster and guinea-pig was unaffected by the four proteinase inhibitors. Surprisingly, and contrary to previous reports in the literature, about 70% of the kininogenase in the cat, rabbit and rat was blocked by soya bean trypsin inhibitor. Lima bean and ovomucoid trypsin inhibitors produced slight inhibition in the rat but were ineffective in the other species.

Trypsin-like proteases were identified in the submaxillary gland extracts of the rat, hamster and cat. The highest activity was present in the rat. Although the trypsin-like activity in the cat was relatively small, it was observed consistently. Preliminary results indicate that similar enzymes are also present in the mouse submaxillary gland. The effect of the four proteinase inhibitors on BAPA hydrolizing enzymes in the rat, cat and hamster was clearly different. Over 80% of the activity in the rat and 20% in the cat was blocked by trasylol and soya bean trypsin inhibitor; much lesser amount but proportionately similar amount of inhibition was obtained with lima bean trypsin inhibitor. None of the inhibitors affected the activity of the enzymes in the hamster.

Whereas kallikrein was identified in the extracts of all six species, high concentrations of amylase were detected only in the guinea-pig. Cat, dog and hamster extracts contained no activity. A very small amount was identified in the rabbit and only a slightly greater amount in the rat.

In preliminary paper electrophoretic experiments attempts to separate the kininogenase, esterase and trypsin-like activities were unsuccessful. Experiments are in progress to separate these activities by chromatography on DEAE-cellulose and sephadex and to

TABLE 1

Submaxillary gland content of secretory enzymes

Species	Kininogenase* Guinea-pig ileum	Kininogenase* Cat jejunum	Esterase † BAEe	Trypsin-like † proteases BAPA	Amylase†
Rat	32,025	20,815	11,090	125	12.0
Guinea-pig	6,575	2,293	43	0	1,436
Cat	3,099	1,346	146	0.06	0
Hamster	2,375	743	7,443	0.8	0
Dog	258	194	39	0	0
Rabbit	135	75	21	0	0.6

Enzyme activity expressed as *synthetic bradykinin equivalents/mg protein tissue extract and as † milliunits/mg protein tissue extract. Mean results of 3 - 6 experiments; each experiment represents the mean of multiple measurements performed on extracts prepared from pooled, freeze-dried submaxillary glands of each of the six species. Zero activity indicates no hydrolysis of substrate by extracts after 20 min incubation in the case of trypsin-like proteases and after 120 min incubation in the case of amylase.

obtain subsequent purification by affinity chromatography and electrophoresis.

DISCUSSION

The kallikreins (E.C.3.4.4.21) are serine proteinases with substrate specificity which overlaps with that of trypsin (E.C.3.4.4.4) and chymotrypsin (E.C.3.4.4.5). Although kallikrein and trypsin belong to the same family of enzymes and are enzymically very effective on synthetic esters of arginine, the submaxillary kallikreins purified so far have no action on BAPA, the typical trypsin substrate. Furthermore, unlike trypsin and chymotrypsin, the submaxillary kallikreins do not hydrolyse any naturally occurring proteins except the kininogens.

The lack of correlation between the kininogenase and esterase activities of kallikrein and their inhibition profiles seems to indicate the presence of additional BAEe hydrolysing enzymes without any kinin releasing activity. These experiments also suggest that either the salivary kallikreins in the different species are not identical enzymes or that a number of different kininogenases, both kallikrein and trypsin-like, occur in these species. Enzymes similar to trypsin have been reported in the rat submaxillary gland. Their presence in the hamster, mouse and cat described in this study suggests a more widespread occurrence. The functional importance of the salivary trypsin-like proteinases is unknown and requires analysis.

Hilton and Lewis (1955, 1956; see Hilton, 1970) have proposed that kallikrein is a local hormone (transmitter) which in the secreting submaxillary gland diffuses from acinar cells into the interstitial tissue space and indirectly produces dilatation of the arterioles. The subcellular and subsequent morphological and tissue distribution studies were designed to ascertain whether knowledge of the intracellular and cellular localization, transport and release of kallikrein would assist in resolving the controversy concerning its physiological role (see Schachter, 1969). These studies in fact have established that kallikrein in submaxillary glands of the mammalian species examined so far is stored intracellularly in organelles which appear to resemble secretory, zymogen granules observed *in situ* in size and histochemical reactivity. Also, whereas both in the 'resting' and in the secreting submaxillary gland images are frequently seen of granules discharging into the acinar and duct lumen, in none of the reports so far have secretory granules been observed crossing the basal plasma membrane and emptying into the interstitial tissue space. These studies have therefore strengthened previous evidence that, at physiological frequencies, submaxillary kallikrein does not initiate or maintain the vasodilatation associated with stimulation of the chorda tympani nerve. On the contrary, since kallikrein is secreted into saliva, these combined studies provide further support for the view that kallikrein has an extracellular, exocrine functional role. The wide variation in kallikrein content of the different mammalian submaxillary glands reported in the present study, 200 fold between the rat and rabbit may reflect a particular species specific physiological role related to its presence in saliva.

We wish to thank the Wellcome Trust for a travel grant, and K.D.B. is grateful to the Medical Research Council for a research grant.

REFERENCES

Bhoola, K.D. (1968). J. Physiol., Lond. 196, 431-445.
Bhoola, K.D. (1969). Biochem. Pharmac. 18, 1252-1254.
Bhoola, K.D. & Dorey, Gundula (1971). Br. J. Pharmac. In Press.
Bhoola, K.D. & Heap, P.F. (1970). J. Physiol., Lond. 210, 421-432.
Bhoola, K.D. & Ogle, C.W. (1966). J. Physiol., Lond. 184, 663-672.
Chang, T.S., Erdös, E.G., Miwa, I., Tague, L.L. & Coalson, Jacqueline (1968). Circulation Res. 23, 507-517.
Erdös, E.G., Tague, L.L. & Miwa, I. (1968). Biochem. Pharmac. 17, 667-674.
Frey, E.K., Draut, H. & Werle, E. (1951). Kallikrein (Padutin) Stuttgart: Ferdinand Enke.
Heap, P.F. & Bhoola, K.D. (1969). J. Anat. 105, 525-532.
Heap, P.F. & Bhoola, K.D. (1970). J. Anat. 107, 115-130.
Heidenhain, R. (1872). Arch. ges. Physiol. 5, 309-318.
Hilton, S.M. (1970). In: Bradykinin and Related Kinins. Advances in experimental medicine and biology, 2, 625-628.
Hilton, S.M. & Lewis, G.P. (1955). J. Physiol., Lond. 129, 253-271.
Hilton, S.M. & Lewis, G.P. (1956). J. Physiol., Lond. 134, 471-483.
Riekkinen, P.J., Ekfors, T.O. & Hopsu, V.K. (1966). Biochim. biophys. Acta. 118, 604-620.
Riekkinen, P.J., Ekfors, T.O., Hollmèn, T. & Hopsu-Havu, V.K. (1967). Enzymologia, 32, 97-109.
Schachter, M. (1969). Physiol. Rev. 49, 509-547.
Werle, E. (1970). In Handbook of Experimental Pharmacology. 25, 1-13. Berlin-Heidelberg-New York: Springer-Verlag.

CHANGES IN KALLIKREIN ACTIVITY OF RAT SUBMANDIBULAR GLAND DURING POST-NATAL DEVELOPMENT

W. T. Beraldo, G. Siqueira, J. A. A. Rodrigues and Conceição R. S. Machado. Departments of Physiology and Morphology, ICB. Universidade Federal M. G. Brasil

Previous papers of this laboratory have shown that rat kallikreins (urinary, pancreatic and salivary) besides their kininogenic activity exhibit the peculiarity of eliciting contraction of rat uterus apparently independent of a substrate in the uterus. As both properties appear at the same zone of the electrophoretic tracing it is suggested that a single substance is responsible for the effects on uterus and plasma. However, certain reagents inhibit only the kininogenic activity leaving the oxytocic unaffected (Beraldo et al, 1966; Araujo et al, 1970 and Siqueira et al, 1970).

The present experiments were undertaken to follow up the appearance of this special kallikrein in the submandibular gland of rats during post-natal development in a further attempt to detect separately the kininogenic and oxytocic activity, taking advantage of the successive and well-defined stages of the development of the glandular structures.

A preliminary account of this study was presented to the XXV International Congress of Physiological Sciences, Munich, 1970.

MATERIAL AND METHODS

Gland extract

Wistar rats of both sexes were used. The animals were anesthetized with pentobarbital sodium (3 mg/100g, ip). One of the submandibular gland was removed, transfered to iced 0.9%. NaCl solution and squeezed to eliminate blood from it. After drying between filter paper, a piece of 10 to 15 mg of the caudal pole of the gland was weighed and ground in a test tube with sand and saline by mean of a glass rod. The macerated tissues were filtered off and the filtrate used for chromatography.

Filtration on Sephadex G-100

The filtration was made through two connected columns of Sephadex G-100 (1x30cm each). 1ml (1mg) of gland extract in saline was passed through the columns, equilibrated and eluted with saline (pH 6), at room temperature; 5ml fractions were collected, at a flow rate of 30ml per hour. The emergence of the activity was indicated by the oxytocic effect of the fractions on the isolated rat uterus and their kininogenic activity when incubated with heated-treated plasma.

Protein determination

The method described by Lowry et al. (1951) was employed.

Isolated rat uterus

Wistar rats weighing 150-200g were used. One horn of the virgin uterus was mounted in 10 ml Jalon solution maintained at 30° C.

Bioassays

The oxytocic activity was assayed on the isolated rat uterus. Previous experiments showed that there is a parallelism between the size of contractions of the uterus elicited by gland extract fractions and those of bradykinin.

The kinin-forming activity was measured on the rat uterus, incubating fractions collected from the Sephadex column with an excess of heated-treated rat plasma, according to Prado, Prado and Brandi (1962). The kinin released at a chosen period of time was compared against a synthetic bradykinin standard. The direct action on the uterus was prevented using amounts of fractions unable to contract the muscle.

Chemicals

Crystalline trypsin (Worthington Biochemical Co.) Trasylol (Bayer) kindly presented by Dr. Klauss Jansen, Leverkusen, Germany. Bradykinin was kindly supplied by Sandoz Wander, Inc. Hanover New Jersey USA and 1-chloro-3-tosyllamido-7-amino-2-heptanone (TLCK) was supplied by Dr. M. Mares-Guia, Minas Gerais, Brazil.

RESULTS

Preliminary experiments demonstrated that addition of submandibular gland extract to the organ bath containing rat uterus, the preparation responded by a contraction after a latent period ranging from 1 to 3 min. On repetition at 2 min intervals, the same amount of extract, after 12 to 15 additions, elicited little or no response. However, when the horn was replaced by its mate from the same uterus, additions of the extract again elicited contraction. Spontaneous recovery of the capacity to produce a new response was not observed even when the preparation was left resting 40 to 60 minutes. A provisional hypothesis which can be put forward is that, besides the oxytocic principle, there is an inhibitor present in the extract. That this is the case is demonstrated by experiments which follow.

Inhibitor present in gland extract

The filtration of gland extract through Sephadex column shows that an inhibitor of the oxytocic principle can be separated in the fractions 40 to 70 ml of the eluted volume (Fig. 1). These fractions when tested on the uterus produce no contraction but show a strong inhibitory effect against the oxytocic principle; the contraction produced by bradykinin is not altered.

Some properties of the fractions of gland extract

Besides the inhibitor, the filtration of gland extract separates fractions presenting powerful oxytocic and kininogenic activities. When incubated with diisopropylfluorophosphate (DFP) both activities are lost in 60 to 180 minutes (Fig. 2). Trasylol also inhibits both activities. These fractions hydrolyze benzoyl-arginine-p-nitralid (BAPA).

Based on these properties it seems that we are dealing with a kinin-forming enzyme which may well be trypsin or a trypsin--like kallikrein.

Trypsin and trypsin-like kallikrein

Gaddum (1953) has shown that the response of isolated rat uterus to trypsin may be obtained repeatedly, and constant effect may be recorded. This effect was also observed in our experiments.

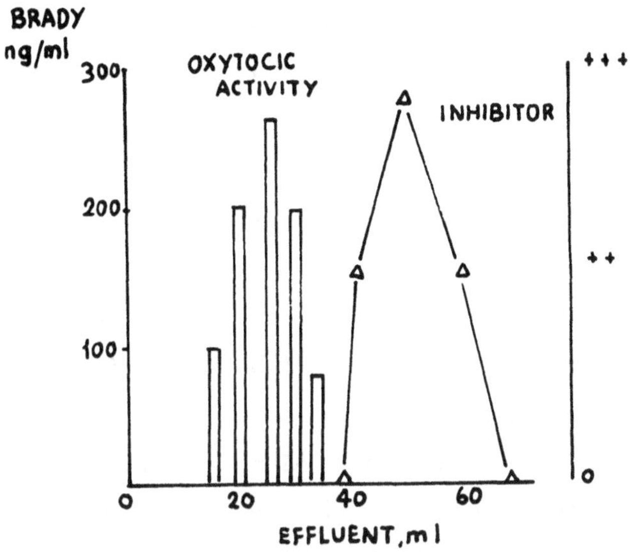

Fig. 1. Separation of oxytocic activity from the inhibitor. Signals (+) represent a semiquantitative avaliation of the inhibitor.

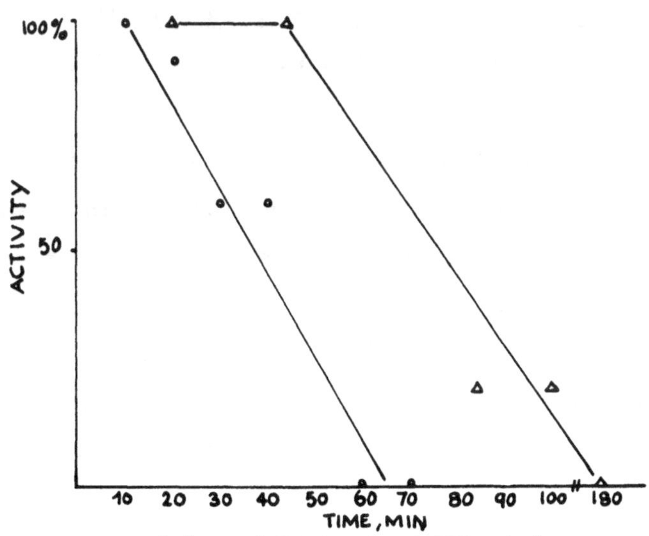

Fig. 2. Time course of the inhibition by DFP of the oxytocic (o) and kininogenic (▲) activities of a pool of fractions obtained from a column of Sephadex G-100.

The presence of proteases in the submandibular gland of rats was reported by Rothschild and Junqueira, (1951) and Sreebny et al. (1955). Riekkinen et al. (1967) demonstrated the presence of trypsin and trypsin-like enzymes in the gland. So, the oxytocic and kininogenic action of the gland extract could be due to the presence of trypsin. However, we found that 0.02 microgram protein of gland fractions caused contraction of the rat uterus equivalent to 10 microgram bovine crystalline trypsin. In addition, 1-chloro-3-tosylamido -7-amino-2-heptanone (TLCK), in conditions that inhibited bovine trypsin, showed without effect on the active fractions of the gland. Mares-Guia and Diniz (1967) demonstrated that rat trypsin is also inhibited by TLCK. These authors pointed out that the lack of reaction between TLCK and rat kallikrein is sufficient to distinguish trypsin-like kallikrein from trypsin.

Kallikrein content and age changes in the structure of the gland

As the submandibular gland of rats shows striking change in its structure during post-natal development (Jacoby and Leeson, 1959) it was decided to measure the oxytocic and kininogenic activities of the gland during the process of development in an attempt to localize the structures related to the production of these activities. Fig. 3 shows that both activities increase parallely with the development of the animal and with the appearance of the acinar cells in the gland (Fig. 4).

At birth, acini are not formed. The gland presents terminal tubules, whose cells show high amount of secretory granules. The intralobular ducts are present. As shows the Fig. 3, in newborn the gland contains no kallikrein at all. So, the granules seen in terminal tubules probably are not associated with the

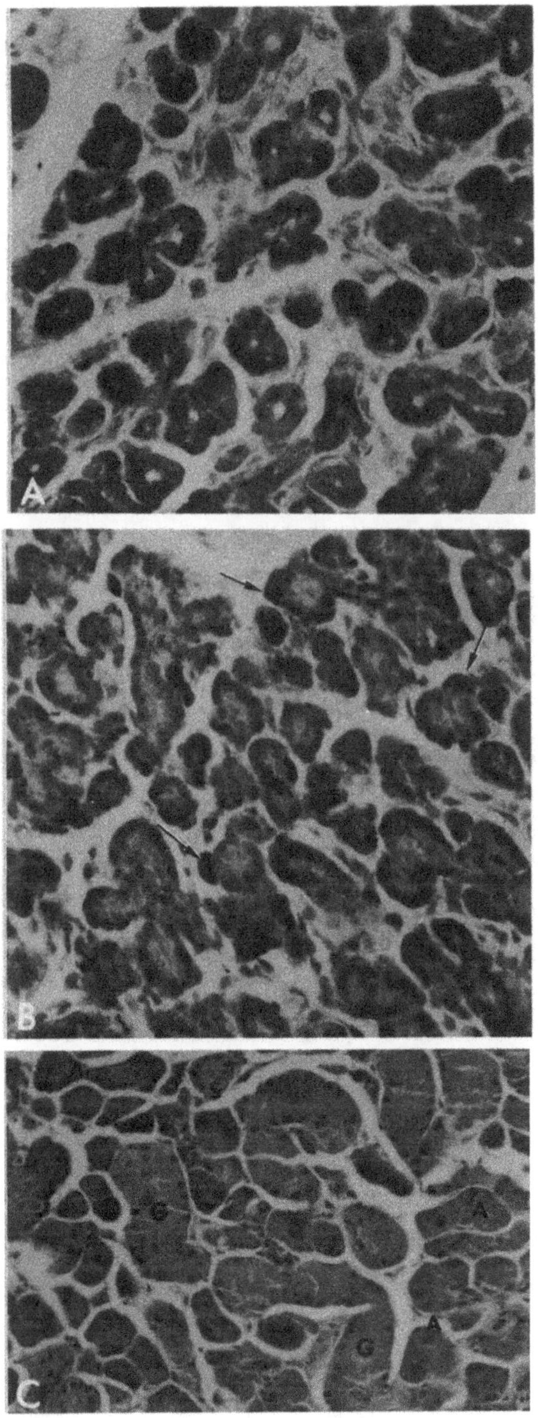

Fig. 4

A. Submandibular gland of 12 hours old rat showing terminal tubules with PAS-postive secretory granules X 270

B. Submandibular gland of 7 days old rat. Note the terminal tubules with small amount of secretory granules and the strongly PAS- positive cells "budding" of them (arrows)

C. Submandibular gland of 150 days old rats showing acini (A) granulated convoluted ducts (G) and striated ducts (S) Hematoxylin eosin staining X 179

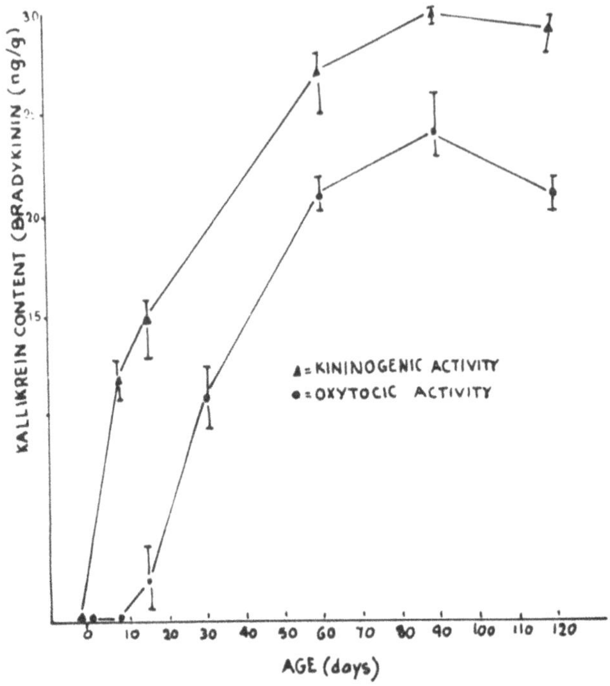

Fig. 3. Kallikrein content (oxytocic and kininogenic activities) of the submandibular gland of rats during development. A pool of fractions was made and used for avaliation of activities. Each point is a mean obtained from 3 to 5 rats ± S.E.M. The vertical bars indicate standard errors of the means. Ordinates: kallikrein content expressed as bradykinin (log-scale).

production of active kallikrein.

During the first week the acinar cells start "budding" off from the terminal tubules. These cells show PAS positive secretory granules from the first week on. In this stage the oxytocic action was not evident but the kinin-forming activity already appeared. The acini reach their complete development in 2 months old animals. In 120 days old rats the convoluted granular tubules are prominent, but the acini are predominant.

It should be noted that there is a correlation between the appearance of acinar cells and the kallikrein activity in the gland, for when the acini dominate the picture there is a correspondent high concentration of kallikrein. So, it is suggested that the granules present in the acinar cells are due to kallikrein.

On the other hand, the participation of structures such as terminal tubes, convoluted granulated ducts and striated ducts, as main source of kallikrein, could be ruled out.

DISCUSSION

The question which arises is whether we are dealing with the same principle or at least two different principles in the gland, one acting directly on the uterus and the other through the releasing of kinin from the plasma.

In a series of experiments using columns of Sephadex it was found that the emergence of both activities took place in same fractions. In addition, there is a parallel increase both in the oxytocic and in the kininogenic activity throughout the developmental process of the animal. The discrepancy observed in 7 days old animals, in which appeared only the kininogenic activity can be explained as due to the smaller sensitivity of uterus to the direct as compared with the indirect action of kallikrein. The complete supression of oxytocic and kininogenic activities by DFP also indicate that a single substance is responsible for both effects.

The possibility that trypsin present in the gland could contribute to the activities of the fractions cannot be accepted, since a protein isolated from the gland presents activity 500 times greater than crystalline trypsin on the rat uterus. Moreover TLCK, in conditions which inhibits trypsin, shows without effect on the oxytocic and kininogenic activities of the glandular extract.

It was found a correlation between the appearance of acinar cells and the kallikrein activity of the gland. This last findings agree with those of Bhoola and Heap (1970) for guinea-pig.

Finally,the inhibitor found in gland extract probably is a protein. It is under study in our laboratory

SUMMARY

It was found in rat submandibular gland a kallikrein presenting besides its kininogenic activity a powerful oxytocic action on the rat uterus. A protein fraction collected from Sephadex G-100 column showed an oxytocic activity 500 times greater than crystalline trypsin. This activity was not inhibited by TLCK, a potent trypsin inhibitor. The concentration of kallikrein in the submandibular gland increases parallely with the development of the animal. There is a correlation between the appearance of acinar cells and the kallikrein activity of the gland. It is suggested that the granules present in the acinar cells are due to kallikrein. Besides the oxytocic and kininogenic activity, it was separated from the column an inhibitor of the oxytocic activity.

We would like to thank Dr. Arinos Magalhães from the Department of Biochemistry for laboratory facilities and Miss Jussara F. Carvalho for technical assistance.

REFERENCES

Araujo, R. L., W. Dias da Silva and W. T. Beraldo. Kininogenic and oxytocic properties of rat pancreatic juice. In Bradykinin and Related Kinins: Cardiovascular, Biochemical, and Neural Actions. Plenum Press, 1970.

Beraldo, W. T., R. L. Araujo and M. Mares-Guia. Oxytocic esterase in rat urine. Am. J. Physiol. 211, 975, 1966.

Gaddum, J. H. The technique of superfusion. Brit. J. Pharmacol. 8, 321, 1953.

Mares-Guia, M. and C. R. Diniz. Studies on the mechanism of rat urinary kallikrein catalysis, and its relation to catalysis by trypsin. Arch. Biochem. and Biophys. 121, 750, 1967.

Prado, E. S., J. L. Prado and C. M. W. Brandi. Further purification and some properties of horse urinary kallikrein. Arch. int. Pharmacodyn. CXXXVII, 358, 1962.

Bhoola, K. D. and P. F. Heap. Properties of kallikrein-containing granules isolated from the submaxillary gland of the guinea--pig. J. Physiol. 210, 421, 1970.

Lowry, O. H., N. J. Rosebrough, A. L. Farr and R. J. Randall. Protein measurement with the Folin phenol reagent. J. Biol. Chem. 193, 265, 1951.

Jacoby, F. and C. R. Leeson. The post-natal development of the rat submaxillary gland. J. Anat. 93, 201, 1959.

Sreebny, L. M., J. Meyer, E. Bachem and J. P. Weinmann. Post--natal changes in proteolytic activity and in the morphology of the submaxillary gland in male and female albino rats. Growth, 19, 57, 1955.

Riekkienem, P. J., T. O. Ekfors, T. Hollmen and V. K. Hopsu-Havu. Purification and physical characteristics of a trypsin--like protease from the rat submandibular gland. Enzymology, 32, 97, 1967.

Rothschild, H. and Junqueira, L. C. V. The possible correlation between cathepsin activity and protein synthesis. Arch. Biochem. and Biophys. 34, 453, 1951.

Siqueira, G., W. Dias da Silva and W. T. Beraldo. A kinin-like forming enzyme in saliva presenting a potent oxytocic activity. In Bradykinin and Related Kinins: Cardiovascular, Biochemical and Neural Actions. Plenum Press, 1970.

EFFECT OF BRADYKININ POTENTIATING PEPTIDE* ON CORONARY CIRCULATION IN CONSCIOUS DOGS**

E. BASSENGE, M. KUCHARCZYK, W. von RESTORFF and E. WERLE
Dept. of Physiology and Dept. of Clinical Chemistry, University of Munich, Germany

Since kinins have a very pronounced dilating effect on the coronary system, it has been proposed (PARRAT 1964) that they may play a major role in the regulation of coronary blood flow. In order to test the hypothesis whether kinins regulate the coronary resistance during reactive or functional hyperemia, recent studies on the role of kinins in myocardial reactive hyperemia (BASSENGE et al. 1969 and 1970) were extended by using a pentapeptide described by FERREIRA et al. (1970), which specifically potentiates the action of bradykinin by inhibiting the enzymes responsible for the cleavage of kinins (FERREIRA et al. 1970). The effect of this bradykinin potentiating pentapeptide (BPP) on myocardial reactive and tachycardia-induced functional hyperemia was tested in the conscious dog, which at rest exhibits a pronounced coronary tone and is thus very sensitive to dilating substances.

METHODS

Dogs (24 - 27 kg) were trained to lie quietly on their left side. Cardiac output and left circumflex coronary flow were measured with electromagnetic flow transducers (KHOURI and GREGG 1961) placed around the

*The pentapeptide was kindly donated by Dr. L.J. GREENE, Upton, N.Y.

**Supported by Deutsche Forschungsgemeinschaft

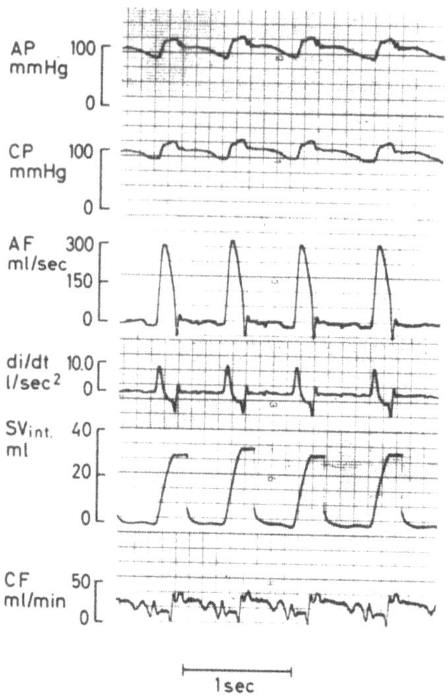

Fig. 1.
Tracings of an experiment under constant infusion of bradykinin potentiating peptide.
Ordinate:
AP: arterial pressure
CP: left circumflex coronary pressure
AF: aortic flow
di/dt: aortic flow acceleration
SV_{int}: integrated stroke volume
CF: left circumflex coronary flow
Abscissa: time

ascending aorta and the origin of the left circumflex coronary artery. Zero coronary flow for various periods of myocardial ischemia could be established with a pneumatic cuff placed distally to the coronary flow probe. Central aortic pressure was recorded with a catheter inserted into the left carotid artery. Tracings of the parameters measured in these experiments are depicted in Fig. 1. - Electrodes for pacing were attached to the left ventricle and the heart paced with a Grass SD 9 stimulator (2 msec, 1.5 times diastolic threshold voltage). A small silicon catheter was inserted into the left circ. coronary artery and brought out at the dog's neck. The catheter was used a) for coronary arterial pressure registrations, b) for injections of bradykinin (2.0 - 10.0 ng), c) for combined injections of bradykinin along with the bradykinin potentiating peptide (BPP) (20 - 100 μg) and d) for constant infusions of BPP (20 - 100 μg/min). The amino acid sequence of the pentapeptide is pyroglutamic acid-Lys-Trp-Ala-Prolin.

RESULTS

The potentiation of the bradykinin effect on coronary flow brought about by intracoronarily administered BPP is shown in Fig. 2. The abscissa shows increasing dosages of bradykinin. The ordinate shows the bradyki-

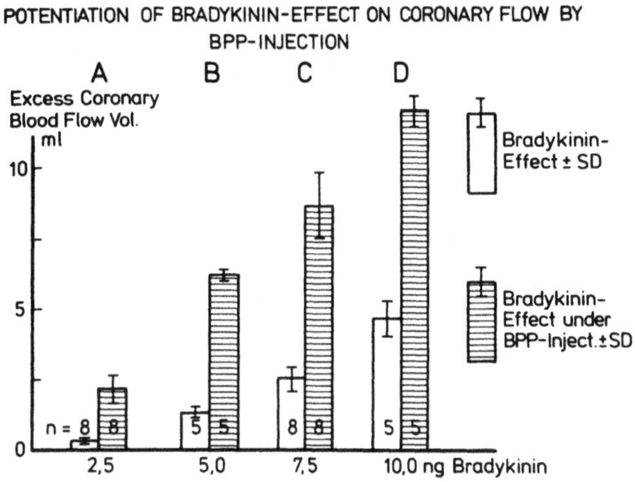

Fig. 2. Ordinate: Excess coronary blood flow volume induced by intracoronarily injected bradykinin (white columns) and the excess flow induced by combined bradikinin-BPP injection (hatched columns). Excess coronary flow represents the area under the coronary flow curve from the beginning to the end of the kinin effect minus the respective area under the control flow level. - Abscissa: Increasing dosages of bradykinin and combined bradykinin-BPP injections. The dosage of injected BPP was by weight in all experiments 10^4 times that of bradykinin. Further details: see text.

nin effect and the effect of combined BPP-bradykinin injection on coronary flow. Increasing dosages of bradykinin result in a more pronounced coronary dilation. The potentiation of the kinin-induced coronary excess blood flow by BPP ranged between 6.5 fold at 2.0 ng of bradykinin and 2.5 fold at 10.0 ng of bradykinin. The potentiating effect of the pentapeptide on the bradykinin induced excess flow decreases with rising doses of bradykinin. The effect of 2.0 ng bradykinin potentiated by BPP is equipotent to the effect of 7.4 ng bradykinin without the potentiating peptide.

Whereas the coronary excess blood flow under bradykinin injection is significantly increased by simultaneous administration of BPP, there is no increase in coronary excess blood flow during reactive hyperemia after 10 sec of ischemia. From the results shown in

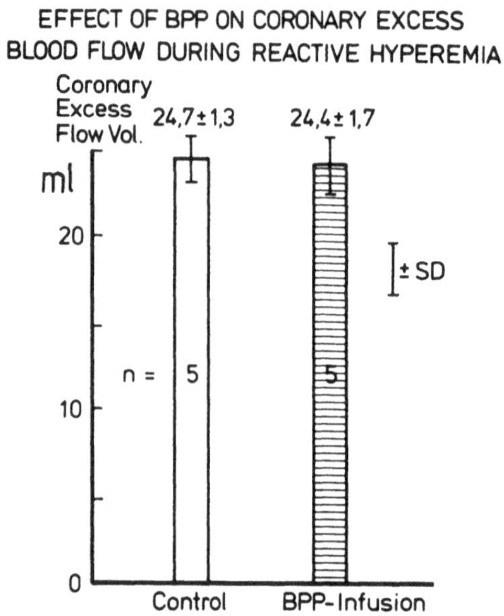

Fig. 3.
White column: Coronary excess blood flow volume (ml) during reactive hyperemia at control.
Hatched column: Coronary excess blood flow during reactive hyperemia under simultaneous intracoronary infusion of 100 µg/min bradykinin potentiating peptide (BPP)
n: number of experiments

Fig. 3 it is obvious that the pentapeptide infusion does not result in a significant rise in coronary excess blood flow during reactive hyperemia. Significant increases in coronary excess flow during reactive hyperemia were possible under our experimental conditions, and were in fact demonstrated following intravenous administration of small dosages of dypiridamole or lidoflazine.

Tachycardia-induced functional myocardial hyperemia was brought about by pacing the resting dogs at 240 beats/min. This resulted in a substantial increase in coronary flow, which is demonstrated in the tracings depicted in Fig. 4. Increasing the heart rate from 94 beats/min to 240 beats/min with the dogs at rest caused an increase in mean and phasic coronary flow from 25 ml/min to approximately 50 ml/min in this experiment. When bradykinin potentiating peptide was infused intracoronarily (100 µg/min), there was no significant increase in coronary flow, either at the control state (heart rate 94/min) or under the condition of tachycardia-induced functional hyperemia (heart rate 240/min). The data of 5 experiments are summarized in Fig. 5. Increasing the resting dog's heart rate by pacing the ventricles at 240 beats/min resulted in an average increase of coronary flow by 65%. This increase was not significantly enhanced by additional infusion of BPP.

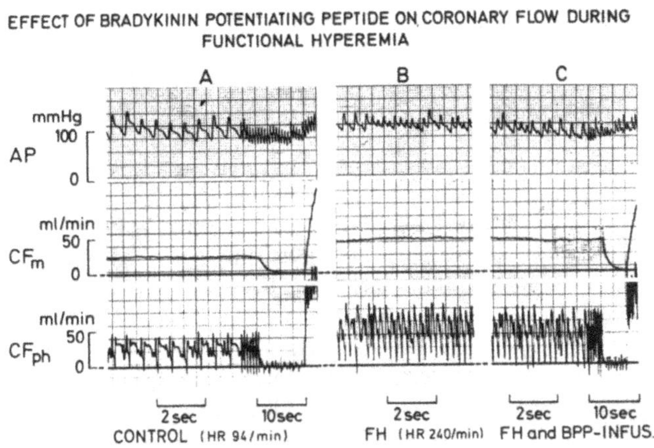

Fig. 4. Ordinate: AP arterial pressure, CF_m and CF_{ph} mean and phasic left coronary flow. Abscissa: time at two different paper speeds. Panel under A depicts the controls (heart rate 94/min), under B tachycardia-induced functional hyperemia (FH), and under C functional hyperemia during simultaneous infusion of bradykinin potentiating peptide (BPP) at 100 μg/min.

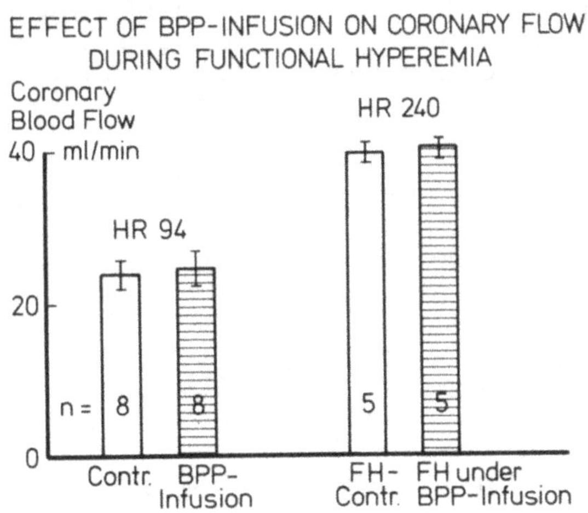

Fig. 5. White columns: Coronary flow of resting dogs at heart rate 94/min and during tachycardia-induced (left ventricular pacing) functional hyperemia (FH-controls). Hatched columns: Coronary flow at respective heart rates under simultaneous infusion of bradykinin potentiating peptide (BPP-Infusion) at 100 μg/min. n: number of experiments.

DISCUSSION

An enhancement of the coronary dilator effect of bradykinin by a factor present in the venom of Bothrops Jararaca was first described by ANTONIO in 1966. Our results demonstrate that the synthetic bradykinin potentiating peptide (BPP) enhances the dog's coronary flow response to exogenous bradykinin when kinins and BPP are administered intracoronarily. BPP seems to be a suitable tool in the identification of reactions brought about by a possible kinin release. In contrast to the potentiation of exogenous bradykinin by BPP there was no significant effect of BPP on myocardial reactive and tachycardia-induced functional hyperemia, which indicates that bradykinin release within the coronary circulation does not contricute significantly to the coronary dilation under these two conditions. PITT et al. in 1969 reported on experiments indicating that intramyocardial or intracoronary liberation of kinins may occur and affect coronary circulation under pathophysiological conditions (pacing stress test in patients suffering from angina pectoris). Local ischemia was shown to be a kininogen depressant (pointing to a kinin release) by BAVAZZANO et al. in 1969. Therefore it may be possible that kinins play a more substantial role in the pathophysiology of the heart (SICUTERI et al. 1970) than they do under normal physiological conditions in adjusting the coronary resistance to match myocardial oxygen demands.

SUMMARY

a) The effect of intracoronarily administered bradykinin on coronary flow was significantly enhanced by simultaneous application of a pentapeptide, which inhibits the enzymes responsible for the cleavage of bradykinin.

b) Myocardial reactive and tachycardia-induced functional hyperemia were not significantly increased by the bradykinin potentiating peptide.

c) No signs of pain in the conscious animals were observed following the intracoronary bradykinin injection.

d) Kinins do not seem to contribute significantly in the coronary regulation under physiological conditions.

REFERENCES

1) ANTONIO, A., in: Int.Symp. on Vasocative Polypeptides, p. 87-90, Edart Livraria Editora Ltda, Sao Paulo 1966

2) BASSENGE, E., E. WERLE, P. WALTER and J. HOLTZ: Advances in Exper.Med.Biol. 8, 141-148 (1970), Plenum Press, New York/London

3) BASSENGE, E., E. WERLE, J. HOLTZ and H. FRITZ: Pharmacological Res.Comm. 1, 136 (1969)

4) BAVAZZANO, A., N. SIDELL, S. MICHELACCI and F. SICUTERI: Advances in Exper.Med.Biol. 8 (1970), Plenum Press, New York/London

5) FERREIRA, S.H., D.S. BARTELT and L.G. GREENE: Biochemistry 9, 2585 (1970)

6) FREY, E.K., H. KRAUT and E. WERLE: Das Kallikrein-Kinin-System und seine Inhibitoren. Ferd. Enke-Verlag, Stuttgart (1968)

7) KHOURI, E.M., and D.E. GREGG: J.Appl.Physiol. 18, 224 (1963)

8) PARRAT, J.R.: Brit.J.Pharmacol. 22, 34-36 (1964)

9) PITT, B., J. MASON, C.R. CONTI and R.W. COLMAN: Pharmacological Res.Comm. 1, 185 (1969)

10) SICUTERI, F., P.L. Del BIANCO and M. FANCCIULLACCI: Advances in Exper.Med.Biol. 9, 315-322 (1970), Plenum Press, New York/London

METABOLIC FATES OF BRADYKININ, ANGIOTENSIN I, ADENINE NUCLEOTIDES AND PROSTAGLANDINS E_1 AND $F_{1\alpha}$ IN THE PULMONARY CIRCULATION

J. W. Ryan, R. S. Niemeyer and D. W. Goodwin

Department of Medicine, University of Miami School of Medicine, Miami, Florida 33152

The lung appears to have a variety of means of eliminating circulating pharmacologic agents. Ether, paraldehyde and volatile sulfhydryl agents (e.g. BAL and 2-mercaptoethanol) pass from blood to the alveolar space. Agents such as acetylcholine can be metabolized by enzymes of the blood circulating through the pulmonary vascular bed. Amines such as norepinephrine and serotonin appear to be taken up by cells and stored for reuse or degradation. Recently, we have postulated the existence of yet another mechanism: degradation of pharmacologic agents by enzymes on or close to the luminal surface of pulmonary endothelial cells (1-8).

Our hypothesis is based on several peculiar features of the immediate metabolic fates of intrinsically-labelled bradykinin, angiotensin I and the adenine nucleotides. Each of these substances disappears during a single circulation through the lungs. Their disappearance is accounted for by enzymic degradation and not by tissue uptake nor transfer to extravascular spaces. Their metabolic products are recovered in almost quantitative yields in the pulmonary venous effluent, neither the parent compounds nor the metabolites being retained by lung. Metabolism of these compounds is selective and does not depend on enzymes of blood or secretion of soluble hydrolase enzymes.

Using cytochemical and kinetic techniques, we have shown that adenosine-5'-monophosphate (5'-AMP) is hydrolyzed by enzymes located within endothelial caveolae intracellulares (pinocytotic vesicles) open to the vascular lumen (7). These enzymes, isolated as a plasma membrane fraction, hydrolyze 5'-AMP to yield the same products formed by intact lungs perfused with 5'-AMP.

The close parallels of the disappearances of bradykinin, angiotensin I and adenosine-5'-triphosphate (ATP) with that of 5'-AMP suggest that these substances are metabolized by similar mechanisms. In the present study we have attempted to refine our understanding of the subcellular location of the relevant metabolic enzymes by examining the kinetics of disappearance of bradykinin, angiotensin I and ATP using a variety of intrinsically-labelled analogs, namely (2,3-[^{14}C]proline)-bradykinin, (8-[^{3}H]phenylalanine)-bradykinin, (10-[^{14}C]leucine)-angiotensin I, (8-[^{14}C]phenylalanine)-angiotensin I, ^{14}C and ^{3}H-labelled-ATP and $^{32}P_{\alpha}$- and $^{32}P_{\gamma}$-ATP. Recently, Vane and colleagues have presented evidence that prostaglandins of the E and F series also may be metabolized but not retained by lung. Therefore we included studies of the metabolic fates of $^{3}H_{5,6}$-prostaglandins E_1 and $F_{1\alpha}$ (9).

MATERIALS AND METHODS

Perfusions were performed as described previously using lungs of Sprague-Dawley rats either _in situ_ or _in vitro_ (7,8). The lungs were perfused with Krebs-Henseleit solution or Locke's solution, aerated with $O_2 + CO_2$ (95:5) or room air, heated at 37°C. Dextran or serum albumin were added in some experiments as osmotic agents. Dextran and albumin had no apparent effect on the metabolic processes. In some experiments, blue dextran (MW>2,000,000) was used as a compound unlikely to leave the vascular space and therefore provided a basis for estimating volumes of distribution and mean transit times (10).

Peptides, nucleotides and nucleosides were purified and identified as described previously, using a combination of gel chromatography, electrophoresis and thin-layer chromatography (1-8). Prostaglandins were identified by established techniques (11).

The radioactive angiotensins were purchased from Schwarz BioResearch. (2,3-[^{14}C]Proline)-bradykinin was a gift from the National Institutes of Health (courtesy of Dr. John Pisano). $^{32}P_{\alpha}$- and $^{32}P_{\gamma}$-ATP were gifts from New England Nuclear Corp. $^{3}H_{5,6}$-Prostaglandins E_1 and $F_{1\alpha}$ were purchased from New England Nuclear Corp.

RESULTS

A. Apparent Mean Transit Times and Volumes of Distribution

In these experiments lungs were perfused with Krebs-Henseleit solution (6 ml/min) until the venous effluent was free of blood

(2-3 min). Then a 100 µl aliquot of blue dextran (5 mg/ml) and labelled vasoactive substance was injected as a bolus into the pulmonary artery. Fractions of venous effluent were collected as quickly as possible (2-2.6 sec/fraction) and were assayed for radioactivity and blue color.

Results of experiments using $^{32}P_{\alpha}$-ATP are shown in Fig. 1. The apparent mean transit times of blue dextran and ^{32}P were identical and averaged about 12 sec. The estimated volumes of distribution of radioactivity and blue dextran were 1.2 ml. Neither of these values were corrected for any artifactual mixing or delay in transit that may have occurred in the venous outflow tract. Although the radioactivity of $^{32}P_{\alpha}$-ATP was not delayed in respect

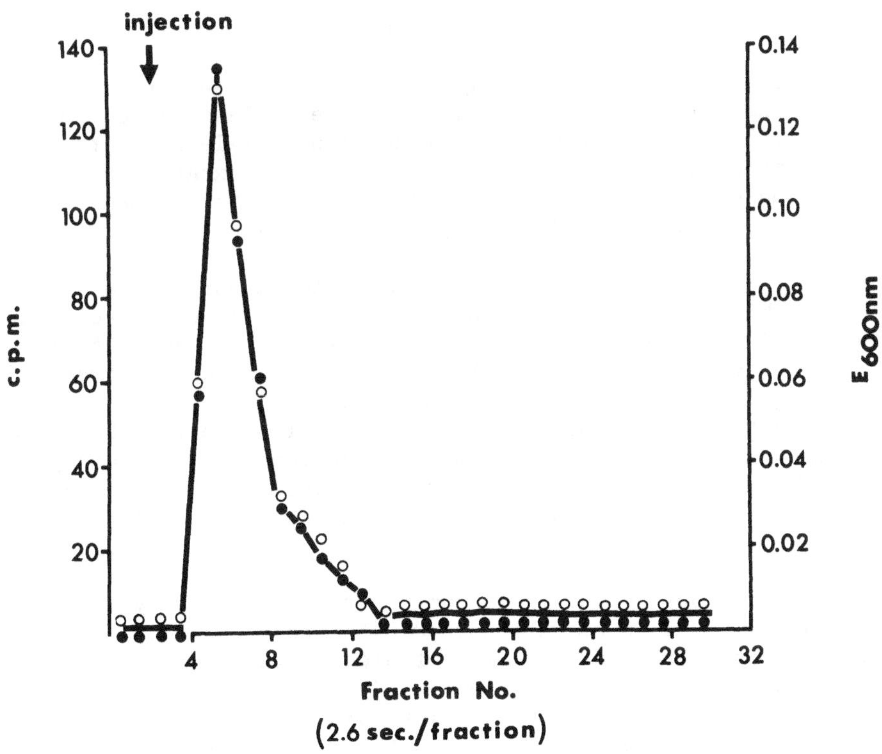

Fig. 1. Mean transit times of blue dextran and radioactivity of $^{32}P_{\alpha}$-ATP through the lungs. A 100 µl aliquot of blue dextran (5 mg/ml) and $^{32}P_{\alpha}$-ATP (20 nCi/ml, 725 mCi/mmole) was injected (see arrow) as quickly as possible into the cannula in the pulmonary artery. Fractions of venous effluent were collected from a cannula in the left atrium. The venous effluent was assayed for blue color (o) and radioactivity (•).

to the transit of blue dextran and had no larger volume of distribution, relatively little (<5%) of the radioactivity emerged as ATP, the remainder of the ^{32}P occurring as 5'-AMP and Pi. Less than 1% of the radioactivity was retained by the lungs.

Similar results were obtained using ATP labelled in the nucleoside moiety with ^{14}C or ^{3}H. Metabolic products included adenosine as well as 5'-AMP. Similarly, the apparent mean transit times and volume of distribution of the radioactivities of (8-[^{3}H]phenylalanine)-bradykinin, (2,3-[^{14}C]proline)-bradykinin, (10-[^{14}C]leucine)-angiotensin I and (8-[^{14}C]phenylalanine)-angiotensin I did not differ from that of blue dextran. Again, none of these compounds emerged intact in the venous effluent. By contrast, $PGF_{1\alpha}$ emerges much later than blue dextran, >10% being retained by the lungs.

B. Steady-State Perfusions

It could be argued that single injections of substances such as the kinins, angiotensin I and the adenine nucleotides do not duplicate the likely circumstance of continuous release of these compounds into the circulation in vivo. Thus we extended the kinetic studies to include continuous perfusion of the vasoactive compounds. Results using (8-[^{14}C]phenylalanine)-angiotensin I and blue dextran are shown in Fig. 2. The appearance and then disappearance of radioactivity in the venous effluent correspond exactly to that of blue dextran, thus indicating identical volumes of distribution (10). The recoveries of blue dextran and radioactivity were 90% and 92% respectively. Less than 1% of either was retained by the lungs. The remainder was not found and may have been adsorbed by tubing or glass or may have been lost during the rapid collection of samples. However, as we have reported previously, none of the radioactivity emerged as angiotensin I (5,8). Approximately 20% was identified as angiotensin II and the remainder as

---

Fig. 2. Continuous perfusion of lungs with (8-[^{14}C]phenylalanine)-angiotensin I. The lungs were pumped with Krebs-Henseleit solution until the venous effluent was free of blood. Then blue dextran (5 mg/ml) and (8-[^{14}C]phenylalanine)-angiotensin I (16 ng/ml, 100 μCi/μmole) were added to the perfusion solution (see bar). Fractions of venous effluent were assayed for blue color (o) and radioactivity (•).

Fig. 3. Continuous perfusion of lungs with $^{3}H_{5,6}$-prostaglandin $F_{1\alpha}$ and blue dextran. The perfusion system was similar to that described in Fig. 2. ^{3}H-Prostaglandin $F_{1\alpha}$ (100 mCi/μmole) was perfused in a final concentration of 0.1 μCi/ml. Venous effluent and the lungs were assayed for radioactivity (•) and blue color (o).

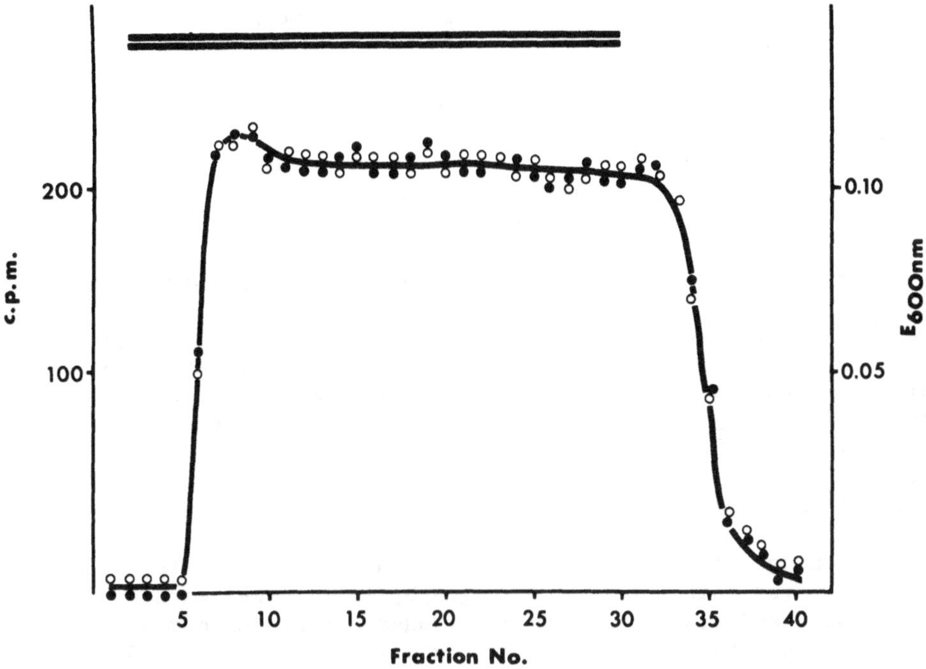
200
100
c.p.m.
0.10
0.05
E600nm
5
10
15
20
25
30
35
40
Fraction No.

Fig. 2

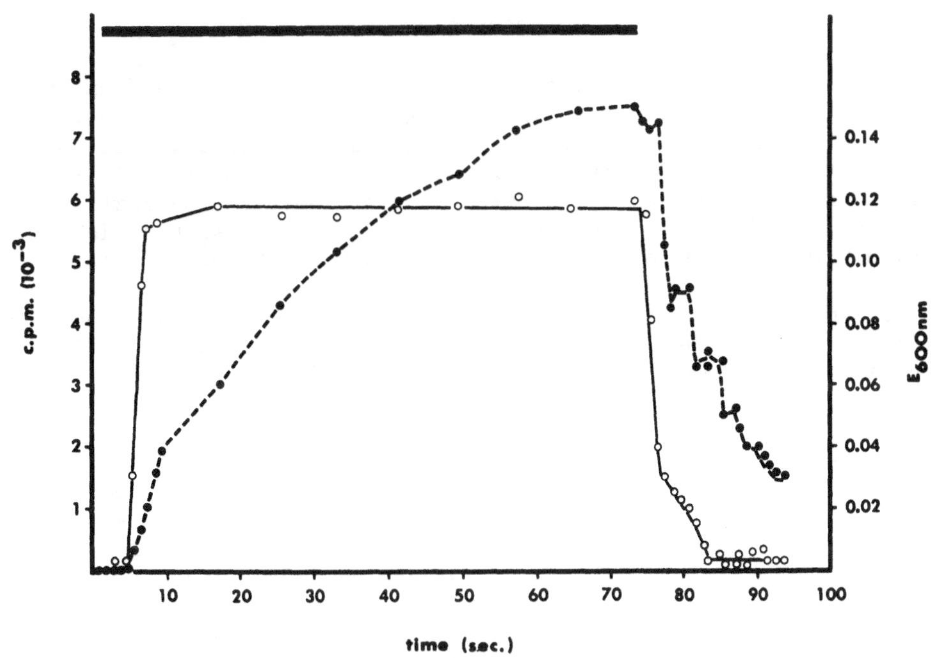
8
7
6
5
4
3
2
1
c.p.m. (10⁻³)
0.14
0.12
0.10
0.08
0.06
0.04
0.02
E600nm
10
20
30
40
50
60
70
80
90
100
time (sec.)

Fig. 3

tetra- , tri- , and dipeptides plus a small amount of free phenylalanine.

Similar results were obtained using $^{32}P_{\alpha}$-ATP, ^{3}H-ATP, (8-[^{3}H]phenylalanine)-bradykinin, and (2,3-[^{14}C]proline)-bradykinin. Nucleotide metabolic products were identical to those described in Section A. Primary radioactive products of bradykinin were Phe-Arg and Pro-Pro (1,2).

Results using $^{3}H_{5,6}$-prostaglandins E_1 and $F_{1\alpha}$ were different. As shown in Fig. 3, the appearance of tritium of prostaglandin $F_{1\alpha}$ in the venous effluent is much delayed in respect to the appearance of blue dextran and reaches its peak concentration several seconds after stopping the infusion. Samples of venous effluent taken at intervals during and after the infusion contain a less polar metabolite (75-80%) and unchanged prostaglandin $F_{1\alpha}$ (20-25%). The lungs retained approximately 20% of the radioactivity, all in the form of the less polar metabolite.

Although we do not, as yet, have mass spectrometric data, results using thin-layer chromatography (silica gel with and without $AgNO_3$) and reactions with acid, base and sodium borohydride indicate that the metabolite of prostaglandin $F_{1\alpha}$ may be 9,11-OH-15-ketoprostanoic acid.

DISCUSSION

Our data confirm that radioactive ATP, bradykinin and angiotensin I are metabolized during a single circulation through the lungs. The volumes of distribution of their radioactivities do not exceed that of the intravascular space, and the mean transit times of the radioactivities are no greater than that of blue dextran, a compound unlikely to leave the vascular lumen. These kinetic data, taken with our previous studies (1-8), support the concept that ATP, bradykinin and angiotensin I, like 5'-AMP (7), are metabolized by enzymes on or close to the luminal surface of pulmonary endothelial cells. Further evidence bearing on this concept and a discussion of its implications will be presented in the following paper (12).

Recently, Vane and collaborators (9) have indicated that prostaglandins of the E and F series may also be metabolized but not retained during circulation through the lungs. Using $^{3}H_{5,6}$-prostaglandins E_1 and $F_{1\alpha}$ we have found this not to be the case. Prostaglandins E_1 and $F_{1\alpha}$ are rapidly removed from the circulation and have a much larger volume of distribution than does blue dextran. At this point, it appears that the parent prostaglandins and their metabolites are released slowly into the venous effluent. The cellular site of their metabolic enzymes is not known.

REFERENCES

1. J. W. Ryan, J. Roblero and J. M. Stewart (1968). Biochem. J., 110, 795.
2. J. W. Ryan, J. Roblero and J. M. Stewart (1970). Adv. Exp. Med. Biol., 8, 263.
3. U. Smith and J. W. Ryan (1970). Adv. Exp. Med. Biol., 8, 249.
4. U. Smith and J. W. Ryan (1971). Chest, 59, 12S.
5. J. W. Ryan, J. M. Stewart, W. P. Leary and J. G. Ledingham (1970). Biochem. J., 120, 221.
6. J. W. Ryan, J. Roblero, J. Stewart and W. P. Leary (1971). Chest, 59, 8S.
7. J. W. Ryan and U. Smith (1971). Trans. Ass. Am. Physcns., in press.
8. J. W. Ryan, R. S. Niemeyer, D. W. Goodwin, U. Smith and J. M. Stewart (1971). Biochem. J., in press.
9. P. J. Piper, J. R. Vane and J. H. Wyllie (1970). Nature, 225, 600.
10. F. P. Chinard, T. Enns and M. F. Nolan (1962). Trans. Ass. Am. Physcns., 75, 253.
11. P. W. Ramwell, et. al. (1968). In Progress in the Chemistry of Fats and Other Lipids, Pergamon Press, Oxford, pp. 231-273.
12. U. Smith and J. W. Ryan (1971). This volume.

ACKNOWLEDGEMENTS

This investigation was supported in part by grants from the U.S. Public Health Service (HE 12843 and HE 14360) and the Council for Tobacco Research, New York.

PULMONARY ENDOTHELIAL CELLS AND THE METABOLISM OF ADENINE. NUCLEOTIDES, KININS AND ANGIOTENSIN I

Una Smith and J. W. Ryan

Department of Medicine, University of Miami and the

Howard Hughes Medical Institute, Miami, Florida 33152

As indicated in previous publications (1-9) and as revealed in the preceding paper (10), it is likely that circulating adenine nucleotides, kinins and angiotensin I are metabolized by enzymes of the luminal surface of the pulmonary endothelium. One of the nucleotides, namely adenosine-5'-monophosphate (AMP), is metabolized by enzymes of caveolae intracellulares open to the vascular lumen (7-9). This enzymic activity is not peculiar to those caveolae open to the vascular lumen, as one can show by reactions performed on fixed blocks of tissue (5,6) that other caveolae and intracellular vesicles are capable of degrading nucleotides. However, in the case of metabolism of circulating nucleotides it appears that only those caveolae open to the lumen have access to the substrate.

The close parallels between the kinetics of metabolism of the kinins and angiotensin I with that of 5'-AMP have already been pointed out (3,7,8) and indicate that the sites of the enzymes may be similar (9,10).

Examination of the luminal surface of pulmonary endothelial cells, especially as they occur in capillaries, shows that it is endowed with special features which make it particularly well-suited to process circulating vasoactive substances of blood. Firstly, its strategic position and the vast surface area which it presents to the blood should provide for rapid and efficient processing of biologically active substances. In addition, the surface area is vastly increased by the endothelial projections (11) and caveolae, many of which open directly to the capillary lumen. The luminal stoma of each caveola is covered by a delicate diaphragm which may in part account for the known specificity of the pulmonary metabolism of kinins (2,3).

Here we report on the metabolism of kinins, angiotensin I and adenine nucleotides by fractions composed of caveolae and plasma membrane of lung, and results of high magnification electron microscopy. The latter is of interest because it provides evidence of substructure of caveolae which may be important in the metabolism of circulating vasoactive substances.

MATERIALS AND METHODS

Techniques for the perfusion (12) and preparation for electron microscopy (6) of rat lungs have been described previously. In order to study details of the endothelial plasma membrane and caveolae at high magnification it was necessary to enhance the contrast of thin sections. Specimens of intact tissue and of pellets were therefore subjected to _en bloc_ staining with 0.5% uranyl acetate prior to embedding. Sections were also stained with uranyl acetate and lead citrate and examined in a Philips EM 300 operated at 60 KV with a liquid nitrogen anticontamination device.

Pellets of plasma membrane and caveolae were prepared from homogenates of lung after reaction of 5'-nucleotidase enzymes with 5'-AMP in the presence of lead nitrate (13). These preparations were incubated at 37°C in 50 mM tris-maleate buffer, pH 7.4, containing 100 mM KCl, 5 mM $MgCl_2$ and 100 mM Na_2HPO_4. Bradykinin, angiotensin I, ATP and AMP (each labelled intrinsically with ^3H, ^{14}C or ^{32}P) were used as substrates. Metabolic products were isolated and identified as described before (1,4,9).

Fig. 1. General survey electron micrograph of the capillary bed of the rat lung. The lung was pumped free of blood prior to perfusion-fixation with 0.25 M glutaraldehyde. The capillaries (*) therefore contain no blood cells. Endothelial cells are indicated by arrows and an endothelial nucleus is illustrated (N). The blood space is separated from the air space (A) by three layers comprising endothelium, basement membrane and epithelium. An alveolar macrophage (AM) is lodged in an air space.

X 30,000

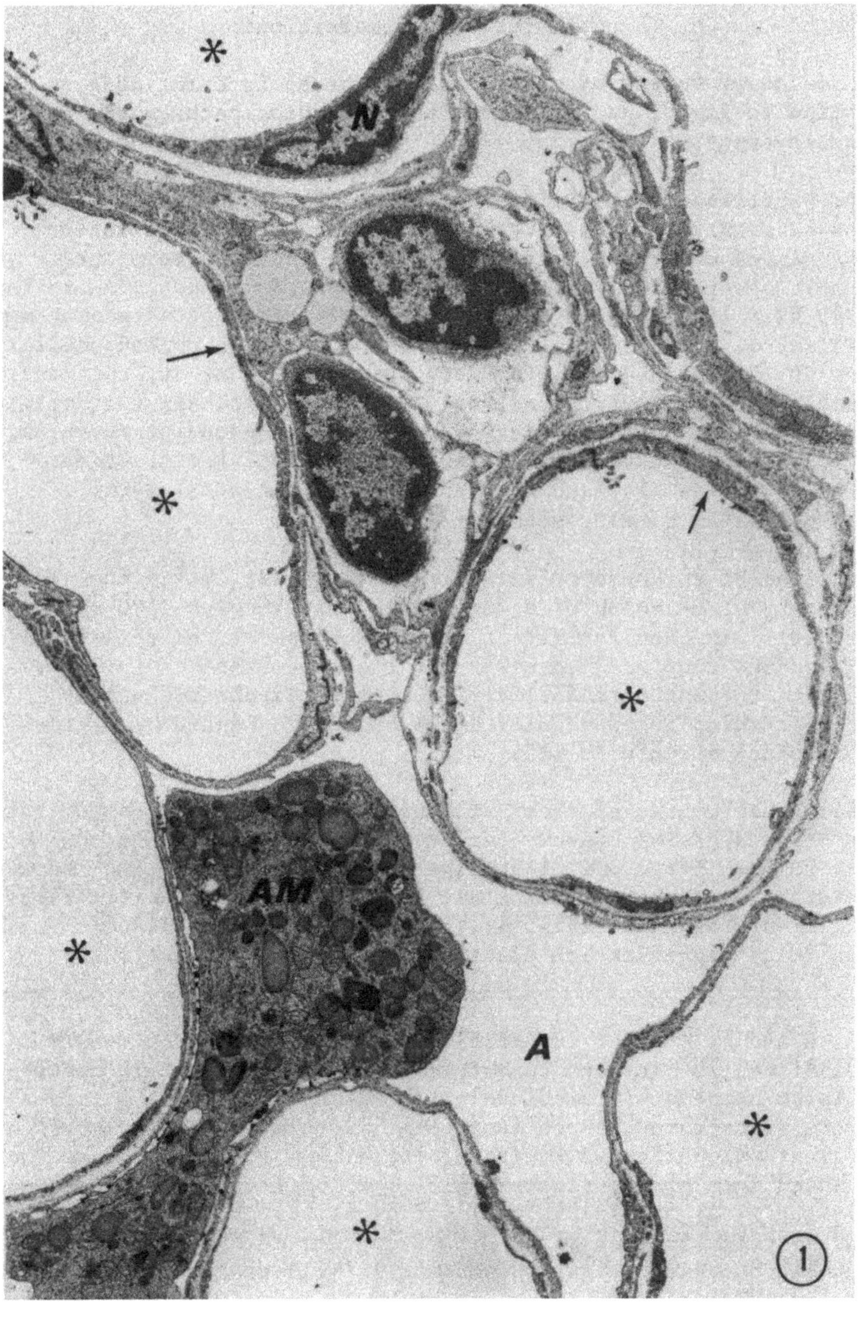
N
AM
A
1

RESULTS

A. Substructural Specializations

A low power survey micrograph of the blood free rat lung is illustrated in Fig. 1 and shows the relationships between the component cell types.

Our experience with the nucleotides (5-9) indicates that caveolae have enzymic properties not shared by undifferentiated areas of plasma membrane (Figs. 9 & 10). The concave membrane of the caveola has regularly-spaced substructural components (approximately 90 Å in diameter) not found on undifferentiated plasma membrane (Figs. 4, 5 & 6). This beading can be seen on endothelial caveolae of lungs fixed intact and in the caveolae of fractions of lung homogenate lightly reacted for 5'-nucleotidase. In the latter preparation the reaction product shows a corresponding preferential deposition. These substructural knobs are reminiscent of mitochondrial particles (14) and may represent a spatial array of enzyme clusters or binding sites.

Sections of the endothelial surface (Figs. 2, 3 & 4) show large numbers of caveolae of a remarkably uniform size (500-900 Å in diameter). This uniformity of size and shape of the caveolae itself bespeaks a special chemistry to their membranes: for instance, a frequent periodicity of membrane components such as lysolecithin which force regular bends in membranes and thus produce spheroidal structures (15).

During the course of these studies we have been impressed that even in regions having a heavy density of caveolae, they do not coalesce to form large vesicles. The uniformity of size and shape is maintained even when caveolae are fused into multiple formations (Fig. 2) or are adjacent (Fig. 4), retaining their individual stomata. High magnification electron microscopy of junctions

---

Figs. 2, 3 and 4. Higher magnification micrographs of pulmonary endothelial cells. In each case the lumen of the vessel is uppermost. Large members of caveolae intracellulares (500-900 Å in diameter) are illustrated, those facing the vascular lumen are covered by a delicate diaphragm each with a central hub (arrows). Caveolae may be fused into groups (asterisk, Fig. 2) or closely adjacent (asterisk, Fig. 4) but retain their original size and shape. Substructural specialization of the inner concave membrane of the caveolae may be evident (arrowhead) and is illustrated in more detail in Figs. 5 and 6.

all at X 140,000

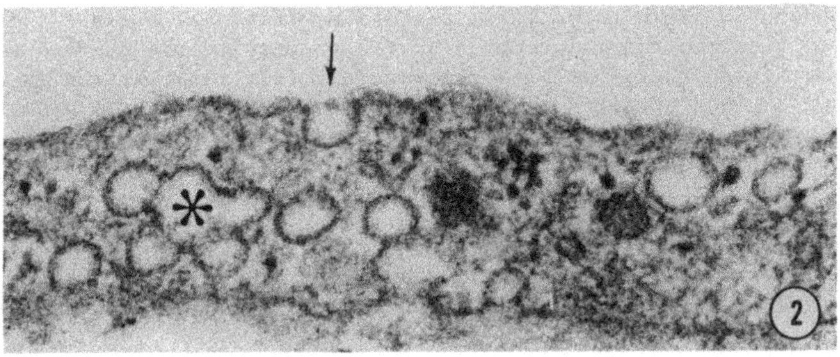
2

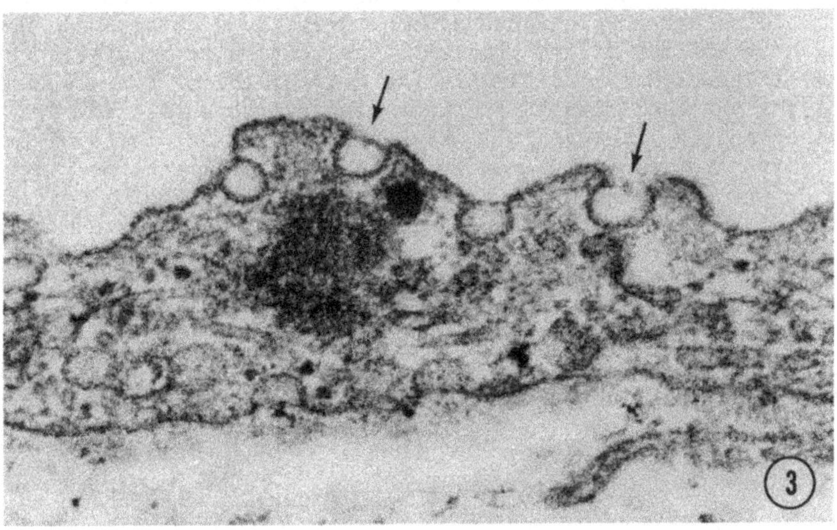
3

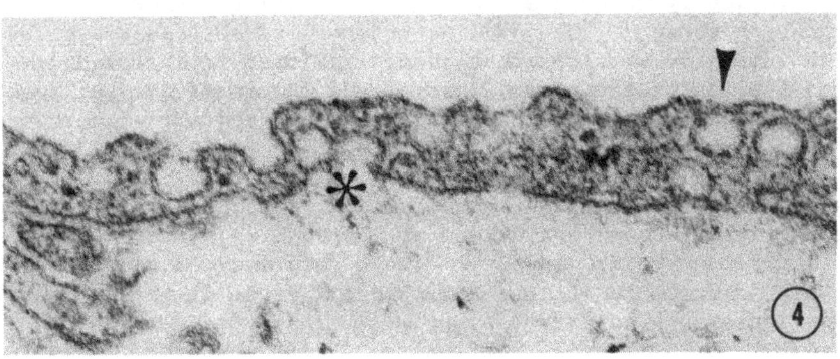
4

between vesicle and plasma membrane and between vesicle and vesicle reveal a new substructural feature. In transverse sections of caveolae a thickened knob is seen at the point of fusion of the plasma membrane, caveola membrane and diaphragm which appears to represent a rim with skeletal properties (Fig. 8). This rim could be critical to the maintenance of the patency of the stomata of caveolae and therefore of surface area. Its true function is not known but may preserve the diaphragm as well as maintain patency of the stoma. The rim persists in isolated fractions of caveolae and appears to contribute to the retention of their characteristic morphology even after homogenization and centrifugation procedures (Fig. 7).

B. Reaction of Vasoactive Substances with Caveola/Plasma Membrane Fractions of Lung

The enzymes which metabolize the polypeptide hormones are not amenable to cytochemical or autoradiographic localization. However, the ability of endothelial caveolae to bind lead phosphate (5-9) (Figs. 9 & 10) has led to the development of a new technique (13) for the isolation of a plasma membrane/caveolae fraction from homogenates of lung (Fig. 7) which allows a direct means of testing the ability of this fraction to metabolize the kinins and angiotensin I. In addition to the preservation of morphological details of both endothelial cells and of the caveolae themselves such as the diaphragm and stomal rim, these pellets retain considerable enzymic activity. The pellets can convert (8-[^{14}C]phenylalanine)-angiotensin I to angiotensin II at rates (at least 0.4 μmole/mg of plasma membrane protein/min) in great excess of that likely to be required of lung in vivo (9,13). Similarly, bradykinin and ATP are metabolized yielding the expected products (1,2,6,7,9,13).

DISCUSSION

Previously we have shown that circulating 5'-AMP is metabolized by enzymes of the caveolae intracellulares open to the vascular lumen (5-9). The kinetic data reviewed in the preceding paper (10) and the findings that plasma membrane/caveolae fractions of lung hydrolyze bradykinin, angiotensin I and ATP provide strong

---

Figs. 5 and 6. Details of substructural components of the caveola intracellularis membrane. Regularly spaced knobs (arrows) occur on the inner concave membrane. The unit membrane construction of the caveola membrane is clearly resolved while the diaphragm consists of a single lamella.

Fig. 5 X 325,000
Fig. 6 X 303,000

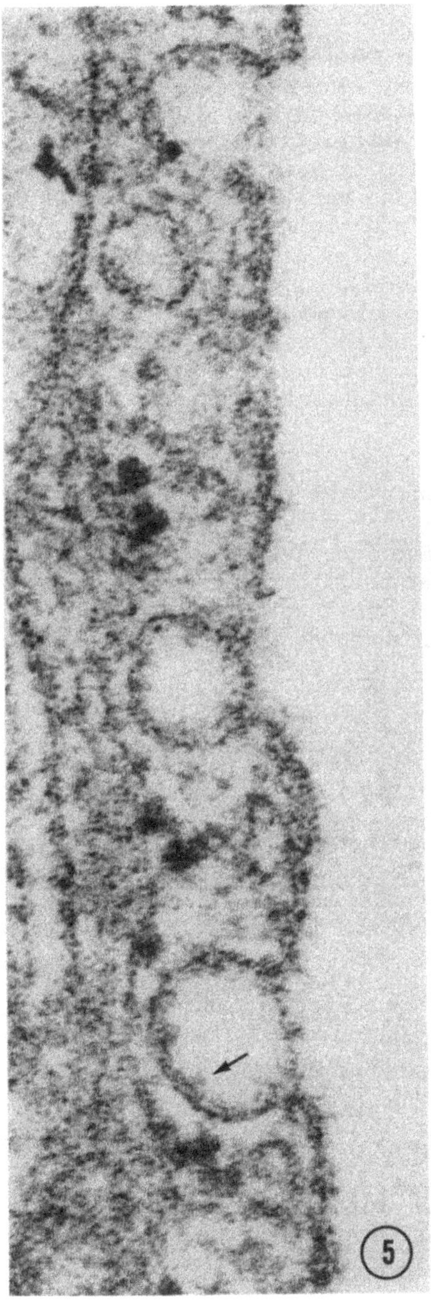
5

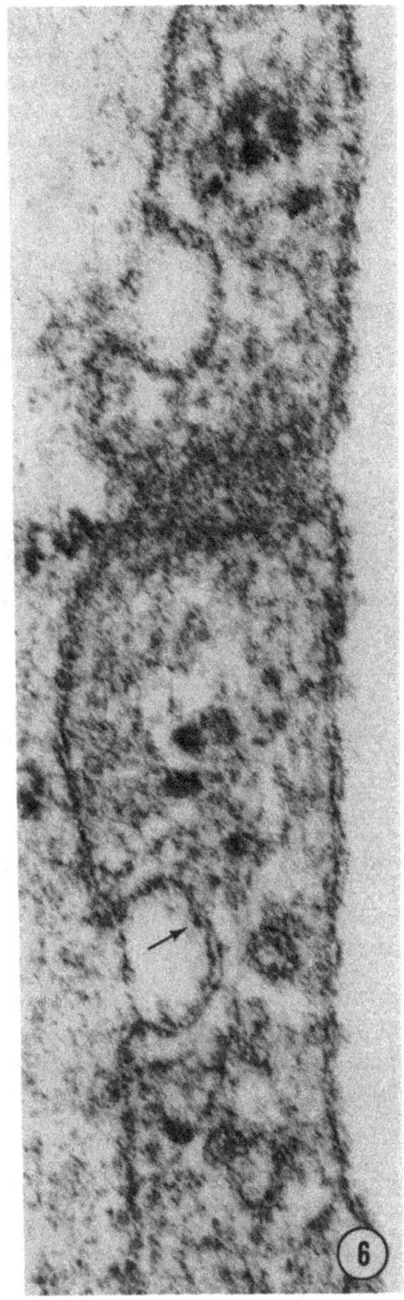
6

support of the concept that each of these vasoactive substances is metabolized by enzymes on or near the luminal surface of pulmonary endothelial cells.

We have examined in detail the endothelial plasma membrane and its caveolae and have found that the concave surface of the caveolae has substructural specializations with a regular periodicity reminiscent of mitochondrial particles. These specializations may be enzyme clusters. The intriguing possibility exists that caveolae contain enzymes, in addition to 5'-nucleotidase, capable of hydrolyzing bradykinin and angiotensin I.

High magnification studies brought up yet another previously undescribed feature of caveolae, namely stomal rims. The thickened knobs seen in transverse sections at the point of fusion between caveolae and plasma membrane or between one caveola and another may represent a skeletal rim possibly maintaining patency of the stoma and integrity of the diaphragm.

The substructural specializations taken with the evidence of specific metabolic activities indicate that caveolae intracellulares (pinocytotic vesicles) have functions other than bulk transport or pinocytosis. As the metabolism of the vasoactive substances discussed in this paper occurs in endothelial cells of lung, our results describe in detail a new nonventilatory function of the lung.

In the cases of the kinins and the nucleotides, metabolism results in biological inactivation, presumably sparing organs on

---

Fig. 7. Cloud of plasma membrane and attached caveolae intracellulares obtained from homogenate of rat lung reacted with AMP and $Pb(NO_3)_2$. The resulting lead phosphate reaction product is localized within the caveolae (arrows) and is responsible for the density of this fraction. X 44,000

Fig. 8. Two caveolae showing the dense knobs (arrows) at the junction of the caveola membrane and diaphragm with the plasma membrane. The knobs may represent transverse sections through a stomal rim possibly with a skeletal function. X 190,000

Figs. 9 and 10. Endothelial cells of rat lungs perfused with AMP and $Pb(NO_3)_2$. The electron dense lead phosphate reaction product (arrows) is restricted to caveolae which face the vascular lumen and indicate the sites of degradation of circulating AMP. The reaction product is clearly localized within the caveolae beneath the diaphragms.

Fig. 9 X 140,000
Fig. 10 X 91,000

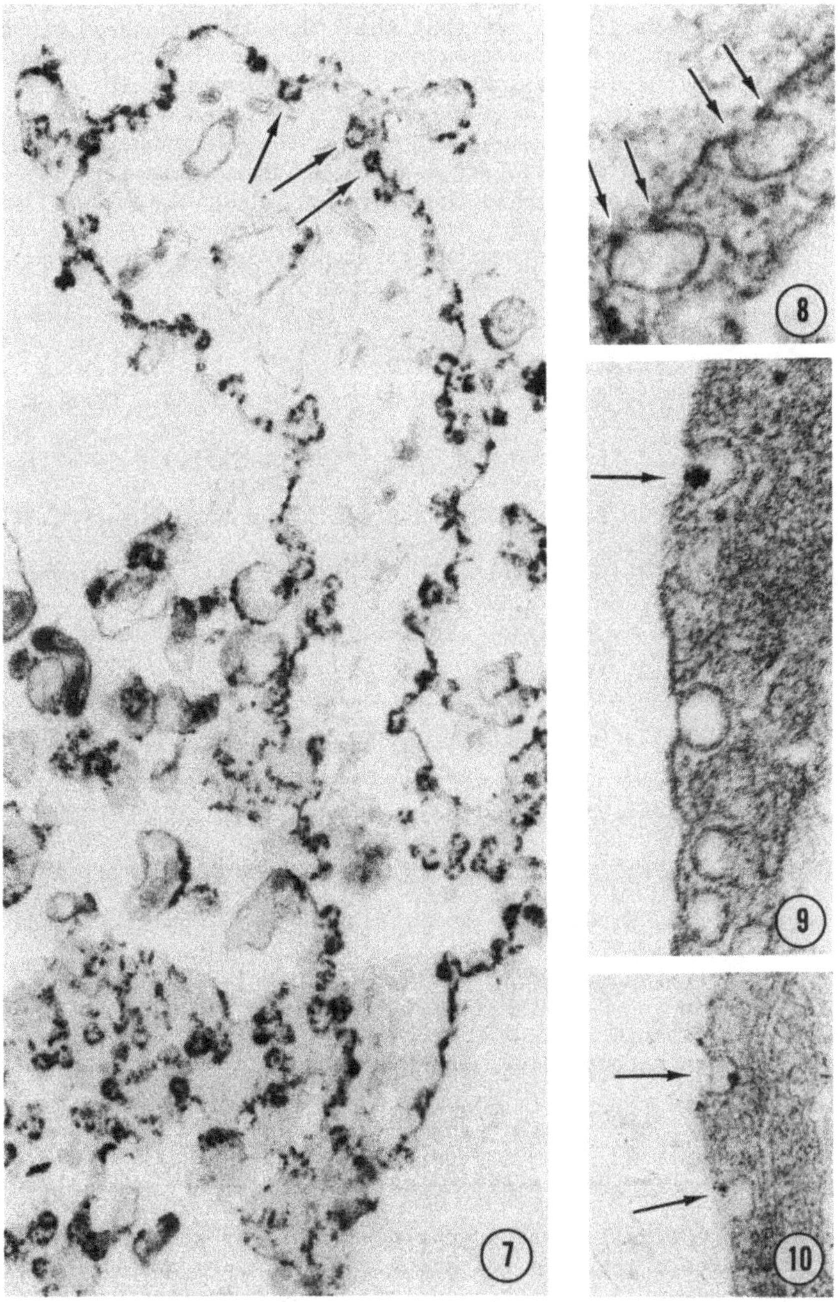
7
8
9
10

the arterial side of the systemic circulation. By contrast, one of the metabolic products of angiotensin I is angiotensin II, a compound which raises blood pressure and stimulates the secretion of aldosterone. Thus it appears that the pulmonary endothelium forms a substance capable of affecting the function of organs at a distance. This situation is analogous to the first description of the endocrine system (16) excepting one major difference. Hormones of an endocrine gland are synthesized within the gland whereas the lung produces angiotensin II by degrading its circulating precursor. Thus we propose that the pulmonary endothelium functions as an intermediate metabolic organ. Whether the endothelium of other vascular beds has a similar capability is not known.

REFERENCES

1. J. W. Ryan, J. Roblero and J. M. Stewart (1968). Biochem. J., 110, 795.
2. J. W. Ryan, J. Roblero and J. M. Stewart (1970). Adv. Exp. Med. Biol., 8, 263.
3. J. W. Ryan, J. Roblero, J. M. Stewart and W. P. Leary (1971). Chest, 59, 8S.
4. J. W. Ryan, J. M. Stewart, W. P. Leary and J. G. Ledingham (1970). Biochem. J., 120, 221.
5. U. Smith and J. W. Ryan (1969). Pharmac. Res. Commun., 1, 197.
6. U. Smith and J. W. Ryan (1970a). Adv. Exp. Med. Biol., 8, 249.
7. U. Smith and J. W. Ryan (1970b). J. Cell Biol., 47, 196a.
8. U. Smith and J. W. Ryan (1971). Chest, 59, 12S.
9. J. W. Ryan and U. Smith (1971). Trans. Ass. Am. Physcns., in press.
10. J. W. Ryan, R. S. Niemeyer and D. W. Goodwin (1971). This volume.
11. U. Smith, J. W. Ryan, D. D. Michie and D. S. Smith (1971). Science, 173, 925.
12. W. P. Leary and U. Smith (1970). Life Sci., 9, 1321.
13. J. W. Ryan and U. Smith (1971). Biochim. Biophys. Acta, 249, 177.
14. A. L. Lehninger (1964). The Mitochondrion. W. A. Benjamin, N. Y.
15. J. A. Lucy (1969). In Lysosomes in Biology and Pathology (ed. J. T. Dingle and H. Fell), Amsterdam, North Holland, p. 313.
16. C. E. Brown-Sequard (1856). Bull. Acad. de Med., 21, 1067.

This investigation was supported in part by grants from the U.S. Public Health Service (HE 12843 and HE 14360) and the Council for Tobacco Research, New York. Figs. 5 and 6 are reproduced from U. Smith and J.W. Ryan, Tissue & Cell, 4, 1972, by permission of the publishers.

BRONCHIAL MOTILITY REGULATION AND BRADYKININ

D. Della Bella, G. Benelli, E. Pajola and P. Valli

Research Laboratories ZAMBON S.p.A.

Bresso-Milan, Italy

Introduction

Several Authors have demonstrated the occurrence of bradykinin in tissues affected by allergic and inflammatory processes: this finding allowed them to postulate the participation of the polypeptide in the pathologic alterations of the vascular permeability and of the smooth muscle motility (Rocha e Silva, 1970).

For instance, in the bronchopulmonary district bradykinin is regarded as responsible for an abnormal influence on the bronchial motor regulation: the observed bronchoconstriction was considered, up to now, as the result of a direct action of the polypeptide on the smooth muscle, as already accepted for acetylcholine, histamine or 5-hydroxytryptamine. Nevertheless, all these agonistic agents share as well the property of stimulating nociceptive, sensitive nervous endings; based on this consideration, in our experiments carried out in order to study the mechanism and the level of the influence of bradykinin on the bronchial smooth muscle, we hypotesized an indirect effect exerted through the stimulation of the sensitive afferent pathway of a nervous reflex arc and the activation of its motor efferences.

The capacity of bradykinin to activate nociceptive afferences has been widely proved (Guzman et. al., 1962; Lim at al., 1962; Keele and Armstrong, 1964; Sicuteri et al., 1965); it is also well known that the bradykinin-induced bronchoconstriction may be strongly increased after the administration of β-blocking agent or after adrenalectomy (Collier et al., 1965).

Experimental results

The experiments were carried out on urethanized guinea-pigs. The intratracheal pressure and the respiratory rate were recorded; the spike activity was registered from: a) the afferent fibres from the lung, in the peripheral trunk of the resected vagus nerve; b) the efferent vagal fibres, in the central trunk of the resected vagus nerve; c) the efferent sympathetic fibres, in the central trunk of the resected splanchnic nerve.
On the left side of Fig. 1 the scheme of the experimental preparation is reported: under these conditions the bioelectric activity of the peripheral trunk of the sectioned right vagus nerve was recorded. This activity is characterized, at rest, by a train of bursts, probably due to the activation of the slowly adapting stretch receptors during the inspiratory phase. The introduction of bradykinin into the jugular vein provokes, if the left vagus

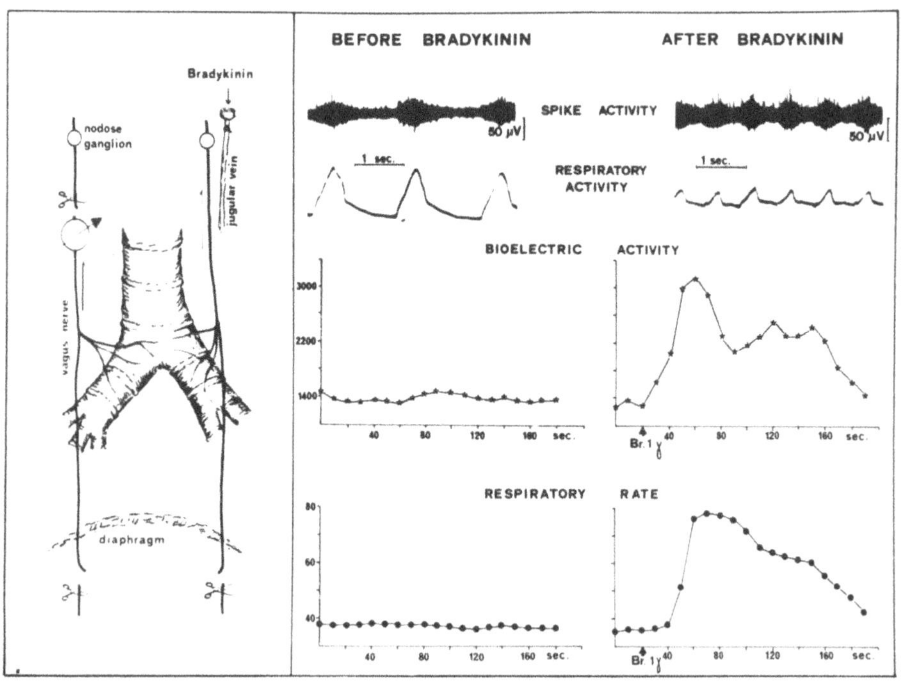

FIG. 1 Urethanized guinea-pig. Left: scheme of the experimental preparation. Right: spike activity, recorded by a digital counter from the peripheral trunk of the right vagus nerve, and respiratory rate before and after intravenous injection of bradykinin. Note the simultaneous increases in both the parameters after bradykinin, made possible by the integrity of the left vagus nerve.

nerve is intact, an evident hyperpnea firstly described by Gjuris et al. (1964) in guinea-pigs and rabbits, and a marked increase in the spike activity, produced in our opinion by the stimulation of both the J and irritant receptors (Paintal, 1969; Widdicombe, 1969), widely present in the bronchopulmonary tissues.
After bilateral vagotomy (Fig. 2) bradykinin introduction still determines a marked increase of the spike activity but the hyperpnea is abolished. Furthermore, as observed by Mills and Widdicombe (1970) in anaphylaxis and histamine-induced bronchospasm and confirmed by ourselves (Della Bella and Ferrari, 1970), the section of either nerve proved effective in significantly reducing also the bronchospastic response to bradykinin.
On the basis of these data we can consider therefore that the integrity of the vagal fibres is essential for the bronchoconstriction, probably allowing the afferent impulses, originating from the peripheral receptors, to reach the centres so as to activate the efferent vagal pathway.

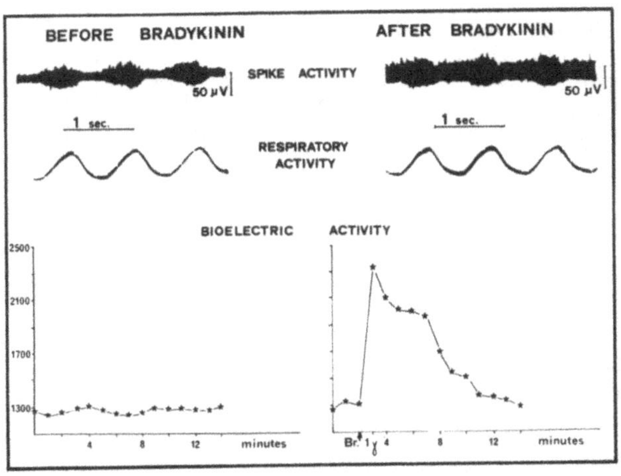

FIG. 2 Urethanized guinea-pig. Spike activity and respiratory rate after bilateral vagotomy, before and after intra venous injection of bradykinin. Note that bradykinin increases the spike activity of the afferent nervous pathway, but does not modify the respiratory rate.

In accordance with our interpretation, the recording of the bioelectric activity from the central end of the acutely resected vagus nerve clearly showed that an evident increase of the spike activity appears (Fig. 3) soon after bradykinin injection.
As regards the distribution of the sensitive neurons, it was demonstrated that they are also located distally to the nodose ganglion, along the peribronchial branches of the vagus nerve (Della Bella and Ferrari, 1970).

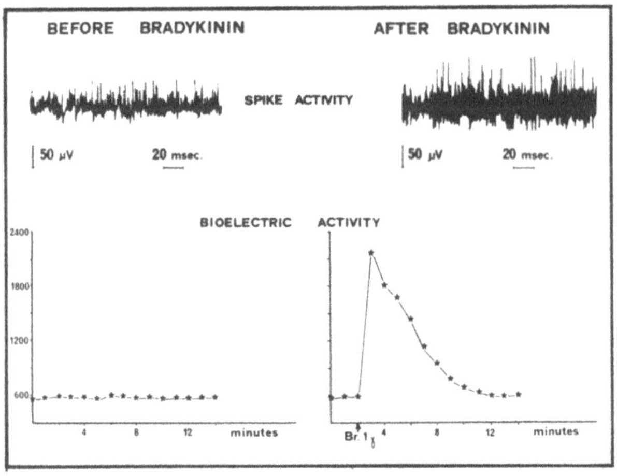

FIG. 3 Urethanized guinea-pig. Spike activity recorded from the central trunk of the right vagus nerve, before and after intravenous injection of bradykinin. Note the increase induced by bradykinin.

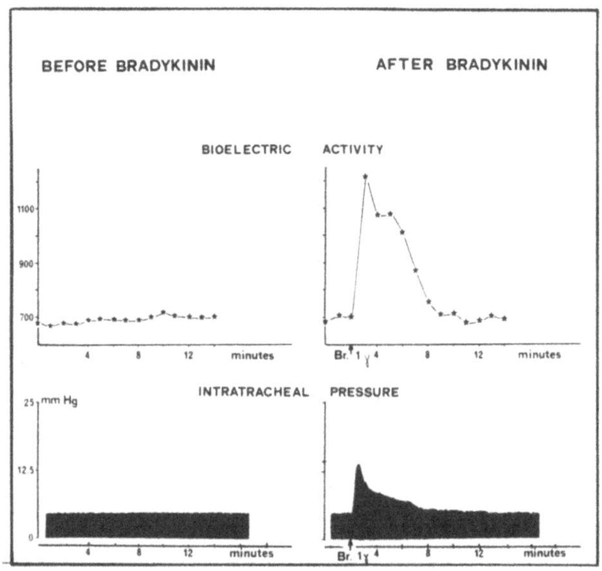

FIG. 4 Urethanized guinea-pig. Bioelectric activity, recorded from the peripheral trunk of the right vagus nerve, degenerated by section carried out 10 days before, and intratracheal pressure before and after intravenous injection of bradykinin. Under resting condition or after bradykinin injection the patterns are similar to those observed by recording from the non-degenerated nerve as shown in Fig. 1.

In fact the spike activity, recorded from the peripheral trunk of the right vagus nerve, whose degeneration was provoked by section carried out distally to the nodose ganglion 10 days before, shows both a resting pattern and changes after bradykinin similar to those observed in the non-degenerated trunk (Fig. 4).
The influence of bradykinin on the sympathetic efferent ways was also investigated: for this purpose, the nervous activity was recorded from the central end of one of the splanchnic nerves after their bilateral resection, before and after vagotomy. Under these conditions we observed a significant increase in the spike activity before the vagal resection, and its disappearance following the bilateral vagotomy (Fig. 5).

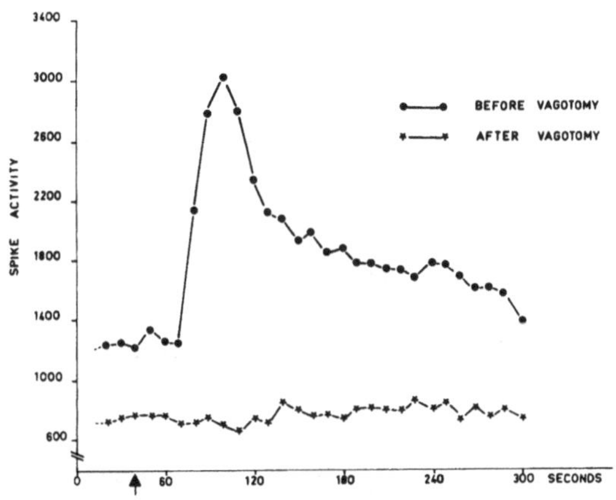

FIG. 5 Urethanized guinea-pig. Effect of intravenous injection of bradykinin (at the arrow) on the spike activity recorded from the central trunk of the sectioned splanchnic nerve, before and after vagotomy. Note the significant increase before the vagal section and its disappearance following the bilateral vagotomy.

Conclusive remarks

- Bradykinin appears to be a potent stimulator of the afferent nervous system in the lung as well as in other districts; because of this property it can also act reflexly, besides as a direct smooth muscle spasmogen. By this indirect mechanism bradykinin proves effective in activating both vagal and sympathetic efferent pathways.
- A vagal activation is mainly responsible for the bronchoconstriction induced by bradykinin. Bilateral vagotomy markedly reduces such motor response, but does not abolish it, a find-

ing probably due to the simultaneous activation by bradykinin of an instrinsic, non cholinergic, nervous system whose presence in the bronchopulmonary district has been recently suggested (Della Bella and Ferrari, 1970).
A sympathetic reflex activation, subserved by both a neural and an humoral way, may be considered a normal component of the nervous response to bradykinin intravenous injection. Its existence has been confirmed in our experiments and its importance in modifying the bronchospastic response to bradykinin has been already shown by Collier et al. (1965) in adrenalectomized or β-blocked animals.

REFERENCES

COLLIER H.O.J., JAMES G.W.L. and PIPER P.J. (1965) - Intensification by adrenalectomy or by β-adrenergic blockade of the bronchoconstrictor action of bradykinin in the guinea-pig. - J. Physiol., 180, 13 P.

DELLA BELLA D. and FERRARI V. (1970) - Nuove ipotesi in tema di regolazione dell'attività motoria bronchiale - Communication at the International Symposium on Inhalation Therapy, Naples, November 2-3rd (in press).

GJURIS V., HEICKE B. and WESTERMANN E. (1964) - Über die Stimulierung der Atmung durch Bradykinin und Kallidin - Arch. Exper. Pathol. Pharmakol., 247, 429.

GUZMAN F., BRAUN C. and LIM R.K.S. (1962) - Visceral pain and the pseudo-affective response to intra-arterial injection of bradykinin and other algesic agents - Arch. Intern. Pharmacodyn., 136, 353.

KEELE C.A. and ARMSTRONG D. (1964) - Substances producing pain and itch - Arnold Publ., London.

LIM R.K.S., LIU C.N., GUZMAN F. and BRAUN C. (1962) - Visceral receptors concerned in visceral pain and the pseudo-affective response to intra-arterial injection of bradykinin and other algesic agents - J. Comp. Neurol., 118, 269.

MILLS J.E. and WIDDICOMBE J.G. (1970) - Role of the vagus nerves in anaphylaxis and histamine-induced bronchoconstrictions in guinea-pigs - Br. J. Pharmacol., 39, 724.

PAINTAL A.S. (1969) - Mechanism of stimulation of type J pulmonary receptors - J. Physiol., 203, 511.

ROCHA e SILVA M. (1970) - Kinin hormones - I.N. Kugelmass Ed., Thomas Publ., Springfield (U.S.A.).

SICUTERI F., FANCIULLACCI M., FRANCHI G. and DEL BIANCO P.L. (1965) - Serotonin-bradykinin potentiation on the pain receptors in man - Life Sc., 4, 309.

WIDDICOMBE J.G. (1969) - Breathing, bronchoconstriction and breathlessness: the role of vagal reflexes in the physiology and pathology of the lungs - Atti Accad. Med. Lombarda, 24, 5.

THE ACTION OF BRADYKININ ON ARTERO-VENOUS HYPERSTOMIA IN MAN

F.Pratesi, C.Corsi, A.Nuti, C.Deidda, M. Tarantelli

Department of Angiology, Main Regional Hospital "Santa Maria Nuova", Florence.

The syndrome of artero-venous hyperstomia of the thigh described by one of us consists of:

1) The presence of an abnormally rapid passage of blood from the arteries to the veins in the district of the deep femoral artery; this passage is effected along channels which differ from both nutritional microvessels and large artero-venous fistulae.

2) The consequent "steal" of blood to the detriment of the superficial femoral artery and to the advantage of the deep femoral artery; this "steal" brings about a circulatory insufficiency in the leg and foot, which reveals itself in a particular type of intermittent claudication.

In 6 subjects suffering from artero-venous hyperstomia of the thigh, we injected bradykinin (10 gamma) into the common femoral artery, and we then carried out:

- rapid seried femoral angiography, respectively before and following the injection;
- oxymetry before and after injection, of samples of blood taken from:
- the femoral vein in the groin, of the same limb,
- the vein of the superficial network of the calf of the same limb,
- a vein in an arm.

RESULTS

1.- With angiographical examination (fig. 1, 2, 3), one notes:

a) the visualisation of the veins of the thigh, which in basal angiography is earlier and more massive than normal, owing to the presence of hyperstomia, becomes, in angiography after bradykinin, even earlier and yet more massive;

b) the visualisation of the veins of the calf, which in basal angiography is normal, becomes, in angiography following bradykinin, earlier and more massive;

c) the speed of penetration of the contrast medium along the superficial femoral artery, which in basal angiography is slower than normal because of the "steal"

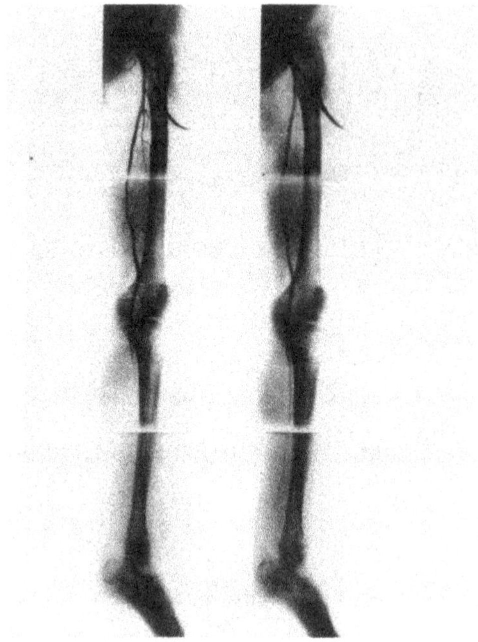

Fig. 1 - Rapid seried femoral angiography in a subject suffering from artero-venous hyperstomia of the thigh. On the left, before: on the right, after - administration of bradykinin (10 gamma) in the femoral artery. Radiogram of 2"

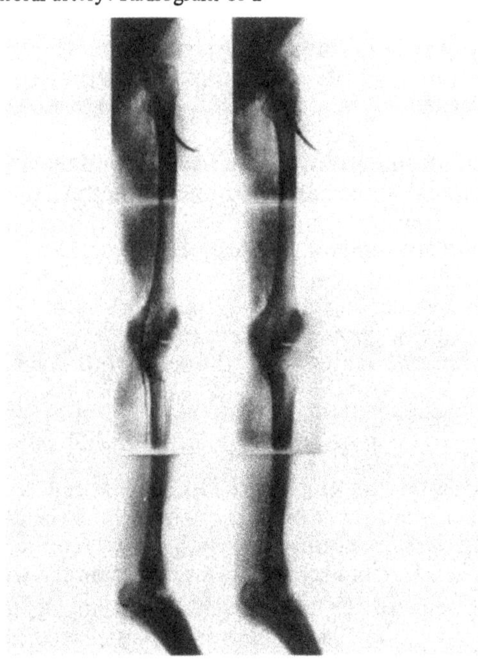

Fig. 2 - As Fig. 1. Radiogram of 4".

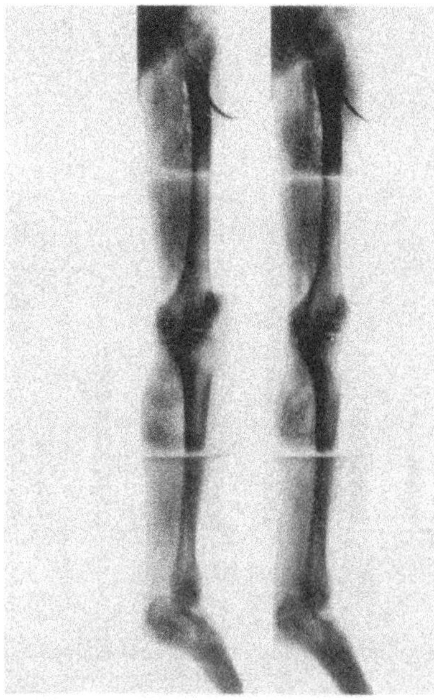

Fig. 3 - As. Fig. 1. Radiogram of 6".

of blood, becomes normal in angiography after bradykinin, because the possibility of a "steal" is removed.

The findings are constant and uniform in all subjects.

2).- With gasimetric examination (fig. 4), one notes:

a) in three subjects, of whom the first of Fig. 4 is the prototype:

- in basal conditions, the femoral artero-venous gradient of the exymetry evaluated by means of a comparison with the artero-venous gradient of the leg and the arm, is less than normal, in accordance with the presence of hyperstomia of the thigh:
- bradykinin accentuates this diminution in the femoral gradient;

b) in one subject (the second of fig. 4):

- in basal conditions the femoral artero-venous gradient of the oxymetry is greater than normal. in contrast with the presence of hyperstomia of the thigh;
- bradykinin brings about a great reduction in the femoral gradient, that is to say, it produces a situation in accordance with the presence of hyperstomia of the thigh.

3.- in one subject. (the third of fig. 4):

- in basal conditions: situation similar to that found in the first subject;
- bradykinin increases the femoral artero-venous gradient in contrast with the presence of hyperstomia.

DEDUCTIONS

1) Bradykinin, in the doses employed, produces an accentuation of artero-venous

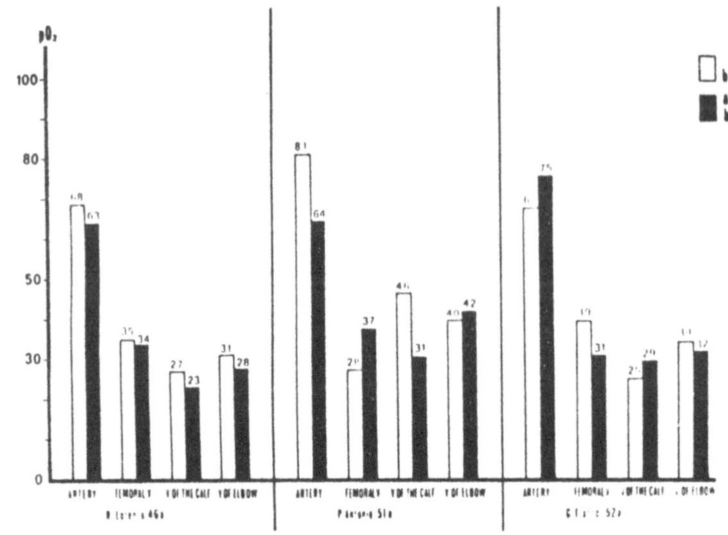

Fig. 4 - Blood pO_2 after bradykinin injected into femoral artery. Artero-venous hyperstomia of the thigh.

hyperstomia of the thigh, and also produces artero-venous hyperstomia of the leg. This deduction may be validly advanced on the grounds of the constancy of angiographical findings.

2) Gasimetric examination confirms this concept. This, however, furnishes less constant results. This diminished constancy may depend upon the fact that the femoral vein in the groin collects both the blood from the district of the deep femoral artery and that from the district of the superficial femoral artery. The interference of a direct or indirect metabolic action of bradykinin is also possible.

3) These deductions, with all the reserve which the extreme difficulty of studying these microhemodynamic phenomena demands, allow us to put forward the working hypothesis that bradykinin may have a particular vasomotor action of dilatatory type, more proximal in respect of the extreme periphery: this action might be of no little importance in the mechanism of its vaso-activity.

SUMMARY

Bradykinin is injected into the femoral artery in men suffering from artero-venous hyperstomia of the thigh.

Before and after injection, the following examinations are carried out:

- rapid seried angiography;
- oxymetry of the arterial and venous blood.

On the basis of the results obtained, we put forward the hypothesis that bradykinin may have a vasomotor action of dilatatory type above the arterioles.

RAT INTESTINAL KALLIKREIN *

I.J. ZEITLIN

WOLFSON GASTROINTESTINAL LABORATORIES, GASTRO-INTESTINAL UNIT, WESTERN GENERAL HOSPITAL, EDINBURGH

INTRODUCTION

Aqueous extracts of mammalian gut wall were shown by Werle and his colleagues (1960) to contain a trypsin-activated hypo-:tensive activity. They called this activity a kallikrein although they presented no direct evidence that kinin release was involved. Amundsen and Nustad (1965) noted that simple saline extracts from intestinal mucosae of rats and rabbits released a smooth muscle stimulant from treated human plasma. However, they did not characterize either the smooth muscle stimulant or the nature of its release. In a recent review, Webster (1970) has suggested that the activities involved in these two studies were not, in fact, kallikreins, but other "endogenous proteases".

At the 1969 Kinin symposium in Florence, I reported that kinin-forming enzyme could be detected in the gut wall of the rat along the whole length of the alimentary tract (Zeitlin, 1970). Current studies (Zeitlin, 1971) have shown that rat intestinal kinin-forming activity may be activated, presumably by autolytic processes, by incubating tissue homogenates in distilled water at room temperature for 18 - 24 hours. Heating such homogenates to 55 ^{o}C for ten minutes caused a similar activation. Both proce-:dures increased the aqueous-extractable activity by about five times.

The smooth muscle stimulant formed by the enzyme was shown to be kinin-like since it contracted the isolated oestrous rat

*This work was supported by an M.R.C. temporary research grant to Dr W. Sircus.

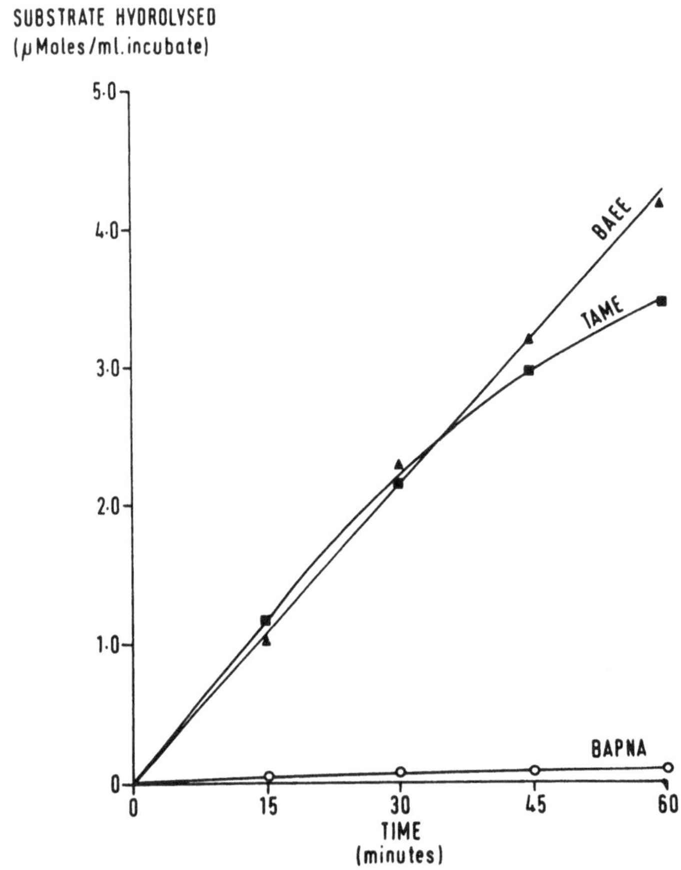

Figure 1

Comparison of the hydrolysis of BAEE (▲), TAME (■) and BAPNA (●) by an activated aqueous extract of rat small intestine (extract from 80 mg. of tissue, per ml. of test incubate). Incubation was at pH 8.5 and 37°C.

uterus and the guinea pig ileum and relaxed the rat duodenum. The indices of discrimination were less than 2 when the kinin was compared with synthetic bradykinin using these three tissues. The pH optimum of the kinin-forming activity was at pH 8.5. The kinin formation was almost completely inhibited (97.4%) by Trasylol and slightly, but not significantly reduced by soybean trypsin inhibitor (SBTI) (7.4%, $P > 0.05$). A specific trypsin inhibitor, tosyl-lysine-chloromethyl ketone (TLCK), had no effect on the kinin-forming activity.

The factors which activate the kinin-forming enzyme also activate arginine esterase activity in the intestinal extracts. Some studies on this esterase activity are now reported.

METHODS

1. Activation of Enzyme

Blood was flushed from rat ileum by intravascular perfusion with 50 ml. of heparinized (10 units/ml.) Krebs-Henseleit solution followed by 50 ml. heparin-free solution. The ileum was removed and the lumen was washed with Krebs-Henseleit. The tissue was then homogenized in distilled water (0.1 gm/ml.). The homogenate was allowed to stand for eighteen hours at room temperature (18 - 22°C). Activation was carried out under toluene to prevent bacterial growth. The homogenate was centrifuged (1 hour, 3000 'g') and the supernatant was filtered through a millipore filter (pore size 0.8 u) to give a clear solution. The extract was then snap frozen and lyophilized for storage at -27°C.

2. Enzyme Estimation

The hydrolysis of BAEE and TAME was followed using the Hestrin method of Brown (1960). Amidase activity was measured as described by Erlanger, Kokowski and Cohen (1961) using benzoyl-arginine-p-nitroanilide (BAPNA) as the substrate. The enzyme estimations were in 0.05 M-barbital buffer at pH 8.5 unless otherwise stated. For the inhibitor studies, inhibitors were incubated with the enzyme for ten minutes before substrate was added.

RESULTS

When the actions of the extract were compared on the synthetic substrates BAEE, TAME and BAPNA, it was found that in molar terms the hydrolysis of BAEE (4.2 µM/hr/ml.) and TAME (3.5 µM/hr/ml.) took place more than forty times faster than that of BAPNA (0.08 µM/hr/ml.) (Figure 1). While BAPNA is a good substrate for plasma kallikrein, it is generally a poor substrate for glandular kallikreins (Schachter, 1969).

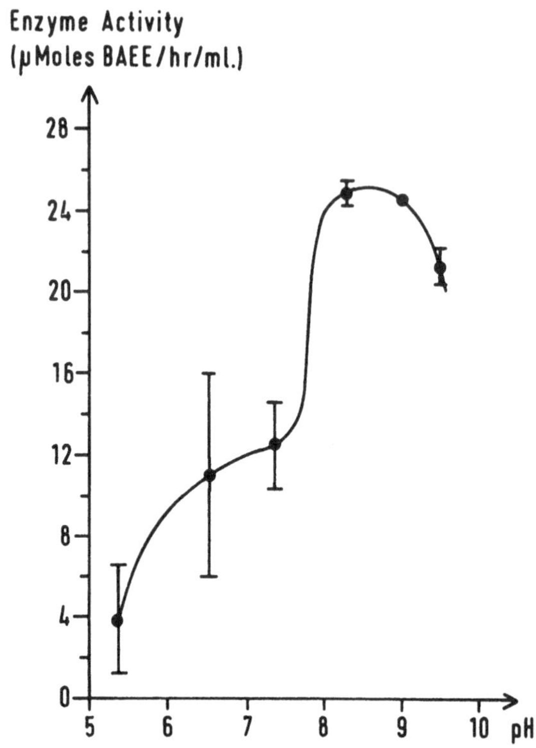

Figure 2

The influence of pH on the BAEE esterase activity of rat small intestine. The results are expressed as mean ± s.d. (N = 6). One ml. incubate contained enzyme from 80 mg. tissue. Incubation was for 1 hour at 37°C.

The pH optimum of the benzoyl arginine esterase activity of the extract, like that of the kinin formation, is at pH 8.5 (Figure 2). The inflection in the pH curve between pH 6 and pH 7.5 is not seen in the curve for kinin formation and is probably due to the presence of more than one arginine esterase. The arginine esterase activity is greatly inhibited (93.4%) by Trasylol (200 KIU/ml.) (Figure 3). The activity is unaffected by ovomucoid trypsin inhibitor (100 µg/ml.) and slightly reduced

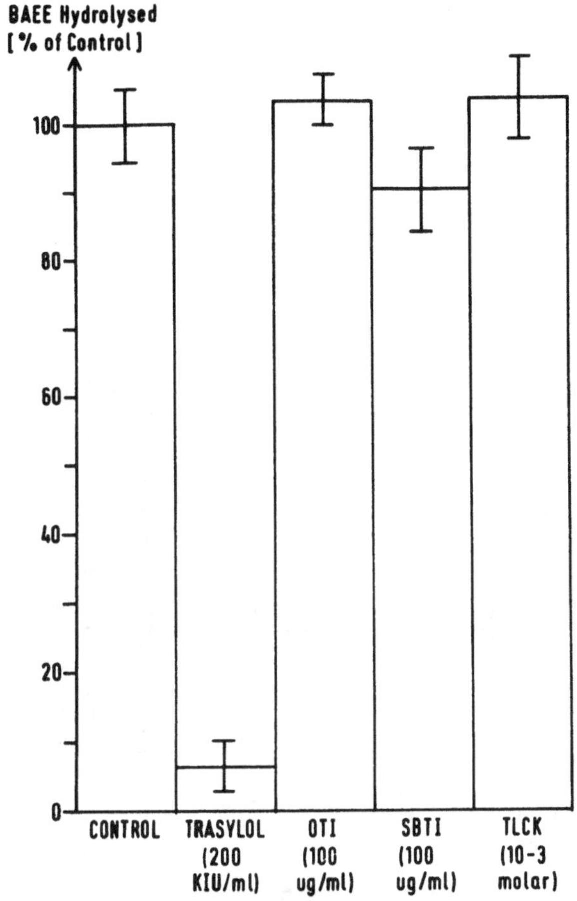

Figure 3

The inhibition of the BAEE esterase activity of rat small intestine by Trasylol, ovomucoid trypsin inhibitor (OTI), soybean trypsin inhibitor (SBTI) and tosyl-lysine-chloromethyl ketone (TLCK). Results are expressed as mean ± s.d. (N = 4).

by soybean trypsin inhibitor (100 μg/ml.) although this change was not statistically significant ($P > 0.05$). It was completely unaltered by the specific trypsin inhibitor, TLCK. The pattern of inhibition of the arginine esterase activity is identical to that of the kinin-forming activity (Zeitlin, 1971).

	TRASYLOL	SBTI	OTI	TLCK
Rat Intestinal Kallikrein	+	0	0	0
Plasmin	+	+	0	
Trypsin	+	+	+	+
Plasma Kallikrein	+	+	0	
*(a) Rat Pancreatic Kallikrein	0	0	0	
*(b) Salivary Kallikrein	+	0	0	
*(c,d) Urinary Kallikrein	+	0	+0	0

Table 1

The table compares the responses of rat intestinal kinin-forming and arginine esterase activities to inhibitors, with the responses of some other kininogenases. +, inhibition; o, no inhibition. Shading indicates responses differing from those of the rat intestinal enzyme.

*(a) Trasylol is inactive on rat pancreatic kallikrein (Vogel and Werle, 1970; (b) Erdos et al, 1968; (c) Nustad, 1970; (d) Mares-Guia et al, 1970.

Table 1 shows a comparison between the inhibition profile of the rat intestinal kinin-forming and arginine esterase activities, and the responses, culled from the literature, of some other kinin-forming enzymes which may be expected as contaminants of rat intestinal extracts. The enzyme is thus distinguished from plasmin, trypsin, plasma kallikrein and rat pancreatic kallikrein. It is not distinguished from rat salivary kallikrein, nor is it distinguished from rat urinary kallikrein although this would not normally be present in intestinal extracts.

CONCLUSION

It is concluded that blood free rat intestine contains a kinin-forming enzyme which is present in the tissue in the form of an inactive precursor. The substance formed by this enzyme is a kinin, very similar to bradykinin. Factors which activate the kinin-forming enzyme also activate arginine esterase activity in the intestinal extracts. The pH optima of both the kinin-forming and esterase activities are at pH 8.5 and the responses of both activities to inhibitors are identical and are different from those of plasmin, trypsin, plasma kallikrein and rat pancreatic kallikrein.

It has been suggested (Webster, 1970) that the activity detected in the gut by earlier workers may be due to endogenous proteases other than tissue kallikrein. The properties described in this paper are those of a kallikrein and support the conclusion that the rat intestinal wall contains a distinct tissue kallikrein.

ACKNOWLEDGEMENTS

The author is grateful to Mr D.R. Keys for technical assistance.

REFERENCES

1. Amundsen, E., Nustad, K. (1965) J. Physiol. 179, 479-488. Kinin-forming and destroying activities of cell homogenates.

2. Brown, M.E. (1960) J. Lab.Clin.Med. 55, 615-624. The colorimetric determination of arginine ester hydrolysis by human sera.

3. Erdös, E.G., Tague, L.L., Miwa, I. (1968) Biochem.Pharmacol. 17, 667-674. Kallikrein in granules of the submaxillary gland.

4. Erlanger, B.F., Kokowski, N., Cohen, W. (1961) Arch.Biochem. Biophys. 95, 271-278. The preparation and properties of two new chromogenic substrates of trypsin.

5. Mares-Guia, M., Silva, E., Diniz, C.R. (1970) in "Bradykinin and Related Kinins", ed.: F. Sicuteri, M. Rocha e Silva, N. Back. New York: Plenum Press, pp. 65-75. Partial purification of rat urinary kininogenase and its reactions with active center reagents of trypsin.

6. Nustad, K. (1970) Br.J.Pharmac. 39, 73-86. The relationship between kidney and urinary kininogenase.

7. Schachter, M. (1969) Physiol.Revs. 49, 509-547. Kallikreins and kinins.

8. Vogel, R., Werle, E. (1970) in "Bradykinin, Kallidin and Kallikrein", ed.: E.G. Erdos, Berlin: Springer, pp. 213-249. Kallikrein inhibitors.

9. Webster, M.E. (1970) in "Bradykinin, Kallidin and Kallikrein", ed.: E.G. Erdos, Berlin: Springer, pp. 131-155. Kallikreins in glandular tissue.

10. Werle, E. (1960) in "Polypeptides which affect Smooth Muscle and Blood Vessels", ed.: M. Schachter, London: Pergamon, pp. 199-209. Kallikrein, kallidin and related substances.

11. Zeitlin, I.J. (1970) in "Bradykinin and Related Kinins", ed.: F. Sicuteri, M. Rocha e Silva, N. Back, New York: Plenum Press, pp. 329-339. Kinin release associated with the gastro-intestinal tract.

12. Zeitlin, I.J. (1971) Br.J.Pharmac., 42, 648-649 P. Pharmacological characterisation of kinin-forming activity of the rat intestinal tract.

GUINEA PIG KININOGEN-KININ SYSTEM IN PREGNANCY AND UNDER HORMONAL INFLUENCE

Glaci R. DaSilva

Department of Pharmacology, Faculty of Medicine of Ribeirão Prêto, USP, São Paulo, Brazil

Evidences on the participation of the kininogen-kinin system in pregnancy and labour have been provided by the work of Armstrong and Stewart (1), Gomes (2), Werle (3), Martinez et al. (4), Erdös et al. (5), Periti and Gasparri (6), Centaro et al. (7), Meirelles (8) and many others, showing the formation of kinins during labour, a fall in bradykininogen (BKg), an increase of the destroyng enzyme (kininase), and of the esterolytic activity, in plasma from pregnant women.

Studies on laboratory animals are scarce; Wiegershausen et al. (9) described increase of BKg during pregnancy in rabbits and rats, with a subsequent fall during labour, and return to normal values in the puerperium.

We decided to take guinea pigs as experimental animals, because of the relatively large period of pregnancy (average 67 days) and higher (than in the rat), values of the BKg level in plasma. We have taken as parameters of investigation BKg levels, the protein content, the esterolytic activity (on TAMe), and the kininolytic activity (kininase) using bradykinin (BK) as substrate.

MATERIAL AND METHODS

The BKg contents were determined by Diniz et al. (10) micromethod, with slight alterations, omitting the final alcohol extraction. After the treatment with trypsin, the volume is adjusted with saline to the required level and the material tested directly upon a trypsin dessensitized piece of guinea pig ileum. The results are given in units of BKg, calculating the units as corresponding to 0.5 µg of synthetic BK.

Total plasma protein (P) was measured by the biuret method (11) and expressed in mg/ml.

The esterolytic activity (EA) was estimated by the variation of pH in an incubate at 37ºC of a fixed amount (50 µmol) of TAMe (tosyl arginine methyl ester) in presence of 20 mg of plasma protein in Tris-HCl as buffer, at pH 7.8 in a final volume of 5.5 ml. The pH determinations were recorded with a titrating potenciometer (TTT_1 Radiometer - Copenhagen) with a single glass electrode (GK 2021^1 B) from the volumes of 0.0067N of NaOH added to keep the pH at the constant level of 7.8. The readings were done at increasing intervals of time (in min) to a zero kinetic reaction. To improve the kinetic of the reaction, a 0.02M of calcium chloride and 0.1M of potassium chloride were added to the incubation mixture. The esterolytic activity is expressed in µmol of TAMe hydrolysed by mg of protein per hour.

The kininase activity (KA) was measured in test samples containing 3.0 µg of BK as substrate, 5.0 µl of plasma and the volume completed to 3.0 ml with Tris-HCl buffer at pH 7.4.

Aliquots were taken every 6 minutes for 18 to 24 minutes, and tested directly upon a piece of guinea pig ileum in a 10 ml chamber, at 37ºC. Kininase activity is indicated as µg of BK inactivated in one minute per mg of plasma protein.

Pregnant guinea pigs were obtained by exposing the females in oestrus to contact with males for a period of 48 hours and then kept in separate cages, for the duration of the experiment. Water and food were provided as usual. To control the hormonal state, vaginal smears were routinely done before and after the initiation of the experiment.

Blood samples were taken by cardiac puncture using a 2.5% solution of sodium oxalate as anticoagulant in the proportion of 0.1 ml per 0.9 ml of blood.

For the experiments with estradiol and aldosterone, immature guinea pigs were used. Estradiol was administered as the benzoate, in peanut oil in the dose of 0.2 mg/animal, for a period of 20 days. During the first 15 days the injections were mainly sub-cutaneous and then in the last 5 days were alternated by the sub-cutaneous and intra-muscular routes according to the scheme proposed by Owman et al. (12). The aldosterone (base) was given by intra-muscular injections in the dosis of 2.0 μg and 4.0 μg daily for 20 days.

MATERIALS

Aldosterone (base) was kindly supplied by Geigy Co., Estradiol benzoate by Shering Co. of Brazil; BRS-640 (Synthetic BK) by Sandoz, Basel; TAMe (tosyl-arginine-methyl-ester)-Sigma; Tris (hydroxy-methyl-aminomethan), Merck; Trypsin, Worthington.

RESULTS AND DISCUSSION

Bradykininogen/Protein Ratios in Normal and Pregnant Guinea Pigs

Table I shows the results obtained with normal guinea pigs (10 animals) in which BKg and plasma protein levels (P) were determined and the ratio BKg/P calculated for control periods of 67 days.

The levels of BKg and protein are kept fairly constant with an oscillation range of 27.0 to 32.0.

TABLE I

BKg AND PROTEIN CONTENTS OF NORMAL ADULT GUINEA PIGS

Days	Protein mg/ml	BKg U/ml	$\frac{\text{BKg} \times 100}{\text{P}}$
1	42.78 ± 0.77*	13.62 ± 0.44	31.83
8	42.70 ± 1.48	11.69 ± 0.31	27.37
16	39.20 ± 1.22	12.52 ± 0.54	31.93
29	39.05 ± 0.88	13.39 ± 0.44	34.28
43	39.71 ± 1.00	12.92 ± 0.44	32.53
67	39.71 ± 1.04	12.84 ± 0.12	32.33

*Mean ± S.E. in 10 animals.

Table II shows the values of BKg and total plasma protein for whole period of pregnancy (67 days) and 12-24 hours of puerperium.

Though a significant fall of BKg was observed after the 54th day of pregnancy, there was also a concomitant fall in total protein. If the ratios BKg x 100/P are calculated as the percentage of BKg per mg of protein, there was no significant changes along the pregnancy period, although a significant increase in the puerperium was observed.

TABLE II

BKg AND PROTEIN IN PREGNANT GUINEA PIGS

Days of pregnancy	Protein mg/ml	BKg U/ml	$\frac{BKg \times 100}{P}$
7	39.48 ± 0.90*	13.10 ± 0.41	33.18
21	39.77 ± 1.55	12.85 ± 0.55	32.31
35	41.35 ± 2.04	13.15 ± 0.70	31.80
47	32.22 ± 0.75	12.47 ± 0.71	38.70
54	29.81 ± 1.28	10.12 ± 0.34	33.94
60	27.61 ± 0.71	8.43 ± 0.49	30.53
12-48 hs, before	26.22 ± 0.80	8.16 ± 0.37	31.12
LABOUR			
12 hs, after	25.19 ± 0.69	10.66 ± 0.40	42.31
24 hs, after	24.41 ± 0.54	13.27 ± 0.56	54.36

*Mean ± S.E. in 8 animals.

For obvious reasons, samples were not taken during labour and so we cannot be sure that a change in such a ratio could have taken place during parturition.

Therefore, a close comparison of our findings with those obtained by clinical workers during labour could not be done. See references 1, 2, 3, 4, 5 and 6.

Variations of Esterolytic (EA) and Kininase Activity (KA) During Pregnancy

As seen in Fig. 1 and Fig. 2, there was no marked variation of the esterolytic activity (upon TAMe) in plasmas taken under normal conditions and over the pregnancy period.

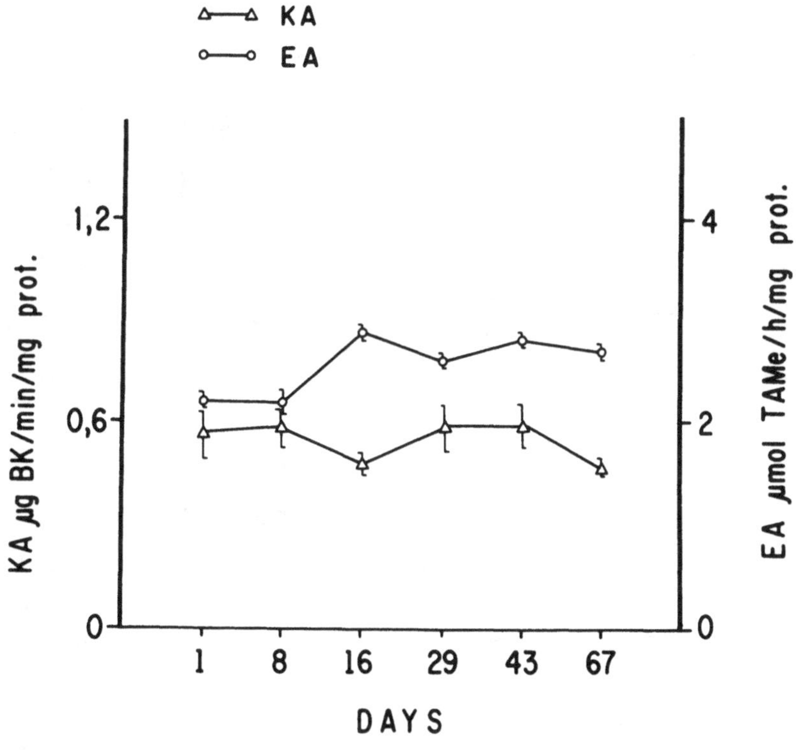

Fig. 1 - Normal values of esterolytic (EA) and kininase (KA) activities in adult female guinea-pigs followed during 67 day.

The KA however, undergoes significant increases attaining a peak at the 47th day of pregnancy, when the activity was found to be increased 4 times of the normal levels.

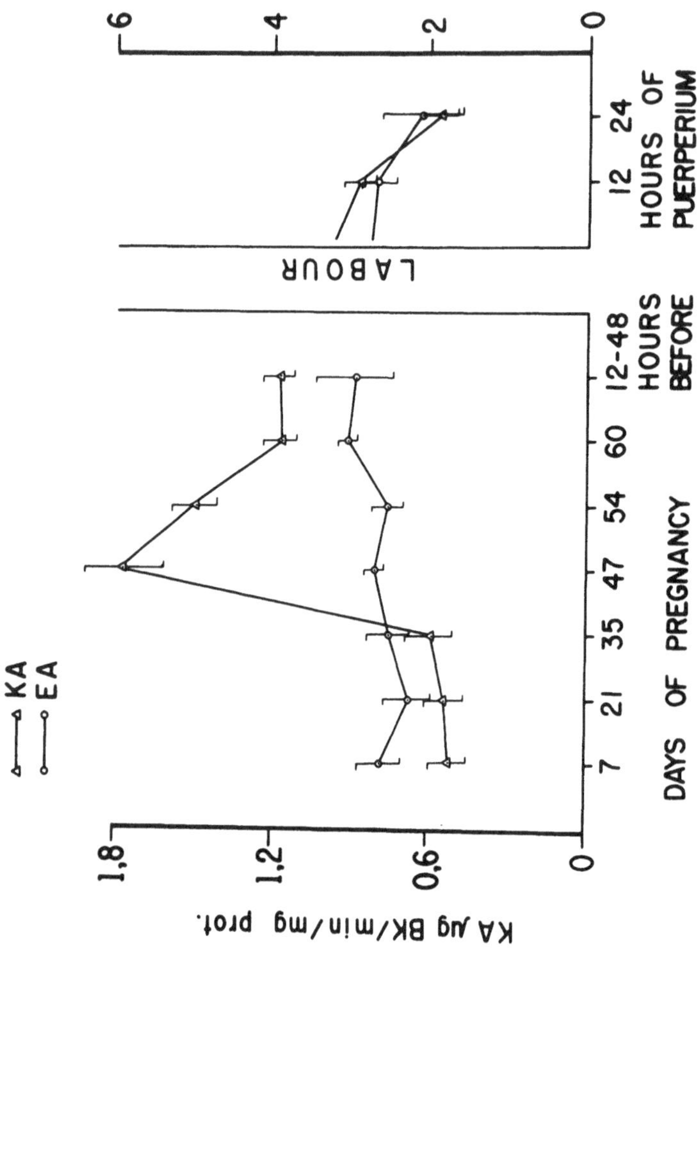

Fig. 2 - Esterolytic (EA) and kininase (KA) activities in guinea-pigs during pregnancy and puerperium. Note that KA values reached a peak at the 47th day of pregnancy and tended to normalize after 24 h after parturition.

After reaching such a peak, KA tends to decrease, though remaining in the average 2.2 times above the normal until 12-48 hours before parturition.

In the puerperium, KA is still above the normal level for the first 12 hours, then tending to normalize after 24 hours from parturition.

Hormonal Influences on BKg Level and Kininase Activity

Estradiol. As shown in Table III, there occurs a significant fall in BKg content with a concomitant decrease of the ratio BKg x 100/P.

The oestrogenic hormone had no effect on KA, as indicated in Table III.

Aldosterone. Fig. 3 shows the effect of chronic treatment of 2.0 µg and 4.0 µg of aldosterone on BKg and total protein (columns to the right) and on KA (full lines at the left). It is to be noted, that a fall in BKg and the increase of KA were already in the range of the maximum effect of aldosterone.

Furthermore, it is noteworthy to say that aldosterone was about 100 times more potent than estradiol to produce a similar fall in BKg. In both cases, the fall in total protein, did not match the fall in BKg, and the ratio BKg x 100/P tend to decrease. This is to be contrasted, with the situation met in pregnancy, when the ratio remained approximately constant. If the fall in BKg in the second half of pregnancy is somehow dependent on the release of endogenous steroidal hormones, it might indicate, that fall in BKg during pregnancy is an independent parallel event with the fall in total protein. If this conclusion will be substantiated by further investigation, the fall in BKg might be a significant event as others have postulated in the evolution of pregnancy (1, 2, 3, 4, 5 and 6).

In Vitro Studies with Estradiol and Aldosterone

In order to see whether the effect of such hormones can be a direct one on the kininogen-kinin system of plasma, we devised "in vitro" experiments, in which 8.0 µg of estradiol benzoate or 0.15 ng of aldosterone, were incubated wich 1.0 ml of plasma and the BKg and KA, estimated after 30 minutes incubation at room temperature.

As seen in Table IV, there was no change either in BKg or KA

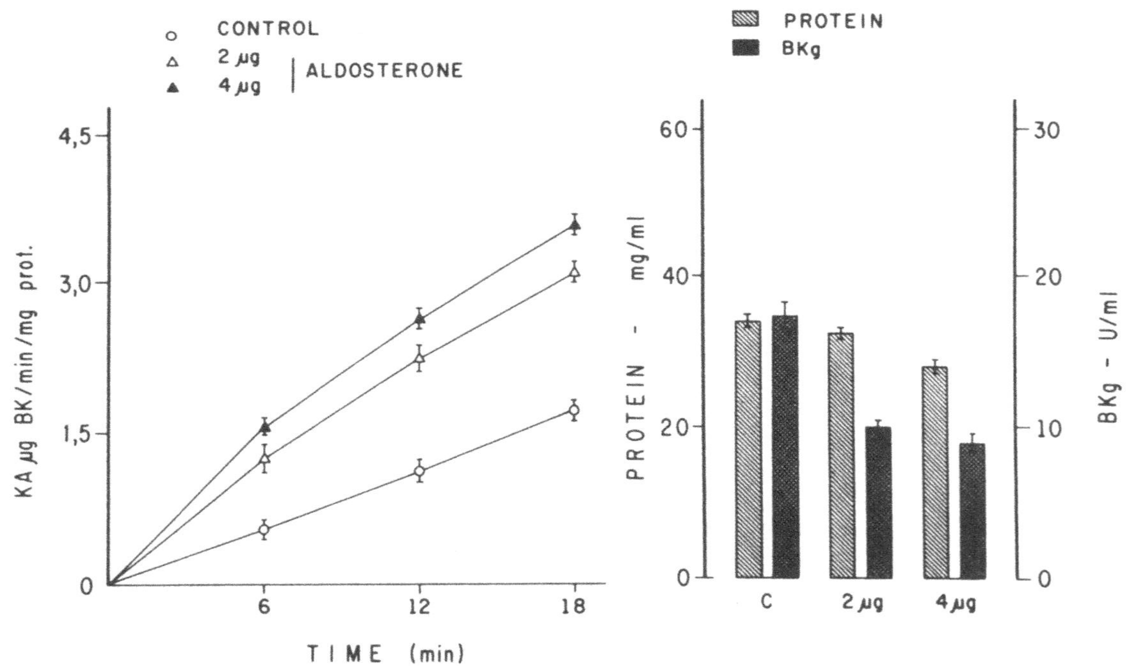

Fig. 3 - Effect of chronic treatment of aldosterone (2.0 and 4.0 µg) on plasmatic BKg and total protein (columns to the right) and KA (full lines at the left).

TABLE III

EFFECTS OF ESTRADIOL IN IMMATURE GUINEA PIGS

	Protein mg/ml	BKg U/ml	$\frac{BKg \times 100}{P}$	BKg Percentual variation	Test "t" of Significancy	KA µg BK/min/mg prot.
Control	31.78±0.21 (10)**	17.27±0.46* (10)	54.34	-	-	0.74 ± 0.05 (10)
Estradiol 0.2 mg***	29.46±0.49 (8)	11.13±0.24 (8)	37.78	-35.60	<0.001	0.67 ± 0.02 (8)

*Mean ± S.E.

**Number of animals in each experiment.

***Estradiol benzoate.

in such conditions.

TABLE IV

"IN VITRO" ESTRADIOL AND ALDOSTERONE ON BKg AND KA OF IMMATURE GUINEA PIG PLASMAS

Hormone	BKg U/ml	KA µg BK/min/mg prot.
-	11.22 ± 0.12 (10)	0.56 ± 0.01 (10)
Estradiol*	11.16 ± 0.23 (5)	-
Aldosterona**	11.04 ± 0.13 (5)	0.60 ± 0.02 (10)

*Estradiol benzoate - 8.0 µg/ml of plasma.

**Aldosterone base in peanut oil - 0.15 ng/ml of plasma.

Between brackets number of animals.

SUMMARY

Using guinea pigs as experimental animals, bradykininogen (BKg), total plasma proteins (P), esterolytic activity (EA) and kininase activity (KA) were measured during pregnancy and puerperium. A significant reduction in BKg was observed that ran parallel with a fall in total protein concentration.

The EA measured using TAMe as substrate remained constant, while KA, using bradykinin as substrate increased sharply to a peak at the 47th day of pregnancy and then declined to the end of pregnancy. The levels of KA were high (from 4 to 2 times the normal) throughout pregnancy returning to normal at puerperium.

To correlate such changes with the discharge of hormones during pregnancy, estradiol and aldosterone were chosen as a first approach to the problem.

Both hormones produced fall in BKg without any significant change in total protein in plasma. Aldosterone was about 100 times more potent than estradiol in producing similar alterations in BKg.

Aldosterone, but not estradiol, increased sharply kininase activity (KA) at doses (2.0 μg and 4.0 μg) that produced nearly maximal effects in the chronic treatment employed.

The conclusion is reached that the fall in BKg as observed in pregnancy, might be independent of the fall in total protein. This would give additional weight to observations of earlier workers in human pregnancy.

Acknowledgements

The author is deeply grateful to Prof. M. Rocha e Silva for helpful advice and criticism; Thanks are also due to Mr. Osmar Vettore for his excellent technical assistance.

This work was supported by "Fundação de Amparo à Pesquisa do Estado de São Paulo" (FAPESP), Grant Biológicas 70/441.

REFERENCES

1. ARMSTRONG, D., STEWART, J.W. (1960) - Nature, 188: 1193.
2. GOMES, F.P. (1958) - Thesis, Faculty of Medicine of Lisbon.
3. WERLE, E. (1960) - In: Polypeptides wich Affect Smooth Muscles and Blood Vessels. Oxford, Pergamon, p. 199.
4. MARTINEZ, A.R., CARVALHO, I.F. and DINIZ, C.R. (1962) - J. Obstet. Gynacol. Brit. Commonwealth, 69: 104.
5. ERDÖS, E.G., WOHLER, I.M., LEVINE, M.I. and WESTERMAN, M.P. (1965) - Clin. chim. Acta, 11: 39.
6. PERITI, P. and GASPARRI, F. (1965) - Biochem. Pharmacol., 14: 1396.
7. CENTARO, A., PERITI, P. and SICUTERI, E. (1963) - Sett. Med., 51: 1.
8. MEIRELLES, R.S. (1967) - Thesis, Faculty of Medicine of Ribeirão Prêto.
9. WIEGERSHAUSEN, B., KLAUSCH, B., HENNIGHANSEN, G. and SOSAT, R. (1968) - Experientia, 24: 1128.

10. DINIZ, C.R. and CARVALHO, I.F. (1963) - Ann. N.Y. Acad. Sci., 104: 77.

11. WEICHSELBAUM. T.E. (1946) - Am. J. Clin. Path. Tech. Bull., 10: 40.

12. OWMAN, Ch., SJÖBERG, N.O., SJÖSTRAND, N.O. and SWEDIN, G. (1970) - Acta Endocrinol., 64: 459.

COMPARATIVE ACTIONS OF KININS OF INCREASING MOLECULAR WEIGHT ON SMOOTH MUSCLE, BLOOD PRESSURE AND VASCULAR PERMEABILITY[(*)]

M. Rocha e Silva, Marina L. Reis and Lyioko Okino

Department of Pharmacology, Faculty of Medicine, USP

Ribeirão Prêto, São Paulo, Brazil

It has been well established in the literature (Stürmer and Berde, 1962; Schröder and Lübke, 1966; Stewart and Woolley, 1967) that the pharmacological actions of bradykinin and its two known natural derivatives (Lys-Bk and Met-Lys-Bk) change if we prolong the chain of amino acids, at the NH_2-terminal, according to the biological preparation utilized, as well as to their resistance to destruction by kininases. For instance, the activity on smooth muscle falls to one half with Lys-Bk and to one third with Met-Lys-Bk. On the other hand, as far as the actions upon blood pressure are concerned, the activity increases by about the same proportion in the reverse direction, in such a way that we have the following sequences:

Rat uterus: Bk > Lys-Bk > Met-Lys-Bk.

Guinea pig ileum: Bk > Lys-Bk > Met-Lys-Bk.

Rabbit B.P.: Met-Lys-Bk > Lys-Bk > Bk.

We have recently analysed (Reis et al., 1971) the actions of a new synthetic polypeptide derived from bradykinin, with 13 amino acid residues:

Gly-Arg-Met-Lys-Bk (GAML-Bk).

(*) Aided in part by Conselho Nacional de Pesquisa (CNPq) and Fundação de Amparo à Pesquisa (FAPESP) of the State of São Paulo.

Authors was considerably smaller than ours. Furthermore, if BAL is given to the animal before the injection of the larger molecular kinins, no change in the ratio can be observed, as shown in Fig. I.

The most striking differences appear when the effects upon vascular permeability (blue test) are compared with the effects upon smooth muscles, and especially the ileum of the guinea pig. This is clearly shown if the indexes of discrimination are calculated. Taking Bk as 1,0, the discrimination index "vascular permeability/ guinea pig ileum" rose to levels of 150 to 300, when the larger kinins Met-Lys-Bk and GAML-Bk, were assayed.

This possibility of discriminating sharply the actions of kinins upon the smooth muscle and on vascular permeability, has a strong bearing on the screaning of mediators for the acute inflammatory reaction or for other physiological effects of the kininogen-kinin system. If we consider a trivial case in which a large kinin such as Met-Lys-Bk or GAML-Bk would appear in biological materials and assayed upon the guinea pig ileum or the rat uterus, as it is usually done in a first screening of activity, and then tested on the skin of the rat, with the blue test, the ratio of 150 to 300 would probably disclassify the material as a kinin, and it would be labelled as a "permeability factor" though the active group in the molecule might still be bradykinin, as in the molecule of GAML-Bk and Met-Lys-Bk. Therefore we might suspect that some "permeability factors" derived from globulins, and that have been described in the literature as independent entities, might be just one of the larger kinins giving such high discrimination indexes when tested by the vascular permeability test against the guinea pig ileum and rat uterus. We know that many permeability factors have precursors localized in the pseudo-globulin fraction, such as those described by Miles and Willhelm (1960). The dependence of such factors from activation of proteases, is strongly suggested by the fact that their activation is prevented by SBI, as the kinin/s generated from the pseudo globulin fraction submitted to heating at pH=2.0 (Rocha e Silva and Holzsacher, 1959; Hamberg, 1962). If the material is dialyed for 3 days at pH=2.0, after precipitation with 50% ammonium sulfate, a material was obtained that was much more active on rat B.P. than Bk or even Met-Lys-Bk, and much less active upon the guinea pig ileum (Ryan and Rocha e Silva, 1971). Therefore we had to conclude that the enzyme that is activated by acid treatment can be differerent whether the material is heated or not heated at

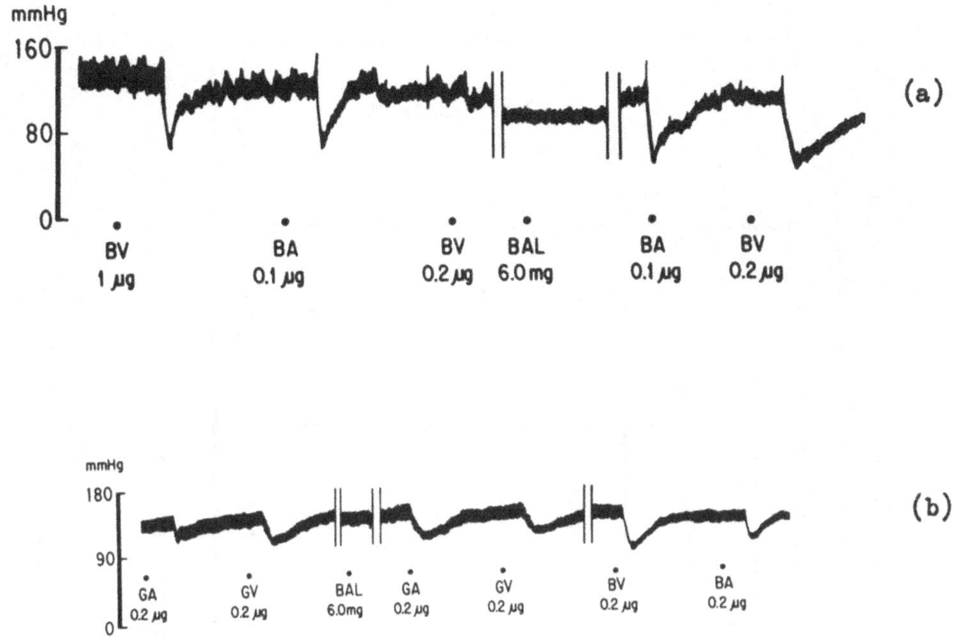

Fig. 1. (a) Rat systemic blood pressure. BV and BA indicate bradykinin by venous or arterial route respectively, before and after the injection of BAL (6.0 mg).
(b) Idem, idem, GA and GV indicate GAML-Bk by arterial or venous route respectively, before and after the injection of BAL (6 mg); BV and BA have the same connotation as in (a).

TABLE I

EQUIPOTENT CONCENTRATIONS (ng/ml or total) OF THE DIFFERENT POLYPEPTIDES IN PRODUCING EFFECTS ON VARIOUS BIOLOGICAL PREPARATIONS

Preparations \ Peptides	Bk	Lys-Bk	Lys-Lys-Bk	Met-Lys-Bk	GAML-Bk
Guinea Pig Ileum	6.25	18.75	25.00	66.0	103.00
Rat Duodenum	0.62	0.28	0.51	1.25	0.86
Rat Uterus	0.25	0.33	0.23	0.70	0.43
Artery B.P.(**)	100.00	112.00	22.8	126.00	100.00
Vein B.P.(**)	1.000.00	400.00	27.00	280.00	120.00
Vascular Permeability(*)	100.00	100.00	25.00	12.00	10.00

(*) The doses (ng) refer to the total amounts injected.

(**) The data referring to Lys-Bk, Met-Lys-Bk and GAML-Bk on BP were recalculated and found slightly different from those presented in Reis et al. (1971), Table I.

This compound was comparatively assayed upon several biological preparations with the 3 peptides described above, and the endecaptide Lys-Lys-Bk, already studied by Stewart et al. (1970). From these experiments it was concluded that all the peptides mentioned above showed a spectrum of activities characteristic of the group of kinins of the bradykinin family, producing fall in blood pressure, stimulation of the guinea pig ileum and rat uterus, relaxation of the rat's duodenum and increased vascular (capillary) permeability. For a characterization of the group see Rocha e Silva (1970), Erdös (1970) and Sicuteri et al. (1970).

We had shown previously (Ryan and Rocha e Silva, 1971) that according to the pretreatment to which acidified ox pseudoglobulin is submitted, either bradykinin or Met-Lys-Bk (see also Elliott and Lewis, 1965), and a larger peptide can be released. Therefore it became important to study the properties of kinins of larger molecular weight which may be released under natural conditions and displaying hypotensive and vascular permeability activities.

Experimental: The actions of Bk, Lys-Bk, Met-Lys-Bk, GAML-Bk and Lys-Lys-Bk were compared in different biological tests, i.e. on the: 1) guinea pig ileum, a) rat uterus, 3) rat duodenum, 4) rat blood pressure, and 5) vascular permeability in rats, and the "indexes of descrimination" (Gaddum, 1955) calculated. Table I, shows the equipotent concentrations (ng/ml or total) of the different polypeptides producing effects on various biological preparations.

Table II shows the ratios of potency of the different peptides upon different preparations, taking bradykinin (= 1,0) as a reference standard (indexes of discrimination).

As regards the data presented in Tables I and II, it is pertiment to stress a few points. It is known that when Bk is given intravenously its effects upon the blood pressure is 7 to 8 times smaller than if given through the artery, in such a way that the ratio artery/vein ≛ 7 or 8 for Bk. This was explained by Ryan et al. (1968) and Stewart et al. (1970) as indicating that Bk is rapidly destroyed in a single passage through the lung by the actions of kininases. If Met-Lys-Bk and Gly-Arg-Met-Lys-Bk (GAML-Bk) are injected intravenously and intraarterially, the ratio "artery/vein" approaches 1,0, indicating that these kinins are more resistant to destruction by kininases than Bk or Lys-Bk. As regards Lys-Lys-Bk the ratio Artery/vein was smaller than 1.00 (= 0.97) in agreement with Stewart et al. (1970), though the ratio found by the latter

TABLE II

DISCRIMINATION INDEXES OF THE DIFFERENT PEPTIDES IN RELATION TO PAIRS OF BIOLOGICAL STRUCTURE

Peptides	INDEXES OF DISCRIMINATION(*)				
	Uterus Rat / Ileum G.P.	Duodenum Rat / Uterus Rat	Duodenum Rat / Ileum G.P.	B.P. (vein) Rat / Ileum G.P.	Vasc. Permeability / Ileum G.P.
Bradykinin	1.00	1.00	1.00	1.00	1.00
Lys-Bk	2.00	2.50	5.20	6.50(**)	2.65
Lys-Lys-Bk	4.20	1.12	4.80	18.0	16.00
Met-Lys-Bk	4.00	1.36	5.40	41.00	111.00
GAML-Bk	9.10	1.30	11.80	142.00	180.00

(*) The indexes of discrimination were calculated by dividing the ratios of potency indicated in Table II, for each pair of biological structures.

(**) This figure has been recalculated and is different from that presented in Reis et al. (1971), Table III.

pH=2.0. The aplitting of different bonds in the bradykininogen sequence might of course explain the results of such experiments.

In conclusion, we have to stress the importance of such findings to explain the participation of bradykinin in physio-pathological phenomena, namely in inflammation and shock, in which the main symptom is increased vascular permeability. We have found that Bk may play a role in acute inflammatory reactions produced by heat or by the injection of sulfated polysaccharides, such as cellulose sulfate, carragena, dextrane, etc. (Rocha e Silva and Antonio, 1960; Rocha e Silva et al., 1968; Rocha e Silva, 1970; Garcia Leme et al., 1970). What has to be stressed is the possibility of participation of kinins with larger molecular weights than bradykinin itself, having a predominant effect on vascular permeability and the systemic blood pressure, being far more resistant than bradykinin to the destructive action of kininases.

Thus, a promising new field has been opened by such studies which might lead to interesting advances for the explanation of the important phenomenon of inflammation in its broader sense.

Acknowledgements: We are indebted to Dr. J.W. Ryan, of the Department of Medicine, University of Maimi, for providing us with the first sample of GAML-Bk, and to Schwartz BioRearch, Orangeburg, New York for subsequent supplies of the material; to Dr. J.M. Stewart of the Department of Biochemistry of the University of Denver Medical School for a supply of Lys-Lys-Bk, and to Dr. L.J. Greene, of the Department of Biology, Brookhaven National Laboratories, for a certified sample of GAML-Bk; to Mr. Hildeberto Caldo for technical assistance.

REFERENCES

ELLIOTT, D.P. and LEWIS, G.P. 1965 - Methionyl-lysyl-bradykinin, a new kinin from ox blood. Biochem. J., 95: 437.

ERDÖS, E.G., Edit. 1970 - Bradykinin, kallidin and Kallikrein. Hdb. exp. Pharmakol., vol. XXV.

GADDUM, H.H. 1955 - In: Polypeptides which stimulate plain muscles. J.H. Gaddum, Edit. E. & S. Livingstone Ltd., Edinburg and London, p. 130.

GARCIA LEME, J., HAMAMURA, L. and ROCHA E SILVA, M. 1970. Brit. J. Pharmacol., 40: 294.

HAMBERG, U. 1962 - Ann. Acad. Scientiarum Fennicae. II - Chimica 113: 1.

MILES, A.A. and WILHELM, D.L. 1960 - In: Polypeptides which affect smooth muscle and blood vessels. M. Schachter, Edit. Pergamon Press, Oxford, p. 309.

REIS, M.L., OKINO, L. and ROCHA E SILVA, M. 1971 - Biochem. Pharmacol., In Press.

ROCHA E SILVA, M. 1970 - Kinin Hormones with Special reference to Bradykinin and related Kinins. Ch. C. Thomas, Springfield, Ill.

ROCHA E SILVA, M. 1970 a - In: Bradykinin and Related Kinins. Sicuteri et al., Edit. Plenum Press, N. York, London, pg. 507.

ROCHA E SILVA, M. and ANTONIO, A. 1960 - Med. Exper. (Basel), 3: 371.

ROCHA E SILVA, M., GARCIA LEME, J. and SOUZA, J.M. 1968 - In: Internat. Symp. on Inflammation, Biochemistry and Drug Interaction (Como, Italy), Bertelli and Houck, Edits., Excerpta Med. Internat. Congress Series, 188: 170.

ROCHA E SILVA, M. and HOLZSACKER, E. 1959 - Arch. Intern. Pharmacodyn., 122: 168.

RYAN, J.W., ROBLERO, J. and STEWART, J.M. 1968 - Biochem. J., 110: 795.

RYAN, J.W. and ROCHA E SILVA, M. 1971 - Biochem. Pharmacol., 20: 459.

SCHRÖDER, E. and LÜBKE, K. 1966 - The Peptides, Vol. II, Academic Press, N. York and London.

SICUTERI, F., ROCHA E SILVA, M. and BACK, N., Edits. 1970 - Bradykinin and Related Kinins. Plenum Press, N. York, London.

STEWART, J.W. and WOOLLEY, D.W. 1967 - In: Internat. Symp. on Vaso-Active Polypeptides: Bradykinin and related Kinins. Rocha e Silva and Rothschild, Edit. Edart, S. Paulo.

STEWART, J.W., ROBLERO, J. and RYAN, J.W. 1970 - In: Bradykinin and Related Kinins. Sicuteri et al., Edit. Plenum Press, N. York, London, pg. 175.

STÜRMER, E. and BERDE, B. 1962 - J. Pharmacol. exp. Ther., 139: 38.

EFFECT OF KININS AND OF KININ-RELEASING ENZYMES ON RAT MAST CELLS

A.M. Rothschild, M.P.O Antonio, J.J. Dias, A. Castania and L.C. Neves
Department of Pharmacology
School of Medicine of Ribeirão Prêto, Brazil

When occurring at the mast cell, the major outcome of the antigen-antibody reaction is the release of histamine (review by Mota, 1966). The formation of bradykinin as a consequence of the antigen-antibody reaction *in vivo* or *in vitro* has been described (Beraldo, 1950; Rothschild, 1968). Trypsin has been shown to release kinin (review by Habermann, 1970); its histamine releasing activity has been reviewed by Rocha e Silva (1955). Chymotrypsin has been described as a kinin releasing agent (Rocha e Silva, 1970), which is also able to stimulate rat mast cell degranulation, an event associated with histamine release (Uvnäs and Antonsson, 1963; Saeki, 1964). A chymotrypsin-like enzyme is activated during antigen-antibody reactions at the mast cell; inhibiton of this enzyme leads to blockade of histamine release (Becker and Austen, 1966). These results seemed to warrant a search for a correlation between bradykinin formation and histamine release. This has been sought for in the present study, in which the effect of kinins, of trypsin and of chymotrypsin on histamine release from isolated rat peritoneal fluid mast cells has been examined.

METHODS AND MATERIALS

The mixed, loose cell population of the peritoneal cavity of the rat, containg 3-5% mast cells, was collected by flushing the abdominal cavity with 10 ml of Krebs-Ringer phosphate buffer. The cells were separated by centrifugation and resuspended in buffer. Incubations were performed in plastic test tubes at 37°C and gentle agitation, for 10 min. Inhibitors were added to the mixture prior to the histamine releasing agent. Final volumes were 1.0 ml. Following incubation, the cells were separated and washed once by centrifugation, suspended in 1 ml of 0.1N HCl, and boiled for 3 min. The solution was neutralized and its histamine content, corresponding to histamine not released by the experimental treatment, estimated by bioassay on the isolated, atropinized guinea pig ileum.

Mast cell degranulation was followed in pieces of rat mesentery incubated under the same conditions and prepared for microscopic examination by standard techniques (Rothschild, 1970). Esterolytic activities of trypsin and chymotrypsin were estimated by the spectrophotometric method of Hummel (1959), using p-tosyl arginine methyl ester (TAME) or benzoyl-L-tyrosine ethyl ester (BTEE) as substrates. The proteolytic activity of chymotrypsin was estimated by incubating 50 ug/ml of the enzyme for 10 min at 37°C, with 2 ml of a 0.25% solution of casein in Krebs-Ringer phosphate buffer, pH 7.4. After adding trichloroacetic acid (final concentration, 6%), the mixture was centrifuzed and the extent of proteolysis estimated from the increased content of 280 nm - absorbing material in the supernatant.

MATERIALS

Trypsin, alpha and beta-chymotrypsin, crystalline, Worthington.

DFP-treated trypsin, 2 x crystallized, Worthington.

Acetyl-trypsin, Mann Research Labs.

Casein, Hammersten, Nutritional Biochemicals.

Bradykinin triacetate, Sandoz.

8-Hydroxyquinoline, p-Toluene sulfonic acid, Benzoyl-L-tyrosine-ethyl-ester, Acetyl-L-tyrosine-ethyl-ester, p-Tosylarginine-methyl-ester, Sigma.

Compound 48/80, Burroughs-Wellcome.

RESULTS

Table I shows that trypsin as well as alpha or beta-chymotrypsin released histamine from isolated rat peritoneal fluid mast cells. p-Toluene sulfonyl-n-arginine-methyl ester (TAME), a specific substrate of trypsin failed to inhibit the action of this enzyme; in contrast, the effect of alpha-chymotrypsin could be inhibited by its specific substrate, N-acetyl-L-tyrosine ethyl ester (ATEE).

This suggested that differently from chymotrypsin, trypsin acts on mast cells in a non-enzymic fashion. This assumption was confirmed by results (Table II), showing that diisopropylfluorophosphate-treated trypsin (DFP-trypsin) had lost its esterolytic activity on TAME but retained its capacity to release histamine. N-acetyl trypsin on the other hand, failed to release histamine, but retained appreciable enzymatic activity on its synthetic substrate.

TABLE I

HISTAMINE RELEASE FROM RAT PERITONEAL FLUID MAST CELLS BY TRYPSIN, ALPHA-AND BETA-CHYMOTRYPSIN; INFLUENCE OF SPECIFIC SYNTHETIC SUBSTRATES ON THE ACTION OF TRYPSIN AND ALPHA-CHYMOTRYPSIN

Additions to medium	Percent histamine released
Trypsin, 200 ug/ml	36 ± 8 (4)
Trypsin + TAME, 2 mM	39 ± 4 (4)
Beta-Chymotrypsin, 50 ug/ml	59 ± 5 (3)
Alpha-Chymotrypsin, 50 ug/ml	59 ± 5 (3)
Alpha-Chymotrypsin + ATEE, 2 mM	38 ± 12 (3)
Alpha-Chymotrypsin + ATEE, 5 mM	0 (3)
Alpha-Chymotrypsin + ATEE, 10 mM	0 (3)

TAME, p-tosyl-arginine methyl ester, ATEE, N-acetyltyrosine ethyl ester.

Figures within parentheses refer to the number of experiments.

p-Toluene sulfonic acid (TSA) readily forms stable salts with amines and amine derivatives (Beilstein, 1928). Table III shows that concentrations of TSA able to block histamine release by chymotrypsin, failed to affect the proteolytic or esterolytic activities of the enzyme. This suggested that TSA inhibited histamine release by reacting with an intermediate, formed during chymotrypsin action on the peritoneal fluid cells. Preliminary results (Rothschild and Oliveira Antonio, 1970), had shown that kinins were able to release histamine and degranulate rat mast cells *in vitro* and that these effects were potentiated by 8-hydroxyquinoline, an inhibitor of kinin-destroying enzymes (Ferreira and Rocha e Silva, 1962). The results presented on Table IV, indicate that TSA was a powerful inhibitor of kinin-induced, 8-HQ-potentiated release of histamine. This inhibition seemed to be fairly specific, since the action of compound 48/80, a synthetic, potent histamine releasing agent was not affected by TSA.

TABLE II

EFFECT OF TRYPSIN, N-ACETYL-TRYPSIN AND DI-ISOPROPYLFLUOROPHOSPHATE (DFP)-TREATED TRYPSIN ON THE RELEASE OF HISTAMINE FROM RAT PERITONEAL FLUID MAST CELLS AND ON THE HYDROLYSIS OF p-TOSYLARGININE-METHYL ESTER (TAME).

Compound	Percent histamine released	Esterolytic activity (U/mg)
Trypsin, 200 ug/ml	23 ± 1 (6)	140
N-Acetyltrypsin, 200 ug/ml	0 (6)	80
Trypsin, 50 ug/ml	23 ± 6 (5)	140
DFP-Trypsin, 50 ug/ml	25 ± 7 (5)	0

TABLE III

EFFECT OF p-TOLUENE SULFONIC ACID (TSA) ON THE HISTAMINE RELEASING, CASEIN HYDROLYSING AND N-BENZOYL-TYROSINE ETHYL ESTER (BTEE) SPLITTING ACTIVITIES OF ALPHA-CHYMOTRYPSIN.

Additions to medium	Percent histamine released	Casein hydrolysed (OD of TCA filtrate)	Esterolytic Activity (U/mg)
Chymotrypsin, 50 ug/ml	59 ± 5	0.17	60
Chymotrypsin + TSA, 2 ug/ml	52 ± 6	0.17	-
Chymotrypsin + TSA, 5 ug/ml	38 ± 11	0.17	63
Chymotrypsin + TSA, 10 ug/ml	0	0.17	60

Each result is the average of three experiments.

TABLE IV

EFFECT OF p-TOLUENE SULFONIC ACID (TSA) ON THE RELEASE OF HISTAMINE FROM RAT PERITONEAL FLUID CELL SUSPENSIONS INDUCED BY BRADYKININ PLUS 8-HYDROXYQUINOLINE (8-HQ) OR BY COMPOUND 48/80.

Additions to medium	Percent histamine released
8-HQ, 400 ug/ml	0
Bradykinin, 0.5 ug/ml	8 ± 5 (5)
Bradykinin + 8-HQ	35 ± 8 (5)
Bradykinin + TSA, 0.5 ug/ml	0 (5)
Bradykinin + 8-HT + TSA	0 (5)
Compound 48/80, 0.1 ug/ml	37 ± 3 (3)
Compound 48/80 + TSA, 10 ug/ml	40 ± 0 (3)

DISCUSSION

Although the results presented in this paper indicate that trypsin can release a fair amount of histamine from rat mast cells, this activity did not seem to be due to proteolytic action. This is borne out by results showing that after blockade of its basic groups by N-acetylation, trypsin lost all its histamine releasing activity but less than 50% of its esterolytic activity on TAME; in contrast, DFP-treated trypsin had no enzymatic activity but retained full ability to release histamine. Trypsin is a basic protein, having an isoionic point of 10.1; the histamine-releasing activity of several non-enzymatic basic protein or peptides has been described (Rothschild, 1966). It is therefore not improbable that trypsin owes its histamine releasing capacity to the presence of basic groupings on its molecule, rather than to the participation of its proteolytically active center. This conclusion would not apply to alpha-chymotrypsin, whose histamine releasing activity could be inhibited by its specific substrate, acetyl-tyrosine ethyl ester and which furthermore, is only a slightly basic protein with an isoionic point of 8.1-8.6; beta-chymotrypsin, which was as active a histamine releaser as the alpha enzyme, is even a predominantly acidic protein having an isoionic point of 4.7. There is thus no reason to believe that the chymotrypsins would be acting by a non-enzymic mechanism.

It has been shown (Uvnäs and Antonsson, 1963; Saeki, 1965), and confirmed by us, that the action of chymotrypsin on mast cells can be blocked by metabolic inhibitors like cyanide, dinitrophenol or anoxia. Since these inhibitors do not affect the proteolytic activity of chymotrypsin (unpublished observation), it is improbable that the histamine releasing effect of the enzyme results from an unspecific proteolytic action leading to the rupture of the mast cell's membrane. Rather, chymotrypsin seems to stimulate the functional mast cell's histamine secretory response, which depends on cell metabolism (Rothschild, 1970), and is induced by basic compounds. It is conceivable that chymotrypsin releases histamine via a basic intermediate, probably a polypeptide, arising after the enzyme's interaction with the mast cell's membrane. The results presented in this paper suggest that bradykinin could be such an intermediate.

REFERENCES

BECKER, E.L. and AUSTEN, K.F. (1966) - Mechanisms of immunologic injury of rat peritoneal mast cells. J. Exp. Med., 124:

BERALDO, W.T. (1950) - Formation of bradykinin in anaphylactic and peptone shock. Am. J. Physiol., 163: 283.

BEILSTEIN (1928) - Handb. Org. Chem., 11: 97.

FERREIRA, S.H. and ROCHA E SILVA, M. (1962) - Potentiation of bradykinin by dimercaptopropanol (BAL). Bioch. Pharmacol., 11: 1123.

HABERMAN, E. (1970) - Kininogens. Handb. exp. Pharmacol., XXV:

HUMMEL, B.C.W. (1959) - A modified spectrophotometric determination of chymotrypsin, trypsin and thrombin. Can. J. Biochem. Physiol., 37: 1393.

MOTA, J. (1966) - Release of Histamine from Mast Cells. Handb. exp. Pharmacol., XVIII: 568.

ROCHA E SILVA, M. (1955) - Liberation of Histamine. In: Histamine; Its Role in Anaphylaxin and Allergy. Ch. Thomas, Springfield.

ROCHA E SILVA, M. (1970) - Kinin Hormones. Ch. Thomas, Springfield.

ROTHSCHILD, A.M. (1966) - Histamine release by basic compounds.

Hand. exp. Pharmac., 18: 386.

ROTHSCHILD, A.M. (1968) - Role of anaphylatoxin and of bradykinin in passive cutaneous anaphylaxis against heterologous precipitating antibody in the rat. Proc. 3rd. Int. Pharm. Congr., 11: 83.

ROTHSCHILD, A.M. (1970) - Mechanisms of histamine release by Compound 48/80. Br. J. Pharmacol., 38: 253.

ROTHSCHILD, A.M. and OLIVEIRA ANTONIO, M. (1970) - Unpublished observations.

SAEKI, K. (1964) - Effects of cp. 48/80, chymotrypsin and antiserum on isolated mast cells under aerobic and anaerobic conditions. Jap. J. Pharmacol., 14: 375.

UVNÄS, B. and ANTONSSON, J. (1963) - Triggering action of phosphatidase A and chymotrypsin on degranulation of rat mesentery mast cells. Bioch. Pharmacol., 12: 867.

This work was aided by Fundação de Amparo à Pesquisa (FAPESP). Foundation for the Support of Science of the State of S. Paulo, Brazil.

THE PROBLEM OF LATENCY AT BRADYKININ AND SOME ANALOGUES

B. Wiegershausen, I. Paegelow, H. Arold, S. Reißmann

Institute of Pharmacology and Toxicology, University of Rostock and Department of Biochemistry, Friedrich Schiller University Jena, German Democratic Republic

The polypeptide bradykinin possesses a strong characteristical slow smooth muscle stimulating activity. In the most cases its action is evident after a defined period of time - its latency. Because of their latency and the character of contraction bradykinin and angiotensin are signified as typically slow reacting substances in contrast to acetylcholine, histamine and serotonin, which are characterized as fast reacting agents. Eledoisin, a tachykinin, is standing between both groups (Wiegershausen and Deptalla 1969).

In this study we attempted to correlate the phenomenon latency of bradykinin acting directly on smooth muscle with one part of the sequence of events in its biological effect at the isolated guinea-pig ileum. In this respect following questions rise:

1. is the latency a part of the diffusion between medium and the receptor sites at the surfaces on smooth muscle cells?
2. do changes in the conformation of the bradykinin molecule and bonding forces at drug receptor interaction influence the latency?
3. is the latency dependent on the activator calcium which mediates drug induced contraction?

METHODS

All experiments were performed at the isolated guinea-pig ileum suspended in normal or modified Tyrode's solution at 37°C. In studies with pH changes we used tris-buffered Tyrode's solution. The pH was varied by 1 n NaOH or by HCl in the range from 5 to 10. In these experiments the solution was oxygenated by pure oxygen. The latency was measured to a constant point of the contraction (5 mm) because the base line was not always exact. At different pH values or different temperatures we compared quantitative identical effects (30 mm-contraction), that is equieffective doses obtained from dose-response curves by interpolation. The action of substances was studied at each pH for 30 minutes at the same ileum. Between each change the gut had a rest of 30 minutes. One experiment began with the lowest, the other with the highest pH value. The same arrangement was used at experiments with different temperatures.

RESULTS

At a contraction of 30 mm bradykinin has a latency of 12.5 to 14 seconds and the isolated ileum reaches its maximum after 32 to 37 seconds.

As known bradykinin is one of the strongest protonized peptides. Its terminal amino acid - arginin - has a pk value of 12.56. Therefore we supposed a connection between this property and its latency, because charged molecules are not able to pass lipid membranes without difficulties in physiological pH also on an isolated organ. But experiments with some bradykinin analogues do not support this conception. Bradykinin modified in position 1 or 9 (change in the intramolecular distances between the protonized guanido groups and the peptide chain) or in position 5 or 8 (change in the conformation of the side chain of the phenylalanin residues) were lower active than bradykinin (unpublished results). But their latency was 30% shorter than it of bradykinin, though all analogues have both guanido groups. When now the latency of an analogue without a strong positive charge is considered, for instance the latency of $Lys^{1,9}$-bradykinin[+], it should be expected a shorter latency. In spite this modification the latency of $Lys^{1,9}$-bradykinin is the same as at the other analogues. In connection with investigations about ionizable groups in the receptor for bradykinin by the method of Ariens (1968) it was generally noticed a dependence of the latency of bradykinin upon the pH. The dependence of bradykinin induced action on the pH has been the object also of a paper by Walaszek and Dyer (1966). He described that a decrease in pH increases the effectiveness of bradykinin and decreases the effectiveness of angiotensin and eledoisin. We could confirm these results. But now it was demonstrated also a change in the latency at different pH values. Bradykinin has its maximal activity at pH 6 and the longest latency at pH 6 and pH 5: 24 to

[+]We are indebted to Dr. Schröder, West-Berlin, for a sample of $Lys^{1,9}$-bradykinin.

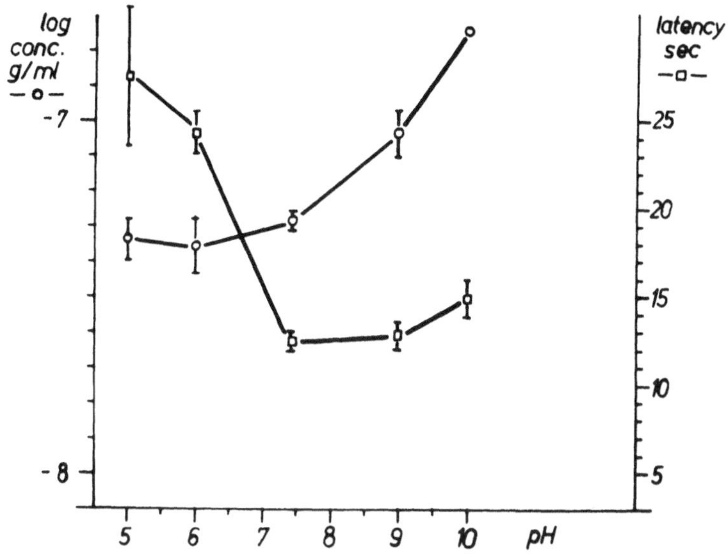

fig. 1 Relation between pH and the contraction induced by bradykinin for constant effects and its latency.

27 seconds in contrast to 12 to 14 seconds at pH 7.4. At a high pH the activity of bradykinin is very low, the latency rises a little (fig. 1).

The latency of the bradykinin analogues studied at the same conditions shows the same course of the curves: the longest latency at pH 5 and 6. By comparison studied angiotensin has its maximal activity at pH 9. Angiotensin is less active at pH 5 in contrast to bradykinin. However, the point of the maximal latency is seen at decreased pH, too (fig. 2). These results were tested as control with the tachykinin eledoisin as well as with the fast acting substances acetylcholine and histamine and also with barium chloride. In all cases the same shift of the curves was observed. The latency was increased at high hydrogen concentration in comparison with the physiological pH. Therefore we can conclude that the latency which is induced by an increase of hydrogen ions is only a non specific charge phenomenon on the membranes, a more difficult possibility for the agent to interact with receptor sites. All compared substances except barium chloride are already strongly protonized at physiological pH. The small differences in the protonization from pH 7.4 to pH 6 or 5 do not explain the steep decline of the slope of the curves from pH 5 or 6 to pH 7.4.

Moreover an attempt has been made to decide, whether the diffusion through the lipid membranes, through the serosa and mucosa of the gut to the receptor sites on the smooth muscle

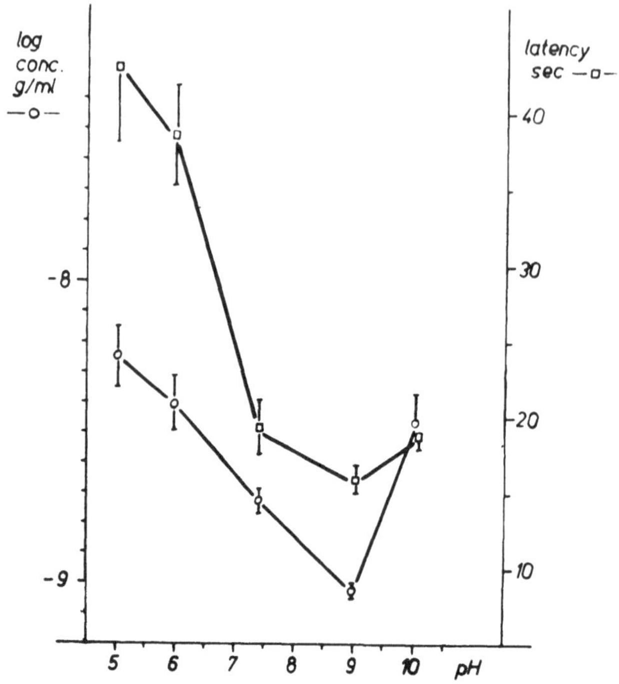

fig. 2 Relation between pH and the contraction induced by angiotensin for constant effects and its latency.

cells can be neglect as a cause for the latency of bradykinin. Therefore the latency of bradykinin was registered in dependence upon the temperature. It could be demonstrated, that the latency is influenced by the temperature being 27° to 42°C in Tyrode's solution. The quotient of the temperature ($Q_{10} = \frac{v_t}{v_t - 10}$) is 1.62 to 1.86 (fig. 3). From this value can be concluded that the latency is partly due to a biochemical process, because the binding of peptides by macromolecules are possibly in this order. A pure diffusion has a lower quotient and a pure chemical reaction is more influenced by variations of the temperature.

Another question is, whether calcium playing an important role at the effect coupling influences the latency of bradykinin induced contraction. Fig. 4 demonstrated the latency measured in calcium-free medium after bradykinin and varied intervals of calcium administration. The latency is 18.5 seconds at simultaneous administration of bradykinin and calcium ions. It is not influenced by the moment of the calcium administration from the begin to the 15. second. But it was not possible to abolish the latency by calcium giving 20 to 30 seconds after bradykinin. About 6 seconds were remaining. But when we now compare the latency of bradykinin in normal Tyrode's solution

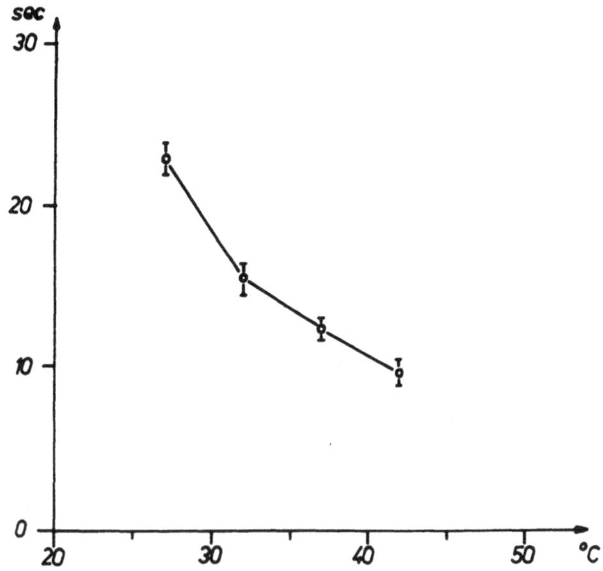

fig. 3 Latency of bradykinin at the isolated guinea-pig ileum in dependence upon the temperature.

(12.4 to 14 sec) with the latency in calcium-free solution at the moment when calcium and bradykinin are given together (18.5 sec), we can find also a difference of 6 seconds. Therefore we suppose that the remaining 40% of the latency, which could not be abolished are caused by the distribution and transference of the calcium in the bath (broken line), that means that the calcium ions do not influence the latency period of the bradykinin induced contraction.

Summarizing it may be concluded from the present results:

1. The well known defined latency of bradykinin is not caused by its strong protonization. Analogues with the same or diminished basicity have both a shorter latency than bradykinin.
2. The pH-induced latency at several substances is not specifical. It can be induced also at nitrogen-free agents, for instance at barium chloride. Therefore we suppose a change of the charge on membrane sites as a cause for the pH-induced latency.
3. By our experiments it can not be excluded that a changed hydrogen concentration can delay the time course of the drug-receptor-interaction by changing of bonding forces or of its conformation.

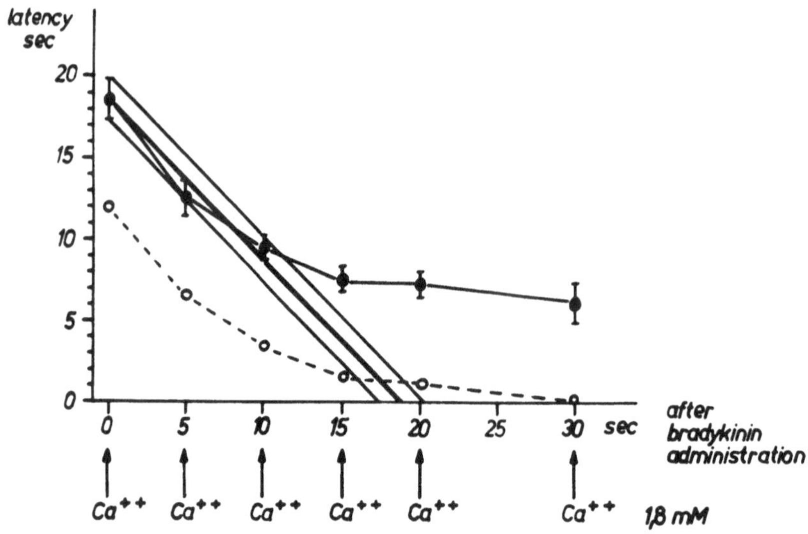

fig. 4 Dependence of latency of bradykinin induced contraction upon calcium administration at the guinea-pig ileum in Ca-free Tyrode's solution (broken line - theoretical curve).

4. The temperature quotient - the Q_{10} - (1.62 to 1.86) refers possibly to a biochemical process as cause of the bradykinin induced latency, that means an interaction between the pharmacon and the receptor or a change in the conformation of it.
5. The calcium ions are without influence on the latency of bradykinin.

So we suppose that the bradykinin induced latency is an event which does not only depend upon the absorption of bradykinin to the receptor sites and does not depend on the stimulus-effect coupling.

REFERENCES

Ariens, E.J.: in "Molecular Pharmacology" vol. I, 1964

Walaszek, E.J., D.C. Dyer: in "Hypotensive Peptides", 1966, p. 329

Wiegershausen, B., H. Deptalla: Arch. int. Pharmacodyn. 182, 1 (1969)

SUPPLEMENTAL INVESTIGATIONS TO THE KININ POTENTIATING ACTION OF SOME LOCAL ANESTHETICS

B. Wiegershausen, G. Rostock and I. Paegelow

Institute of Pharmacology and Toxicology, University of Rostock, German Democratic Republic

Previously we have reported (Wiegershausen and Deptalla 1969a and b; Wiegershausen 1970) that some local anesthetics increase the action of bradykinin both on the isolated hindlimb of the cat and on the isolated guinea-pig ileum.

Procaine potentiates the vasoconstriction on the isolated hindlimb and the contraction on the guinea-pig ileum. Tetracaine only increases the response of bradykinin after its removal, that is sensitization. Cocaine, however, induces different effects as potentiation and as sensitization on both models (Wiegershausen 1970). These results received under our experimental conditions suggest a similar structure of bradykinin-receptors in the visceral and vascular smooth muscles and are contrary to other authors (Edery and Grunfeld 1969), who discussed different structures of bradykinin-receptors.

It is known (Edery 1964, 1965; Edery and Grunfeld 1969) that the proteolytic enzyme chymotrypsin also sensitizates the guinea-pig ileum to bradykinin. Supplemental investigations should show whether differences exist between the phenomena potentiation and sensitization by local anesthetics and chymotrypsin. From this reason we combined the sensitization induced by chymotrypsin with the potentiating action of the local anesthetics.

The figure 1 demonstrates the different influence of procaine on the bradykinin induced contraction. 10^{-4} g/ml procaine which does not cause a measurable contraction reduces the action of bradykinin on the normal ileum. But 4.10^{-4} g/ml procaine potentiates the contraction induced by bradykinin. After pretreatment with 10^{-4} g/ml chymotrypsin procaine potentiates

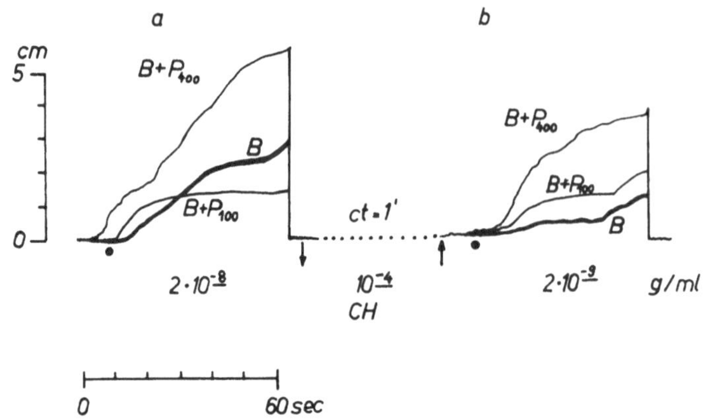

fig. 1 The potentiation of bradykinin-contraction by procaine before (a) and after (b) sensitization with chymotrypsin. B = bradykinin; P_{100} = procaine 10^{-4} g/ml; P_{400} = procaine 4.10^{-4} g/ml; CH = chymotrypsin; ct = time of contact.

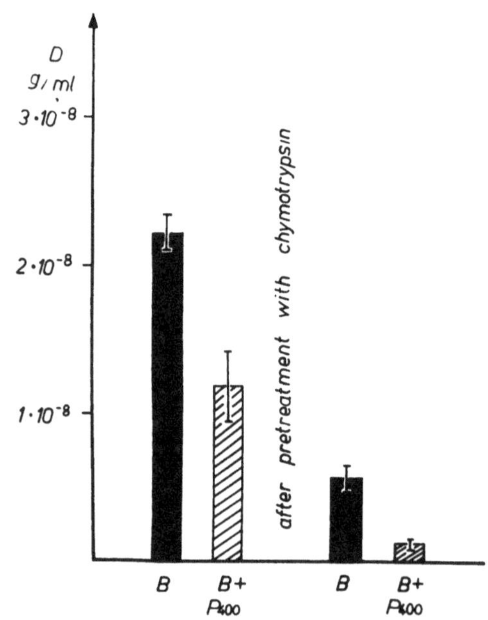

fig. 2 The influence of procaine and chymotrypsin at equiactive concentrations of bradykinin.
B = bradykinin; P_{400} = procaine 4.10^{-4} g/ml.

in both concentrations. However the potentiating action of 10^{-4} g/ml procaine is lower than the action of 4.10^{-4} g/ml procaine.

Our results demonstrate that the procaine induced potentiation is dependent upon the concentration after sensitization with chymotrypsin.

In figure 2 we have compared these results with equiactive concentrations of bradykinin taking from dosis-response-curves by interpolation at the height of 30 mm-contraction. The statistical calculations show that the action of bradykinin significantly increases 1. by procaine (4.10^{-4} g/ml), 2. by chymotrypsin and 3. mostly pronounced by procaine (4.10^{-4}g/ml) after sensitization with chymotrypsin.

By combination of the potentiating action of procaine with the sensitization by chymotrypsin the sensibility of the guinea-pig ileum is increased (fig. 3). The sensibility at bradykinin in comparison with the normal ileum is about twenty-fold higher that it will be possible to determine concentrations of bradykinin lower than 5.10^{-10} g/ml.

Tetracaine and cocaine sensitizate the isolated guinea-pig ileum at bradykinin. Pretreatment with tetracaine (fig. 4) and cocaine (fig. 5) produces an additional increase of the potentiation by procaine in comparison to the control.

It is known by Feinstein and coworkers (Feinstein 1966; Feinstein et al. 1968; 1970) that the local anesthetics release calcium ions from intracellular stores and inhibit the uptake

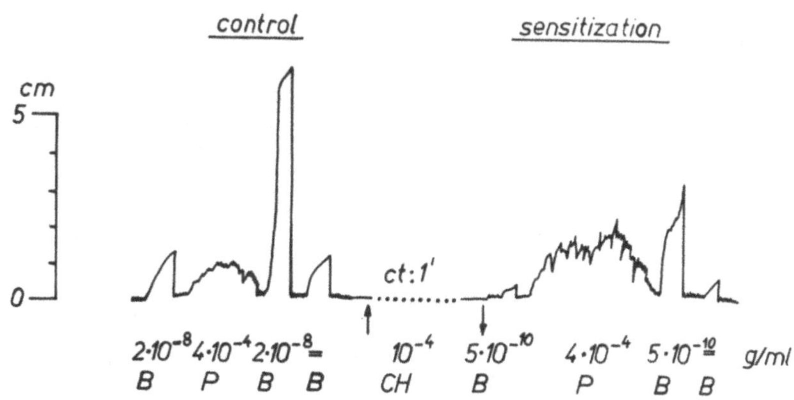

fig. 3 The potentiation of bradykinin-contraction by procaine before and after sensitization with chymotrypsin. B = bradykinin; P = procaine ; CH = chymotrypsin; ct = time of contact.

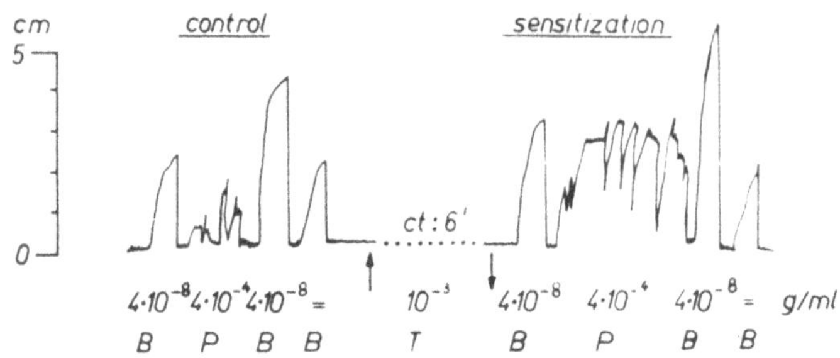

fig. 4 The potentiation of bradykinin-contraction by procaine before and after sensitization with tetracaine.
B = bradykinin; P = procaine; T = tetracaine; ct = time of contact.

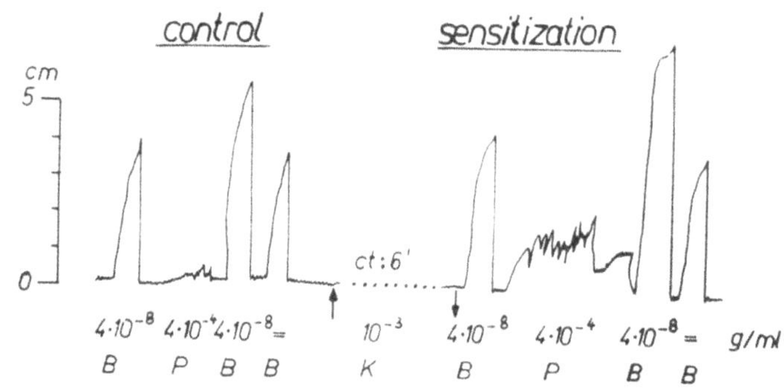

fig. 5 The potentiation of bradykinin-contraction by procaine before and after sensitization with cocaine.
B = bradykinin; P = procaine; K = cocaine; ct = time of contact.

of calcium ions through the cell membrane. These results explain only the localanesthetic action but not the potentiating action. Since Bianchi and Bolton (1967) could show the contribution of calcium ions on the caffeine induced contraction by experiments on skeletal muscle, we investigated the influence of the ionic concentrations on the potentiation and sensitization.

The calcium concentration in the Tyrode solution was reduced to 0,36 mM; 0,18 mM and 0,09 mM calcium chloride. Calcium chloride diminishes significantly only in concentrations lower than 0,18 mM the procaine induced potentiation (level of error $\alpha < 0{,}01$). On the contrary the sensitization by tetracaine, cocaine and chymotrypsin does not vary in dependence upon the calcium concentration.

In Tyrode's solution containing 10 mM potassium chloride we have also observed the potentiating action of procaine, but the effect is significantly lower than in normal solution (level of error $\alpha < 0{,}05$). Therefore the potentiation is influenced by partial depolarization, but the sensitization is not influenced.

We know from our experiments that the contracting action of procaine is diminished by decrease of the bath temperature. Therefore we investigated potentiation and sensitization at 25° C. At this temperature both phenomena are not influenced.

The results are summarized in table 1. We can demonstrate that the phenomenon potentiation under our conditions depends

modifications from	potentiation	sensitization
$[Ca^{++}]\downarrow$	+	—
$[K^{+}]\uparrow$	+	—
temperature ↓	—	—

tab. 1 The influence of different ionic concentrations and temperature on the bradykinin-contraction.
+ = depend upon; - = independ upon.

upon calcium and potassium ions but it does not depend upon a decrease of temperature to 25° C. The sensitization is not influenced by these changes.

At this moment we can not give an answer to the question whether sensitization and potentiation are different phenomena of the same or different mechanisms. In previous studies we have tried to exclude an inhibition of kininase activity as possible course of the potentiation (unpublished). We could not detect any inhibition of human plasma kininase by local anesthetics. The kininase of guinea-pig ileum homogenates was not investigated. It is known that chymotrypsin inhibits the kininase both of the guinea-pig ileum homogenates and the kininase of the intact guinea-pig ileum. But this mechanism can not be responsible for the sensitization of the guinea-pig ileum to bradykinin, because by Edery and Grunfeld (1969) gluthation potentiates the bradykinin induced contraction after pretreatment with chymotrypsin, too.

From our experiments we must suppose that the potentiating and stabilizing action of local anesthetics depend upon different calcium-stores in the smooth muscle cells, as we know from several authors that different calcium-stores play an important role in sites of action of some agonists (Altura, Altura 1970). Electrophysiological investigations, relation of the sodium/potassium exchange and the role of the calcium ions on the pharmacomechanical coupling are capable to give exacter conceptions.

REFERENCES

Altura, B.M., B.T. Altura: Proc. Soc. exp. Biol. Med. 135, 739 (1970)

Bianchi, C.P., T.C. Bolton: J. Pharmacol. exp. Ther. 157, 388 (1967)

Edery, H.: Brit. J. Pharmacol. 22, 371 (1964)

Edery, H.: Brit. J. Pharmacol. 24, 485 (1965)

Edery, H., Y. Grunfeld: Brit. J. Pharmacol. 35, 51 (1969)

Feinstein, M.B.: J. Pharmacol. exp. Ther. 152, 516 (1966)

Feinstein, M.B., M. Paimre, M. Lee: Trans. N. Y. Acad. Sci. 30, 1073 (1968)

Feinstein, M.B., M. Paimre: Fed. Proc. 28, 1643 (1970)

Wiegershausen, B., H. Deptalla: Arch. int. Pharmacodyn. 177, 278 (1969)

Wiegershausen, B., H. Deptalla: Arch. int. Pharmacodyn. 182, 1 (1969)

Wiegershausen, B.: in: Bradykinin and Related Kinins, Plenum Press 437 (1970)

COMPARATIVE ACTION OF PEPTIDES ON THE GALL BLADDER AND THE SPHINCTER OF ODDI

F. Lembeck and H. Juan

Institute of Pharmacology

University of Graz, Austria

INTRODUCTION

Several peptides contract the gall bladder in the guinea pig, cat, dog, chicken and in man: Cholecystokinin-Pankreozymin (CCK-PZ) (Ivy and Oldberg, 1928) and its C-terminal octapeptide (CTO) (Mutt and Jorpes,1968; Rubin et al., 1969; Ondetti et al., 1970; Hedner,1970), caerulein (Bertaccini et al., 1968, 1969, 1970; Agosti et al., 1971), substance P (SP) and a physalaemin-like octapeptide (P 30) (Starke et al., 1968). A concommitant relaxation of the sphincter of Oddi by CCK-PZ and CTO was reported in the cat (Hedner, 1970), and by caerulein in the guinea pig (Bertaccini et al., 1968), dog and in man (Lin and Spray, 1969), but a spasmogenic effect on the isolated ductus choledochus of the calf (Agosti et al., 1971).

The purpose of the present study was to compare the actions of peptides on the biliary system with their effects on salivation and blood pressure in several species. Since many of the latter effects have already been investigated they could mainly be compiled from the literature; only supplementary experiments were performed.

METHODS

1. <u>Animals and Anaesthesia.</u> a) Guinea pigs, 350 - 450 g. Urethane, 1.2 g/kg i.p. b) rabbits, 2 - 3 kg.

Urethane 1.25 g/kg i.p. c) cats, 2.5 - 3.5 kg. Urethane 0.5 g/kg and chloralose 40 mg/kg i.p. d) rats (Wistar, own breed), 300 - 400 g. Urethane, 1.3 g/kg i.p. e) mongrel dogs, 16 - 24 kg. Induction with cyclohexylamine (Ketalar[R]), 20 mg/kg i.m. and continued with pentobarbital i.v. f) chickens, 1 - 1.3 kg. Phenobarbital, 180 mg/kg i.m.

2. Contraction of the Gall Bladder was measured in the guinea pig, cat, rabbit, dog and chicken. The gall bladder was separated from the surrounding tissue. Damage of the small blood vessels which supply the gall bladder was carefully avoided. The ductus choledochus and the ductus hepaticus were ligated. A cannula was tied into the fundus of the gall bladder and connected to a vertical glass tube filled with saline (20 cm high, 1.2 mm ø, 0.2 ml volume), into which the fluid could move during the contraction of the gall bladder. Via a T-piece, a Statham element was connected to the system. The whole system was set under an initial pressure of 10 - 17 cm H_2O by connection to a Mariotte's bottle.

3. Perfusion of the Sphincter of Oddi. Distal to the ductus hepaticus a cannula was inserted into the ductus choledochus and connected to a Mariotte's bottle filled with saline. The pressure was adjusted to yield a flow of 0.01 - 0.2 ml/min (see results). At a distance of a few cm from the tip of the cannula a Statham element and a vertical tube (20 cm high, 1.2 mm ø, 0.2 ml volume) were connected to the cannula. According to Bernoulli's law, the level of saline in the vertical tube was somewhat lower than in the Mariotte's bottle when the flow in the cannula was fast, but reached the level in the Mariotte's bottle when it stopped. The pressure in the inflow system reflected the spontaneous as well as the evoced movements of the ductus choledochus or the sphincter of Oddi and also revealed immediately obstructions in the outflow system as caused occasionally by a plug of mucus. The outflow from the duodenum was lead through a cannula to a cross-connection about 30 cm below the table. Its branches were connected to a Statham element, to a vertical tube (30 cm high, 1.2 mm ø, 0.3 ml volume) and to an electromagnetic outflow valve. Fluid coming from the duodenum rose in the vertical tube and this rise was recorded by the Statham element. At fixed time intervals the valve was opened and released the fluid from the tube. The recorded level in the vertical tube could therefore serve as a measure of the outflow/time. By this method exact estimations of outflow volumes as small as 2 - 5 ul/time unit could be

recorded.

4. Salivation. Only the influence of SP and caerulein on the salivation of guinea pigs was estimated by adoption of the method already used in rats by Lembeck und Starke (1968 a).

5. Blood Pressure. In guinea pigs and rats the blood pressure was recorded in the carotid artery, in cats, rabbits, dogs and chickens the femoral artery was connected to a Condon mercury manometer.

6. Substances. Caerulein (F.I. 6934) and physalaemin (purity: 80 %): Dr. A. Anastasi, Farmitalia, Milano. CTO (Asp-Tyr(SO_3H)-Met-Gly-Trp-Met-Asp-Phe-NH_2): Dr. M. Ondetti, Squibb Research Institute, New Brunswick, N.J. We thank Dr. Anastasi and Dr. Ondetti for the kind supply of these substances. Bradykinin (Sandoz, Basel); substance P, a standard preparation from bovine gut by Fa. Sanabo, Vienna, 13.25 units/mg; for a limited number of experiments a substance P preparation from brain ("Fa") according to the method of Lembeck und Starke (1968) was used. To avoid glass adsorption and to protect from bacterial contamination the peptides were dissolved and diluted in 0.9 % saline solution containing 200 mg/l p-toluene sulfonic acid and 25 mg/l Cialit[R] (Ethyl-mercuriathiobenzazol-5-carbonic sodium: Lembeck et al., 1968 a).

RESULTS

1. Gall Bladder Contraction

In the guinea pig, cat, rabbit, dog and chicken, SP, caerulein and CTO were equally effective, whereas physalaemin was much less active. A few further studies were performed in the guinea pig only: the action of SP was unchanged after atropine (0.1 - 1.0 mg/kg). The action of bradykinin and histamine was weaker than that of the forementioned peptides. The actions of acetylcholine and serotonin were too inconsistant to establish a dose-response-curve. In three experiments "Fa" was compared with the usual SP preparation; its activity was found to be one third to one half of SP on the gall bladder of the guinea pig and the dog when amounts equieffective on the guinea pig ileum were injected. The order of sensitivity to the first three peptides in different species was: dog $>$ rabbit $>$ chicken $>$ guinea pig

$>$ cat (see tab. 1).

2. Perfusion of the Sphincter of Oddi

SP, caerulein and CTO were equally effective in the guinea pig, cat, rabbit, dog and chicken with respect to the increase in flow through the sphincter of Oddi, whereas physalaemin was much less active in each of the five species. The order of sensitivity was: dog $>$ rabbit $>$ chicken $>$ guinea pig $>$ cat. In the cat, but not in the rat a pronounced spontaneous motility was always present. The increase in flow was accompanied by an increased motility in the ductus choledochus or the sphincter of Oddi. Occasionally, the increase in flow was preceded by a short initial reduction of flow in the cat and the chicken when high doses were injected.

In the rat the results were opposite. All the peptides elicitated a decrease in flow in the following order: SP $>$ caerulein $>$ CTO $\gg$ physalaemin. In some experiments on the guinea pig a catheter was inserted from the duodenum through the sphincter of Oddi into the ductus choledochus to find out if the relaxation was restricted only to the sphincter. In these experiments SP, CTO, caerulein (doses of 20 - 160 units/kg or ng/kg) and physalaemin (doses of 80 - 640 ng/kg) decreased flow thus indicating that only the sphincter region relaxes, whereas the duct contracts (tab. 1).

3. Salivatory Response

This investigation revealed a very high sensitivity to SP in the guinea pig (tab. 1). As compiled from the literature the rank order of sensitivity on the salivatory response to SP was: guinea pig $>$ chicken $>$ dog $>$ rat; no effect was observed in the rabbit and cat (Lembeck und Starke, 1968 a). Caerulein did not stimulate the salivary glands of the guinea pig in doses up to 3000 ng/kg.

4. Blood Pressure

The threshold dose necessary to lower the blood pressure in the guinea pig was comparable to that in the rat. The sensitivity in various species decreased in the order: rabbit, dog $>$ rat, guinea pig $>$ cat (tab. 1). An opposite reaction to SP and the other tachykinins has been found in the chicken which responds by a rise in blood pressure (Lembeck et al., 1968 a).

Table 1: THRESHOLD DOSES OF SP (units/kg)

	contract. g.bladder	increase of flow through sph.Oddi	salivation	decrease of blood pressure
GUIN.PIG	1	10	10-30	1-3
CAT submand.	3	20	- a)	10-200
RABBIT	.1-.3	1	- a)	.1-.3
DOG parotis	.1	.3-1	30-100 b)	.1-.3
RAT	0	5-10 x)	300-600 a)	1
CHICKEN	.3	1	50 c)	1-3 xx)

a) Lembeck und Starke, 1968 a   x) decrease !
b) Lembeck und Hettich, 1969   xx) increase !
c) Lembeck et al., 1968 a   - no effect

Table 2: APPROXIMATE EQUIVALENTS TO 100 UNITS SP

100 units SP correspond to the following ng physalaemin,caerulein,CTO and mg "Fa"

	contract. of g.pig ileum	contraction of gall bladder in				
		G.PIG	CAT	RABBIT	DOG	CHICK.
physalaemin	132(123-141)	2300	2800	550	800	650
caerulein	510(465-555)	70	65	80	85	75
CTO	480(440-520)	110	95	95	115	90
"Fa"	2.8	8.4			5.5	

5. Some Additional Findings

SP and physalaemin increased the rate of respiration in guinea pigs and rabbits as also was shown earlier in the dog (Lembeck et al., 1968) whereas caerulein and CTO did not have this effect. Since the respiratory movements impared the registration of the movements of the gall bladder, the animals were usually kept under artificial respiration. No tachyphylaxis could be observed during the course of the experiments. Moreover, a slight increase in sensitivity seemed to be prominent. Sometimes very large doses of peptides resulted in somewhat smaller effects than preceding lower doses; it might be that the contraction of the duct surmounts and obscures the action on the sphincter.

DISCUSSION

Some earlier investigations failed to demonstrate an action of SP on the gall bladder in the guinea pig (Hultman, 1955) and in the dog (Pernow, 1953); we found that the demonstration of such an effect depended on careful avoidance of bleeding from the supplying vessels. A gall bladder contraction by SP was found in the dog by Starke et al., 1968.

This investigation was an attempt to correlate the effect of SP on the gall bladder to its action on blood pressure and salivation in several species. It also includes a comparison with the potency of other substances acting on the gall bladder like caerulein and the C-terminal octapeptide of CCK-PZ (CTO) (tab. 2).

SP was highly specific with respect to its effect on the gall bladder of the cat. The threshold dose which affected blood pressure was at least 100 times higher than the threshold dose for the gall bladder, and salivation was not influenced by 10.000 times higher doses.

Compared with the salivatory action, the gall bladder contraction (in the rat: the contraction of the ductus choledochus) was already obtained with much smaller doses. The smallest difference between the action on the biliary system and on salivation was found in the guinea pig, a much larger one in the dog - and the salivary glands of rabbits and cats did not respond at all to SP.

In the cat and possibly in the guinea pig the action of SP on the biliary system is the most prominent in vivo effect. In the dog the rat and the rabbit it is the fall of blood pressure. Such quantitative differences are quite important in the assessment of a physiological role to a certain effect.

Quantitative comparison among peptides showed that 1 unit SP possesses the activity of 1 ng of caerulein or CTO on the gall bladder contraction and sphincter Oddi relaxation in guinea pigs, rabbits, cats, dogs and chickens (tab. 2). These three peptides were found to be much more active than physalaemin in all species. The effect of bradykinin, histamine, acetylcholine and serotonin - although only investigated in the guinea pig - were smaller and rather inconsistent in the case of the latter two substances. One unit SP corresponds to about 5 ng caerulein or CTO on the guinea pig ileum, but 1 ng caerulein or CTO on the gall bladder. This shows again the very selective action of caerulein or CTO on the gall bladder.

On the ileum 1.3 ng physalaemin were as active as 1 unit SP. But on the gall bladder 5.5, 23, 8.0, 6.6 resp. 28 ng (rabbit, guinea pig, dog, chicken, resp. cat) physalaemin had the same activity as 1 unit SP. SP, therefore, has a much more pronounced action on the gall bladder than physalaemin - an unusual finding, because almost all other quantitative comparisons between these two peptides revealed identical potencies (Bertaccini et al., 1965).

The value of the SP preparation used in the present experiments is limited by the fact that it had been prepared from bovine gut and was not highly purified. Therefore a contamination of this preparation with CCK-PZ had to be taken into account. Due to the limited availability, only a few experiments could be carried out with "Fa", a highly purified SP preparation from brain (Lembeck und Starke, 1968). This "Fa" was about one third as active on the gall bladder of the guinea pig and about one half as active in the dog (based on its potency on the ileum). This means that the usual SP preparation must indeed have been contaminated with other gall bladder contracting agents, but also that SP itself has a higher gall bladder contracting action than physalaemin in the tested animals.

It has been discussed repeatedly whether the sphincter of Oddi relaxes actively, or whether it follows passively any contraction of the gall bladder or the ductus choledochus. Our perfusion experiments clearly indicate a direct relaxation in all cases. The activity of substances on the sphincter of Oddi was weaker than their action on the gall bladder. Only Lin and Spray (1969) found in the dog that the sphincter relaxant dose of caerulein was lower than its gall bladder contracting dose. The relaxation was solely within the sphincter region, since measurements with a cannula inserted from the duodenum into the ductus choledochus showed that the duct contracts upon higher doses of the peptides. Spontaneous motility of the duct as judged from changes in the inflow pressure was found to be low in the rat and high in the cat where the movements did not correlate with those of the duodenum.

In contrast to the other species, in the rat, which does not have a gall bladder, a reduction of flow was seen, even with small doses. This finding indicates that in the rat the function of the missing gall bladder is to some extent performed by the ductus choledochus, and further that there is no sphincter mechanism in this species. The isolated terminal bile duct of the calf was also contracted by caerulein (Agosti et al., 1971).

The structure of substance P which was recently published by Chang et al. (1971) reveals a certain similarity in sequence between the carboxyl terminal sequence of SP (1) and physalaemin (2), caerulein (3) and CTO (4):

1) Arg-Pro-Lys-Pro-Glu-Glu-Phe-Phe-Gly-Leu-Met-NH_2

2) Pyr-Ala-Asp-Pro-Asn-Lys-Phe-Tyr-Gly-Leu-Met-NH_2

3) $^+(SO_3H)$ Pyr-Glu-Asp-Tyr$^+$-Thr-Gly-Trp-Met-Asp-Phe-NH_2

4) $^+(SO_3H)$ -Asp-Tyr$^+$-Met-Gly-Trp-Met-Asp-Phe-NH_2

SUMMARY

The comparative effectiveness of substance P, physalaemin, caerulein and C-terminal octapeptide of CCK-PZ on gall bladder and sphincter of Oddi was investigated in guinea pigs, cats, rabbits, rats, dogs and chickens. In all species, except the rat, SP showed a gall bladder contraction in the range of 0.1 - 3.0 units/kg i.v. and an increase (rat: decrease) of flow through the sphincter of Oddi. The effect was produced by doses lower than those stimulating the salivation and equal or much lower (cat) than the threshold doses which affected the blood pressure. The activity of SP on the gall bladder was higher than that of physalaemin and in the range of caerulein or the C-terminal octapeptide of CCK-PZ.

REFERENCES

Agosti, A., P. Mantovani and L. Mori: Action of caerulein and related substances on the sphincter of Oddi. Naunyn-Schmiedebergs Arch.Pharmak. 268, 114 (1971).

Bertaccini, G., G. Ballarini, A. Agosti and G. Zannetti: Action of caerulein on the biliary system of the dog. Arch.int.Pharmacodyn. 183, 261 (1970).

- T. Braibanti and F. Uva: Cholecystokinetic activity of the new peptide caerulein in man. Gastroenterology 56, 862 (1969).
- J.M. Cei and V. Erspamer: Occurence of physalaemin in extracts of the skin of Physalaemus fuscumaculatus and its pharmacological actions on extravascular smooth muscle. Brit.J.Pharmacol. 25, 363 (1965).
- G. de Caro, R. Endean, V. Erspamer and M. Impicciatore: The action of caerulein on the smooth muscle of the gastrointestinal tract and the gall bladder. Brit.J. Pharmacol. 34, 291 (1968).

Chang, M.M., S.E. Leeman and H.D. Niall: Amino-acid sequence of substance P. Nature New Biology 232, 86 (1971).

Hedner, P.: Effect of the C-terminal octapeptide of cholecystokinin on guinea pig ileum and gall bladder in vitro. Acta physiol.scand. 78, 232 (1970).

Hultmann, E.H.: The relation between cholecystokinin and substance P. Acta chem.scand. 9, 1042 (1955).

Ivy, A.C. and E. Oldberg: A hormone mechanism for gall bladder contraction and evacuation. Amer.J.Physiol. 86, 599 (1928).

Lembeck, F., F. Geipert und K. Starke: Sialogene Wirkung von Peptiden bei Hühnern. Naunyn-Schmiedebergs Arch. Pharmak.exp.Path. 261, 422 (1968 a).

- and R. Hettich: Comparative study of the effects of substance P on blood pressure, salivatory functions and intestinal motility. Naunyn-Schmiedebergs Arch. Pharmak. 265, 216 (1969).
- und K. Starke: Behandlung von Substanz-P-Präparaten mit Aluminiumoxid. Naunyn-Schmiedebergs Arch.Pharmak. exp.Path. 259, 307 (1968).
- -: Substanz P und Speichelsekretion. Naunyn-Schmiedebergs Arch.Pharmak.exp.Path. 259, 375 (1968 a).
- - und U. Weiss: Sialogene Wirkung von Substanz P und einem physaläminartigen Octapeptid bei Infusion in verschiedene Blutgefäße. Naunyn-Schmiedebergs Arch. Pharmak.exp.Path. 261, 329 (1968).

Lin, T.M. and G.F. Spray: Effect of pentagastrin, cholecystokinin, caerulein and glucagon on the choledochal resistance and bile flow of conscious dog. Gastroenterology 56, 117 (1969).

Mutt, V. and J.E. Jorpes: Chemistry and physiology of cholecystokinin-pancreozymin. Proc.Int.Union Physiol. Sci. VI. 193 (1968).

Ondetti, M.A., B. Rubin, S.L. Engel, J. Pluščec and J.T. Sheehan: Cholecystokinin-Pancreozymin. Amer.J.Digest. Diseases 15, 149 (1970).

Pernow, B.: Studies on substance P. Acta physiol.scand. 29, suppl. 105 (1953).

Rubin, B., S.L. Engel, E.D. Grigas, E. Yiacas and B.I. Gold: Gallbladder contracting activities of octapeptide analogs of the C-terminal sequence of cholecystokinin (CCK). The Pharmacologist 11, 277 (1969).

Starke, K., F. Lembeck, W. Lorenz und U. Weiss: Gallen- und Pankreassekretion unter Substanz P und einem Physalaemin-Derivat. Naunyn-Schmiedebergs Arch.Pharmak.exp.Path. 260, 269 (1968).

INHIBITORY ACTIVITY OF GLUCAGON* ON CAERULEIN EXOCRINE STIMULATION, INDEPENDENT OF HYPERGLYCEMIA

A.G. GLÄSSER, L. DORIGOTTI, and G. VAGHI

Farmitalia, Istituto Ricerche

20146 Milano (Italy)

Introduction

Glucagon may influence the exocrine pancreatic secretion stimulated by gastrointestinal hormones or by cholinergic drugs both in the conscious (1, 2, 3) and in the anaesthetized dogs (4, 5).

According to Necheles (5) glucagon produces a biphasic modification of pancreatic secretion in animals under secretin or mecholyl infusion : first a transient increase in secretory rate followed immediately after by long-lasting depression. On the contrary Dyck et al.(1) found that almost physiological doses of glucagon have only an inhibitory action on pancreatic volume flow and enzyme concentration in dogs with a pancreatic fistula and stimulated by secretin or by a combination of secretin plus cholecystokinin. To explain their results the authors postulated either a competition of glucagon for common drug receptors or a vasoconstriction with a consequent reduction in blood flow to the pancreas. For Nakajima and Nagee (2) the inhibition of pancreatic secretion might be due to the well-konwn hyperglycemic action of glucagon. On the contrary, the vascular effect of glucagon appears to be totally

---

* Cristalline glucagon probably about 90 % pure. Contains about 10 % of a component resembling mono-desamido glucagon which has nearly the same glycogenolytic potency as glucagon. Insulin content : 12 milliunits per mg or less.

We thank Dr. I. Slater from the Lilly Research Labs., Indianapolis (USA) for these informations and for the generous gift of glucagon.

independent of changes in blood sugar concentrations (5).

Caerulein, a cholecystokinin-like peptide, besides stmulating the exocrine pancreas (7), causes a release of insulin (8) and glucagon (9) from the endocrine gland. Infusions of caerulein as low as 2 ng/kg/min were found to increase three times the glucagon concentrations (measured by radio-immunoassay) in the venous outflow from the pancreas.

Therefore, since glucagon can be released not only by cholecystokinin, as shown by Unger (10), but also during an administration of caerulein, we have investigated the influences of glucagon on two important pancreatic effects of caerulein : increase in water secretion from exocrine pancreas (7) and functional vasodilatation within the gland (11). To convalidate our results, some control experiments with glucose and insulin were performed too.

Methods

Beagle dogs of either sex, weighing 8 to 12 kg were anaesthetized with uretane chloralose and blood flow changes were measured uncruently by a Nycotron electromagnetic flowmeter, as described previously (11). Exocrine pancreatic secretion was counted by means of a photoelectric Palmer drop recorder from the incannulated main pancreatic duct. All signals were recorded by a galvanometric Honeywell Visicorder 1508.

Glucose content in the arterial blood was determined repeatedly by the method of Hoffman (12) before and after administration of caerulein, glucagon, and dextrose. All drugs were injected into a systemic vein rapidly or infused slowly by a pump.

Results

1. Effects of caerulein, glucagon, and dextrose on pancreatic, splanchnic, and femoral vascular beds.

Caerulein, as reported in our previous papers (11, 13, 14), caused a prompt blood flow increase through the caudal pancreaticoduodenal artery, the responses being satisfactory dose-related in the range of doses from 2 to 20 ng/kg i.v. Blood pressure or heart rate were not affected by these doses. While pancreatic blood flow was doubled or even increased three times, flow through the superior mesenteric artery and through the femoral artery was only slightly and irregularly affected (the positive or negative variations being less than 20 %.

Injections or slow intravenous infusion of glucagon into a vein produced strong and relatively long-lasting increases of blood flows not only in the caudal pancreaticoduodenal artery but also, contrary to caerulein, in the superior mesenteric artery and in the jejunal branches of this vessel. Blood flow through the femoral artery was usually reduced.

The flow responses of the splanchnic arteries were satisfactory dose-dependent from 0.3 to μg/kg and maximum responses were higher than those obtained with caerulein. Also latency of action, total duration as well as time to reach peak effects were longer than those of caerulein (fig. 1 and 2). While mean and diastolic pressures were regularly reduced by glucagon, pulse pressure was increased occasionally when the systolic pressure was left unchanged or showed a slight elevation.

Dextrose (0.5 g/kg) caused a moderate increase in blood flow through the superior mesenteric and the caudal pancreaticoduodenal arteries only in the cases of rapid intravenous injections of concentrated solutions.

Insulin, on the contrary, was entirely without cardiovascular effects in similar control experiments.

Fig. 1

The vascular resistance (PRU) of the caudal pancreaticoduodenal artery (P.D.A.) is markedly reduced by caerulein, while those of the femoral (F.A.) and of a jejunal artery (J.A.) are left unchanged.

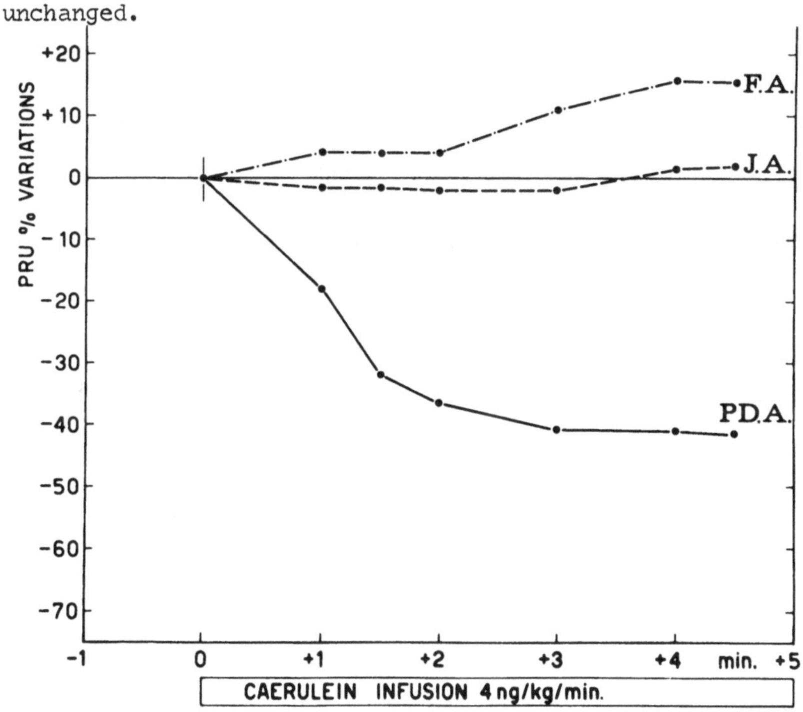

Fig. 2

Glucagon, contrary to caerulein, reduces the vascular resistances of both the pancreatic (P.D.A.) and the intestinal (J.A.) vascular district (vasodilatation) while that of the femoral district (F.A.) is increased (vasoconstriction).

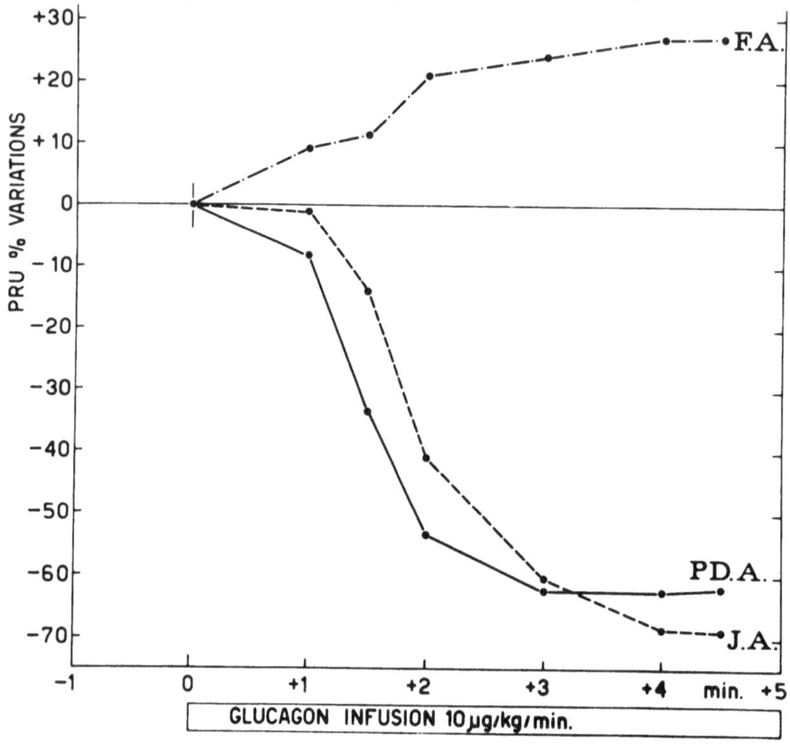

2. Interference of glucagon on the exocrine stimulating action of caerulein.

Glucagon, contrary to caerulein does not possess a stimulating action on the resting pancreatic secretion neither at a single intravenous dose (0.3 to 10 µg/kg) nor when infused slowly at a dose as high as 10 µg/kg/min. The gland in secretory activity during an infusion of caerulein (2 to 4 ng/kg/min) was, however, strongly affected by glucagon. Single doses from 1 to 10 µg/kg i.v. of glucagon injected into a systemic vein had a dual opposite effect on exocrine secretory rate. There was an initial transient increase followed by a long-lasting (10-30 min) depression of secretion (fig. 3).

In accordance whith these results slow intravenous infusions of 1 and 10 µg/kg/min of glucagon blocked the stimulating activity of single low doses of caerulein to an extent that caeru-

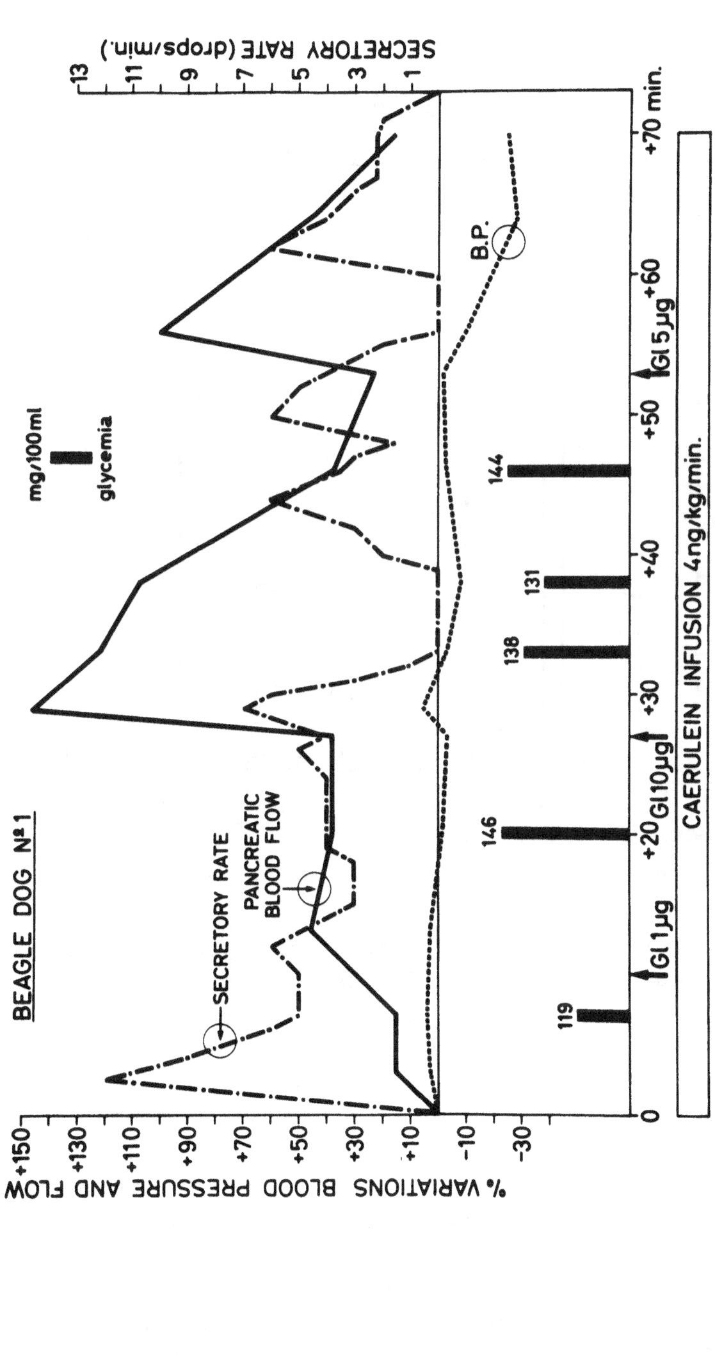

Fig. 3

Per cent variations in pancreatic blood flow, secretory rate and glycemia produced by glucagon (Gl) µg/kg during caerulein infusion.

Fig. 4
Exocrine secretion in response to increasing nanogram/kg doses of caerulein before and during glucagon infusion.

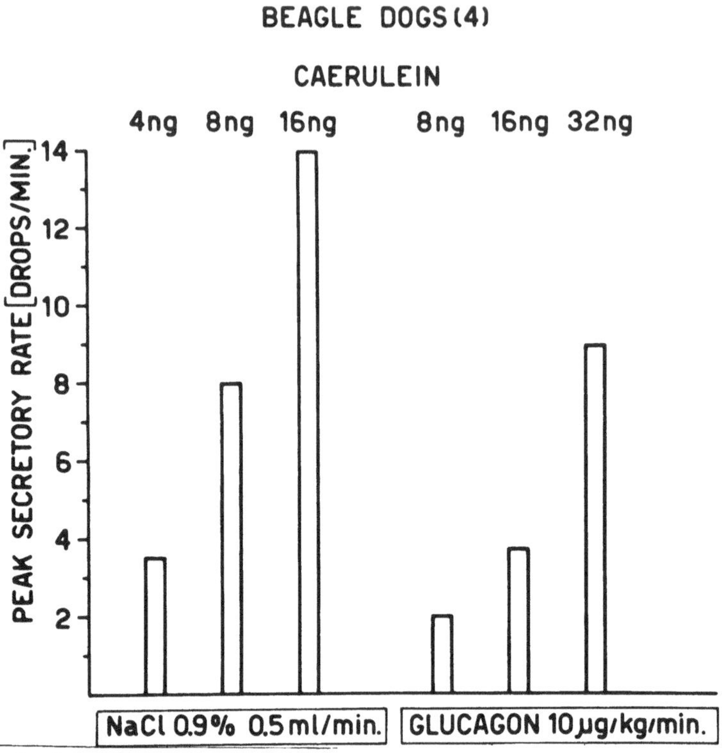

lein had to be injected in doses 2-4 times higher to restore the original activity (fig. 4).

3. Effect of high blood sugar levels on pancreatic exocrine secretion.

It is well know that intravenous glucagon and dextrose increase the content of glucose in the blood of dogs. In our experiments too, glucose levels were rised by single systemic intravenous doses of glucagon as low as 0.3-1 µg/kg or by slow intravenous infusions of 1 and 10 µg/kg/min (glycemia + 60 to 128 %). The responses were however, variable and scarcely dose--related. In particular, glycemia could not be further increased, at least in some animals, when already high as a consequence of previous administrations of glucagon (fig. 3). Blood glucose levels comparable to those produced by glucagon were produced by injections of dextrose, generally 0.5 g/kg i.v. The hyperglycemia observed after intravenous dextrose had no effect on the exocrine secretory rate stimulated by caerulein. In contrast, glucagon nullified secretion for relatively long periods of time in presence of a comparable hyperglycemia. A transient and modest decrease of the stimulating effect of caerulein was obtained, however, in experiments similar to those shown in fig. 4 when single doses of caerulein were injected after dextrose: the first response to caerulein was slightly inhibited, while the second, five-ten minutes later, was already normal.

4. Inhibition of caerulein by glucagon in absence of glycemic variations.

Experiments of the type shown in fig. 3 in which animals received multiple doses of glucagon show that glucagon can inhibit caerulein without affecting practically blood glucose levels.

Conclusion

Our results in the anaesthetized dog show that glucagon is able to alter the pancreatic stimulating activity of caerulein at doses which are active on blood glucose concentrations and on splanchnic blood flows, but not too distant from those considered to be physiological (1).

Immediately after systemic intravenous administration of glucagon, secretory rate of the pancreas under caerulein stimulation is transiently accelerated and then depressed for longer periods of time. The last response is surely independent of blood sugar levels since glucagon is effective even in absence of any variation of glycemia and injections of dextrose do not duplicate the effects of glucagon in spite of a comparable hyperglycemia. Nor is a reduction in pancreatic blood flow the cause of the depression of secretion produced by glucagon, since this peptide increases blood flow to the pancreas in resting conditions and even when pancreatic vessels are already dilatated by caerulein.

The initial acceleration of secretory rate can not be attributed to a vasodilatation in the pancreas favouring the penetration of more molecules of caerulein into the gland since rapid intravenous injections of glucose may increase in some experiments pancreatic blood flow without accelerating pancreatic secretion under caerulein stimulation.

A competition of glucagon for a common receptor in the pancreas could be a better explanation for ours and Dyck's (1) results. In this case, however, we hawe to admit that widely different structures such as those of secretin, caerulein and even cholinergic drugs should act on the same receptor, contrary to the belief that only chemically related antagonists may have a common tissue receptor.

The possibility that the effects of glucagon may result from an activation of the enzyme adenylcyclase must be considered, but can not be discussed on the basis of our results.

References

1. Dyck W.P., Rudick J., Hoexta B., and Janowitz H.D. : Influence of glucagon on pancreatic exocrine secretion. Gastroenterology : 56, 531-537, 1969.
2. Nakajima S. and Nagee D.F. : Inhibition of exocrine pancreatic secretion by glucagon and D-glucose given intravenously. Can. J. Physiol. Pharmacol. : 48, 299-305, 1970.
3. Sweeting J.C. : The effect of glucagon on bile and pancreatic juice. Gastroenterology : 54, 1276, 1968.
4. Walker L. and Necheles H. : Effect of glucagon on pancreatic secretion. Amer. J. Physiol. : 187, 638, 1956.
5. Necheles H. : Effects of glucagon on external secretion of the pancreas. Amer. J. Physiol. : 191, 595-597, 1957.
6. Tibblin S., Kock N.S., and Shentz W.S. : Dissociation of the hyperglycemic and vascular effects of glucagon. Surgery : 67, 816-825, 1970.
7. Erspamer V. : Progress report : Caerulein. Gut : 11, 79-87, 1970.
8. Bertaccini G., De Caro G., and Melchiorri P. : The effects of caerulein on insulin secretion in anaesthetized dogs. Brit. J. Pharmac. : 40, 78-85, 1970.
9. De Caro G., Improta G., and Melchiorri P. : Effect of caerulein on glucagon secretion in the dog. Experientia : 26, 1145-1146, 1970.
10. Unger R.H., Ketter H., Duprè J., and Eisentrant A.M. : The effects of secretin, pancreozymin, and gastrin on insulin and glucagon secretion in anaesthetized dogs. J. Clin. Invest. : 46, 630-645, 1967.
11. Dorigotti L. and Glässer A.H. : Comparative effects of caerulein, pancreozymin, and secretin on pancreatic blood flow. Experientia : 24, 806-807, 1968.
12. Hoffman W.S. : A rapid photoelectric method for determination of glucose in blood and urine. J. Biol. Chem. : 120, 51-55, 1937.

13. Glässer A.H. and Dorigotti L. : Vasodilating action of caerulein and I.C.I. 50,123 on the canine pancreatic vascular bed. In "Pharmacology of Hormonal Polypeptides and Proteins"(Eds. Back W., Martini, L., and Paoletti R.) Plenum Press, New York, pp. 541-543, 1968.

14. Dorigotti L., Glässer A.H. : Relationhip between vasodilator and pancreatic stimulating properties of some caerulein-like peptides. In "Prostaglandins, Peptides and Amines" (Eds. Mantegazza P. and Horton E.W.) Academic Press, London, pp. 181-183, 1969.

BRADYKININ ANTAGONISM BY BIFLAVONYLS FROM GINKGO BILOBA L. AND CUPRESSUS TORULOSA

A. S. Ramaswamy and N. Basu

Department of Pharmacology, All-India Institute of Medical Sciences, Ansari Nagar, New Delhi-16 (India)

A large number of plant products containing the C_6 - C_3 - C_6, skeleton were grouped under the term flavonoids, are widely distributed in the plant materials. The recent discovery of a new class of natural products related to the flavonoids among the Gymnosperms were unexpected. They have been designated as biflavonyls to indicate that their structure contain two flavonoid units. (Baker and Ollis, 1961, Murti et al., 1967, Ramaswamy, 1970). The chemistry of the biflavonyls used in the study are given in Fig. 1 and 2.

MATERIALS AND METHODS

Intestinal Smooth Muscle: Isolated segments of the guinea pig ileum preparations were suspended in a 5 ml bath of Tyrode solution, at 37°C, gassed with oxygen. The contractions induced by agonists were added, recorded on a smoked drum with a frontal writing lever. In the beginning of the assay, bradykinin was tested to obtain repeatedly uniform responses. During the experiment, flavonoids were added to be in contact for 3 minutes, and bradykinin was added to elicit contraction. The dose ranges used varied from 10 µg to 50 µg per ml., and higher concentrations were used when needed. The drugs were dissolved in propyline glycol and further dilutions were done with normal saline. After each contraction the preparation was rested for 30 minutes. A total of six experiments were done for each dosage range used. Synthetic bradykinin (BRS 640, Sandoz, Basel, Switzerland) was used as the

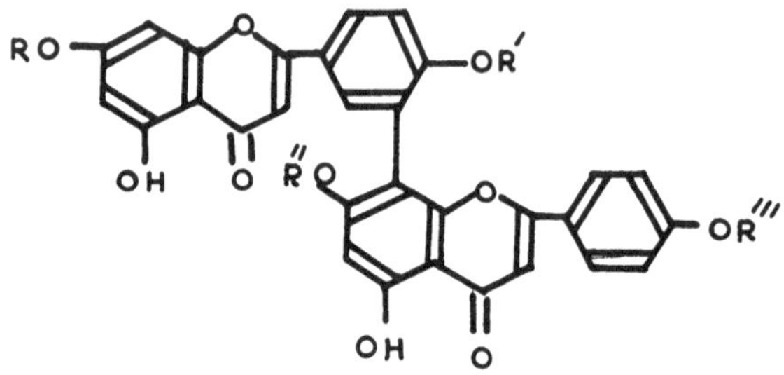

	R	R^{I}	R^{II}	R^{III}
1 (a) Ginkgetin	Me	Me	H	H
1 (b) Iso-ginkgetin	H	Me	H	Me
1 (c) Amentoflavone	H	H	H	H

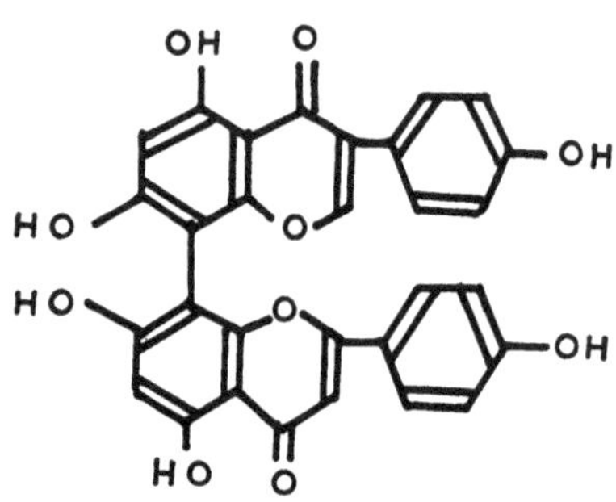

II

Cupressuflavone

Chemical structure of Biflavonyls

spasmogen (Burn, 1952). The biflavonyls, Ginkgo biflavonoid(s) (1a, b) and cupressu - flavone (II), Amentoflavone (1c) were used in this study.

RESULTS

The results obtained in the study are given in Table I.

Table I

Antagonism of bradykinin by bi-flavonyls*

No.	Common Name	ED_{50} and Range µg/ml
1	Amentoflavone	200 - 500
2	Cupressuflavone	50 - 100
3	Ginkgo biflavonyl mixture	10 - 50

*Test object: Guinea pig ileum

DISCUSSION

From this study, it appears that the bis-flavonoids are more potent than the simple flavonoids. (Leme and Walaszek 1967, Natarajan et al., 1970). Among the biflavonyls, Ginkgo biflavonyl mixture is more active than cupressuflavone and amentoflavone. Probably the permeability enhancing action of bradykinin and the spatial structure occupancy of the bradykinin receptor are both blocked by bis-flavonoids, than by the monomeric flavonoids. However, further detailed study of relevant compounds in this series are necessary before arriving at the conclusion.

SUMMARY

Biflavonyls from Ginkgo biloba, Amentoflavone, and Ginkgetin and Iso-Gingketin mixture, and cupressuflavone, from C. torulosa have been studied for the anti-bradykinin action on guinea-pig ileal preparations. Anti-bradykinin action of biflavonyls seems to be superior to that of monomeric flavonoids.

ACKNOWLEDGEMENT

Authors' grateful thanks are due to Drs. R. B. Arora, C. Dwarakanath and T. R. Sheshadri, F. R. S., for keen interest in this work. The gift of bradykinin from Dr. A. Carletti, Sandoz, Basel, is hereby duly acknowledged. This work was in part supported by grants from the Indian Council of Medical Research, New Delhi (India).

REFERENCES

Burn, J. H., Practical Pharmacology, Blackwell Scientific Publications, Oxford (1952).

Baker, W. and Ollis, D., Recent developments in the chemistry of natural phenolic compounds, (ed. D. Ollis) Pergamon Press, London (1961), p. 152.

Leme, J. G. and E. J. Walaszek, Pharmacologist. 9, 242 (1967).

Murti, V. V. S., Raman, P. V., and Sheshadri, T. R., Tetrahedron 23, 397 (1967).

Natarajan, S., Murti, V. V. S., Sheshadri, T. R., and Ramaswamy, A. S., Curr. Sci., 39, 533 (1970).

Ramaswamy, A. S., Jour. Ind. Med. Assoc., 55, 163 (1970).

PATHOLOGY

PLASMA KININOGEN AND ARTERIOVENOUS RATIO IN LATE PREGNANCY TOXEMIA

V. Levi d'Ancona, S. Severi and P. Periti

Clinica Ostetrica e Ginecologica dell'Università di Firenze, Firenze (Italy)

In a previous communication (Periti and Gasparri, 1965) we reported that the blood of a pregnant woman in her third trimester shows a significant difference between the kininogen contents of the arterial and of the venous plasma. The mean of this difference is over 1000 μg/ml, in favour of the former.

We later reported (Gasparri and Periti, 1969) the data referring to a small number of cases of late pregnancy toxemias, in which the mean values of the arterial and venous kininogen did not differ significantly. We have now controlled the behaviour of arteriovenous kininogen in a larger number of toxemia cases, carrying out assays on samples kept at 38°C from the time the blood was drawn until use.

There were 20 cases of toxemia, characterized by edema, proteinuria and hypertension (EPH-Gestose), and 12 healthy women in the third trimester of pregnancy with a normal course. On the average (see figure), the two groups of cases do not show differences in the arterial plasma kininogen level (7.1 ± 1.5 μg for the toxemic cases, 7.1 ± 1.5 μg for the normal ones). The kininogen contents in the venous blood are different instead (6.6 ± 1.5 μg and 5.5 ± 1.7 μg, respectively). The arteriovenous ratio is therefore much higher in the cases of normal pregnancy (7.1/5.5). In other words, the difference between the mean kininogen value in the arterial blood and in the venous blood is

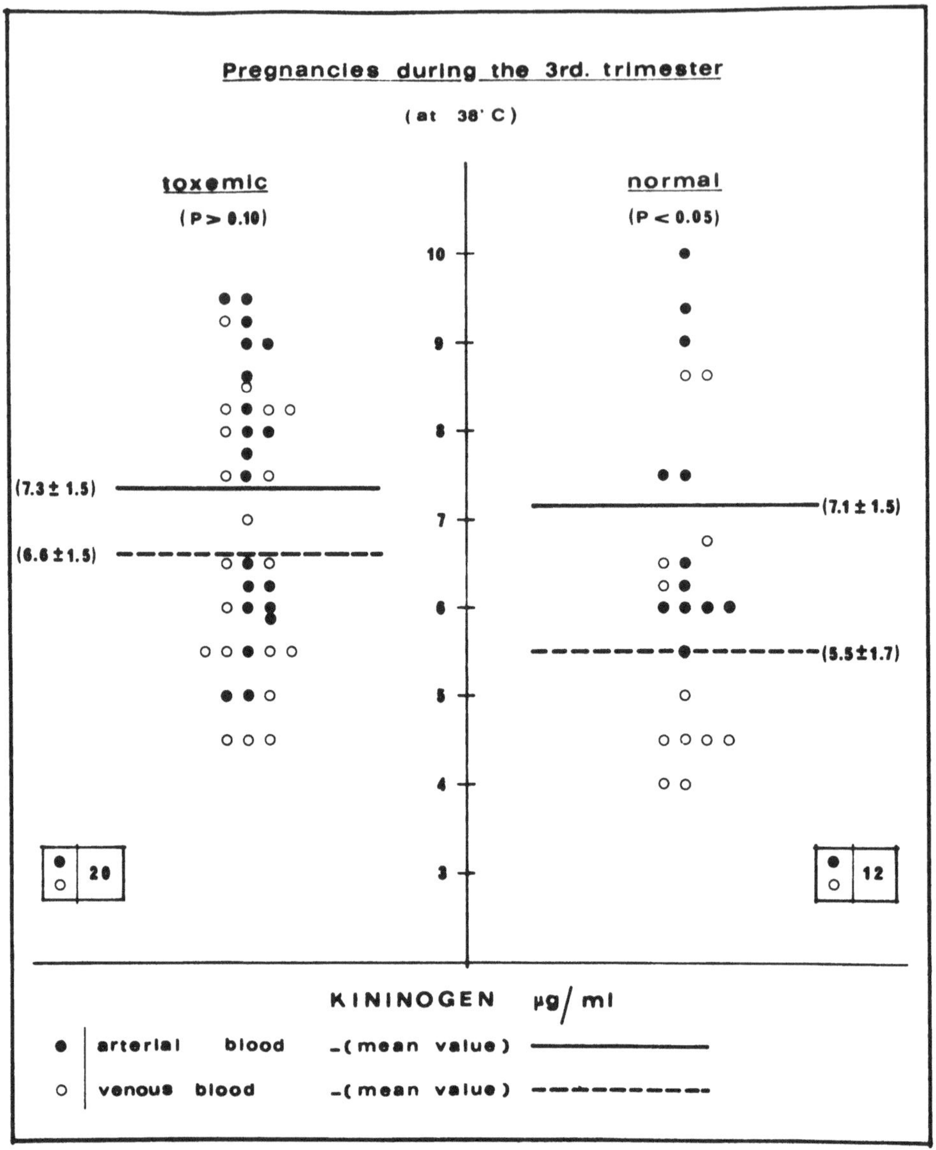

significant (P < 0.05 according to Student's t) for the cases of normal pregnancies, but not for toxemic pregnancies. Presuming that the blood is subjected to kininogenase activation in the peripheral vascular bed and that this activation is reflected by the kininogen depletion, we could therefore reach the conclusion that in the case of toxemia this activation is reduced. We cannot yet explain the significant difference between the mean values of the arterial and of the venous plasma kininogen during the third trimester of normal pregnancies. We can therefore only advance the simple hypothesis that in cases of toxemia the damage of the tissues expressed by the edema and microcirculatory pathology characterizing the syndrome may, in some way, be tied to the missing or reduced peripheral depletion of plasma kininogen.

REFERENCES

Gasparri F. and Periti P. in Die Spätgestose, Sonderdruk page 119, Schwabe & Co, Verlag, Basel, Stuttgart, 1969.

Periti P. and Gasparri F. in Hypotensive Peptides, page 536, Springer-Verlag, New York Inc., 1966.

FETAL DISTRESS AND PLASMA KININOGEN IN THE NEWBORN

S. Severi, M. Paradiso and P. Periti

Clinica Ostetrica e Ginecologica dell'Università di Firenze

Firenze (Italy)

Recent researches on the possible role of kinins as mediators of circulatory changes in the newborn infant (Melmon and co-workers, 1968) have led us to re-examine the behaviour of kininogen in the newborn during the first hours (24 hrs) of life in relation to a possible condition of distress of the newborn, general distress involving respiratory and circulatory functions.

Kininogen assay was carried out on blood samples drawn and kept at 38°C until plasma was used. The results can be seen in figure 1.

There were 15 newborn in distress and 23 normal. All the pregnancies were at term. We know that in the newborn kininogen level in the venous blood is about a third of the value found in the mother at the moment of delivery (Periti and co-workers, 1962; Periti and Gasparri, 1965; Melmon and co-workers, 1968). We do not know, instead, the amount of kininogen in fetal blood before the time of birth, before that is the circulatory changes occurring when the lungs start to function.

The mean values found for the venous blood kininogen were 2.1 μg/ml in the group of the 15 newborn in distress and 2.5 μg/ml in the group of the 23 healthy infants born at term. The difference is not significant ($P > 0.05$ according to Student's t). We should bear in mind that in the adult a condition of stress or shock (Sicuteri and co-workers, 1965) is accompanied by a depletion of

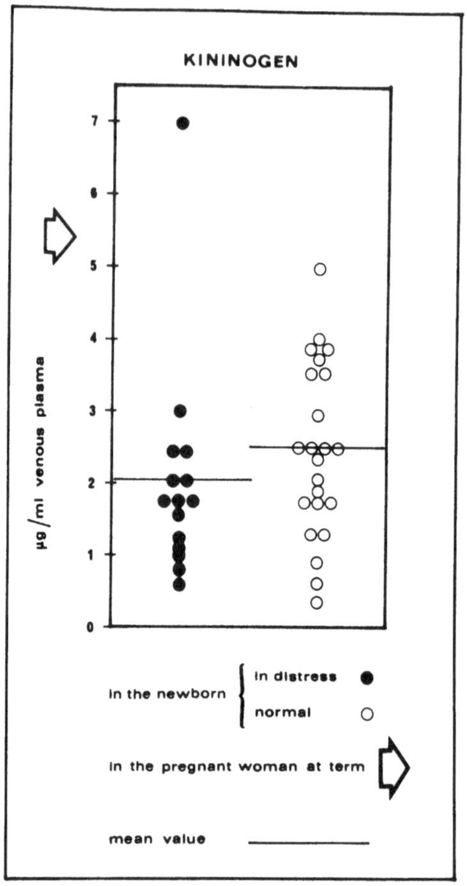

Figure 1.

kininogen and that kininogen in the newborn does not decrease following exposure of the plasma to glass, unlike kininogen in the adult (Melmon and co-workers, 1968). These considerations prompt us to advance the suggestion, on the grounds of our data, that in the newborn kininogen is probably a substrate which is sensitive to particular activators, as schematically shown in figure 2.

This figure shows the main factors which can be considered responsible for the activation of fetal blood kininogen in the newborn. These are (i) the release of fetal granulocytes at the first moments of breathing, (ii) the increase in blood PO_2 and (iii) the cooling of Wharton's jelly in the umbilical cord, and therefore of cord blood, as soon as the fetus has passed through the genital way.

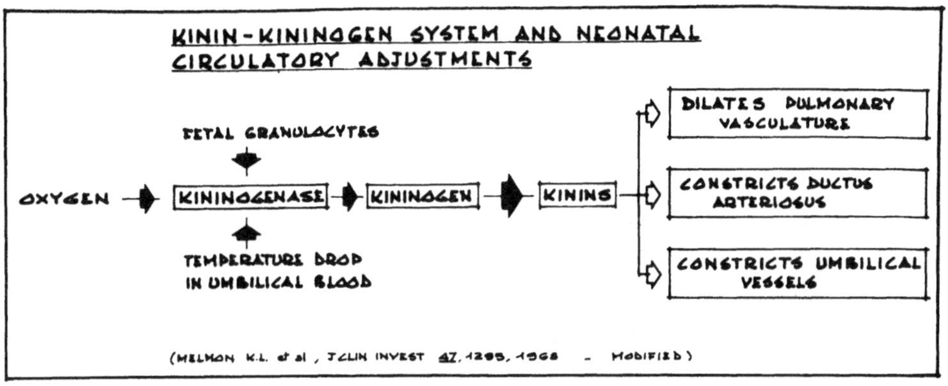

Figure 2.

It is perhaps the same type of kininogen as the type we find in the adult forming system I (that is, kininogenase I - kininogen I). And it is perhaps similar to the only system present in the blood of the dog, lacking, that is, kininogenase II and kininogen II, which form system II, parallel to system I in man, the only system in the rat (Vogt, 1965). System I, in fact, is not activated on exposure to glass, it is activated by acidification, similarly to what was recently shown for the plasma of the newborn infant (Melmon and co-workers, 1968).

Concluding, in the newborn, blood kininogen, at least in the first 24 hours of extrauterine life, does not seem to be affected by conditions of distress, which in the adult, instead, are accompanied by a more or less marked depletion of the precursor of kinins. Still to be ascertained is whether the fetus and the newborn infant really have only system I at their disposal or whether the other component, represented by system II in the adult, is consumed during labour or during the first minutes of life. On the other hand, infants delivered by cæsarian section do not show kininogen levels of blood which are different, both quantitatively and qualitatively, from those of infants born spontaneously. In the newborn, during the first days of life there is a slow, progressive increase in kininogen level, until on the 5th. day values close to those of the adult are reached (Periti and co-workers, 1963).

REFERENCES

Melmon K.L. et al.: J. Clin. Invest. 47, 1295, 1968.

Periti P. et al. : Boll. Soc. Ital. Biol. Sper. 38, 672, 1962.

Periti P. et al. : Settim. Med. 51, supp. 1, 70, 1963.

Periti P. and Gasparri F.: in Hypotensive Peptides, page 536, Springer-Verlag, New York Inc., 1966.

Sicuteri F. et al.: in Hypotensive Peptides, page 522, Springer-Verlag, New York Inc. 1966

Vogt W. : in Hypotensive Peptides, page 185, Springer-Verlag, New York Inc., 1966.

CHANGES IN KININOGEN CONTENT OF SERUM AND SOME ORGANS DURING INJURY AND INFLAMMATION IN RATS

H.P.Zach and E.Werle

Institute of Clinical Chemistry and Clinical Biochemistry, University of Munich, Germany

We have explored some components of the kallikrein-kinin-system after injury and inflammation in rats. Prekallikrein, kininogen and kininase-activity in serum were estimated by following methods: prekallikrein was activated by trypsin (1 mg/ml) and tested against pancreas kallikrein on the blood pressure of anaesthetized dogs. To estimate serum kininogen heat-acid-denatured serum was incubated with trypsin (2mg/ml, 25 min at 37°C); then the samples were chromatographed on Amberlite IRP-64 and the eluats were tested against synthetic bradykinin on the rat uterus. Sometimes the amounts of kinin-activity of the eluats were controlled on the blood-pressure of anaesthetized dogs. Only such smooth muscle activity was regarded as a kinin activity which was inactivated by incubation with chymotrypsin, carboxypeptidase B and native serum. To estimate kininase-activity serum was diluted with tris-buffer and incubated with bradykinin. Then that moment was determined, at which 50% of added bradykinin were inactivated.

All declarations about prekallikrein and kininogen content or kininase-activity were related to the protein content of the samples. All investigations were done in female rats, weighing about 250 g. The rats were stunned by a beat on their head and arterial and venous blood was collected by neckdissection.

The first time we saw a connection between tissue injury and kallikrein-kinin-system was when we explor-

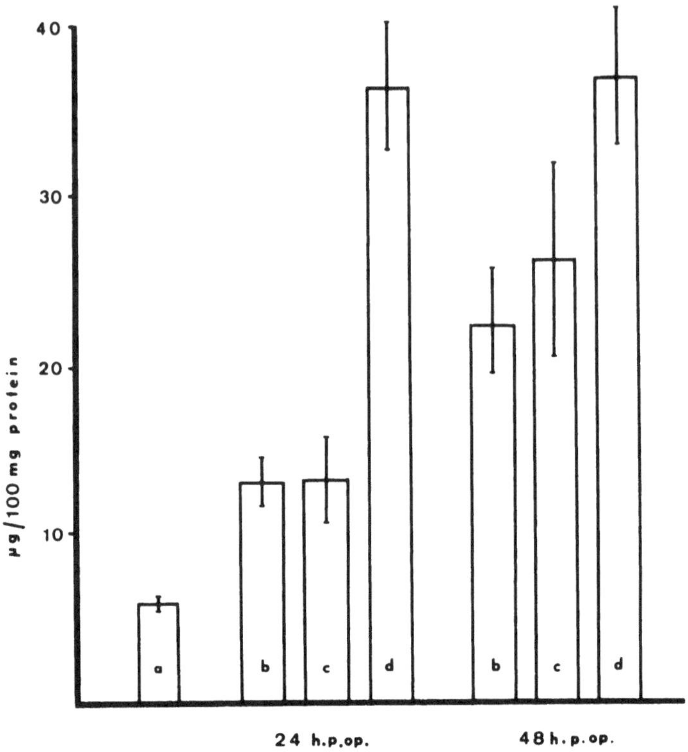

Fig.1: Kininogen in serum of rats given in µg trypsin-liberated bradykinin/100 mg protein ($\bar{x} \pm s$, n = 3)
a = untreated rats
b = 24 and 48 hours after laparatomy
c = 24 and 48 hours after adrenalectomy
d = 24 and 48 hours after nephrectomy.

ed the influence of a total nephrectomy on serum kininogen level. Figure 1 shows the kininogen content of serum 24 and 48 hours after laparatomy and adrenalectomy or nephrectomy on both sides. In comparison to non operated rats 24 and 48 hours after laparatomy or adrenalectomy the kininogen content has increased about the 2 or 3 fold. 24 hours after extirpation of both kidneys kininogen has even increased about 500 or 600%.

We were induced by the fact that even a laparatomy is able to influence the kininogen content of the serum to watch these changes for a longer period. In figure 2 changes in kininogen content and protein concentration of the serum are shown in the course of

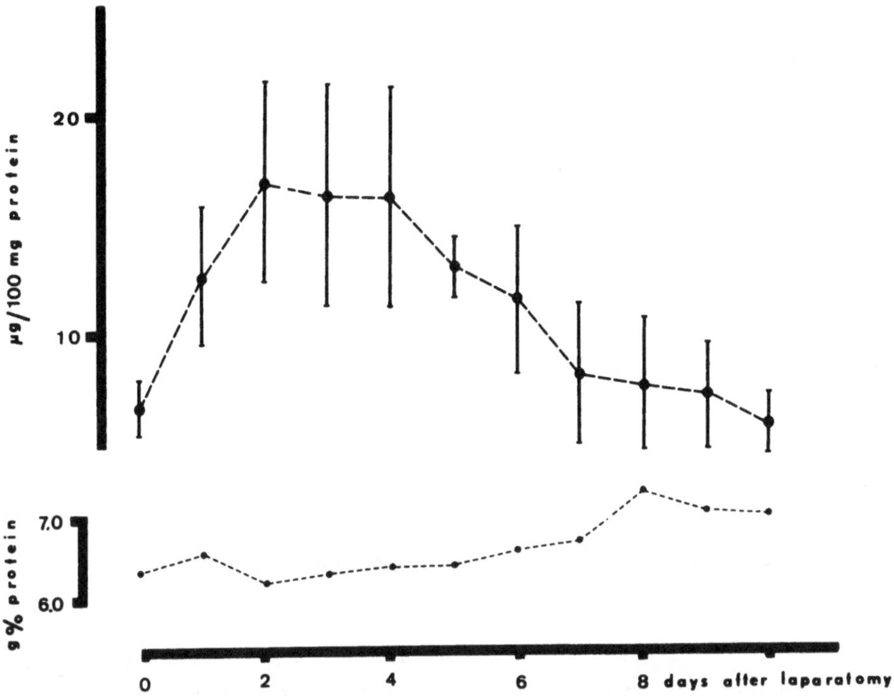

Fig.2: Kininogen in serum (in µg of trypsin-liberated bradykinin/100 mg protein) and serum protein concentration (in g%) after laparatomy in rats ($\bar{x} \pm s$, n = 5).

10 days after laparatomy. There is a strong increase of serum kininogen up to the 3 fold within the first 2 days after laparatomy and a slow decrease till the 8th or 9th day when the kininogen level shows normal values again. When the kininogen level is augmented the concentration of serum proteins is slightly diminished; in the course of wound-healing kininogen level and protein content become normal.

The tissue damage produced by laparatomy seems to be responsible for the increase of serum kininogen. Neither narcose nor an increased liberation of suprarenal hormones, which probably can be expected as a consequence of the operation stress, influence the kininogen level. We were able to state that neither narcose nor application of high doses (10 mg/kg for 10 days) of cortisol were able to produce a change in kininogen content. On the other hand adrenalectomized rats reacted with an augmented kininogen level on tissue injury.

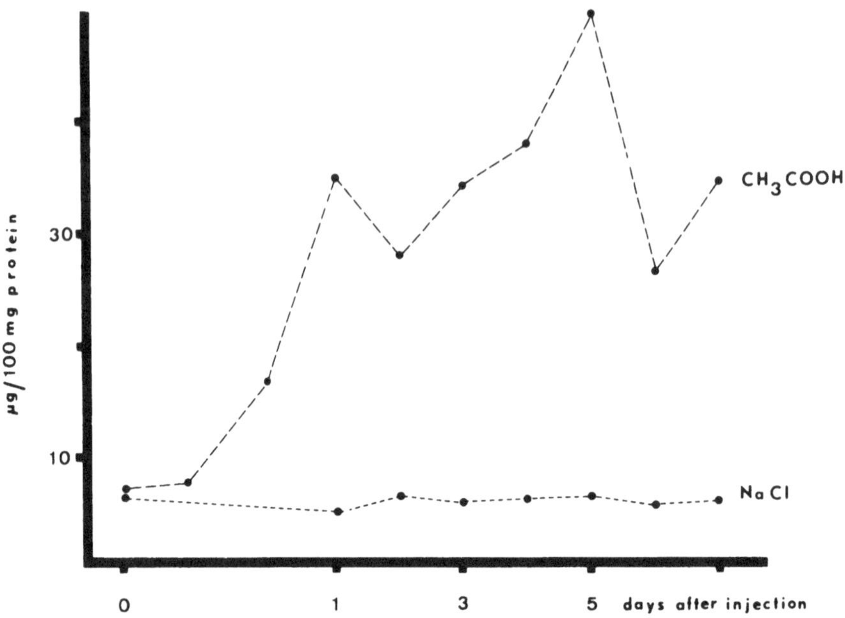

Fig.3: Kininogen in serum of rats (in µg trypsin-liberated bradykinin/100 mg protein) after i.m. injection of 0,2 ml CH_3COOH or 0,2 ml NaCl-solution ($\bar{x}$, n = 2).

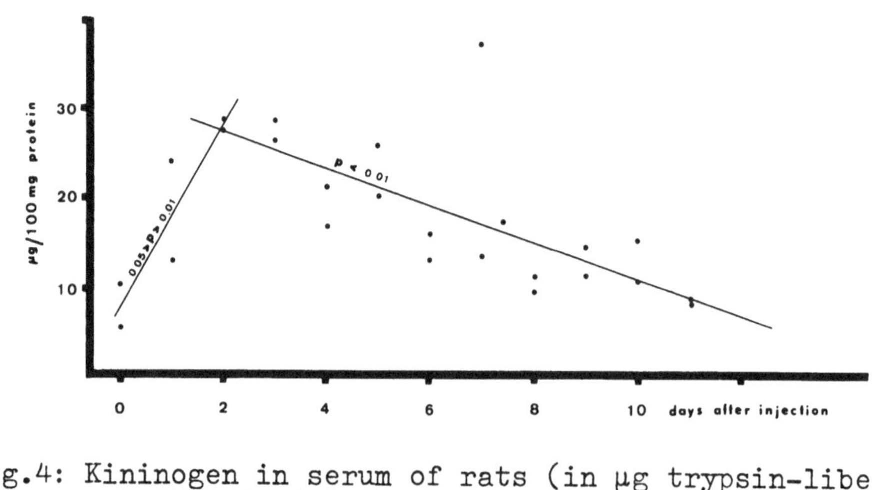

Fig.4: Kininogen in serum of rats (in µg trypsin-liberated bradykinin/100 mg protein) after injection of 0,05 ml oleum crotonis (n = 2).

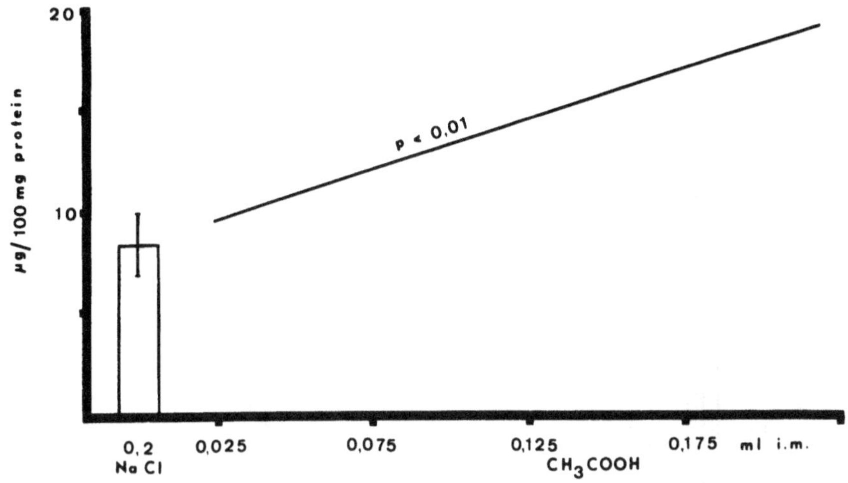

Fig.5: Kininogen in serum of rats (in µg trypsin-liberated bradykinin/100 mg protein) 24 hours after i.m. injection of NaCl-solution and CH_3COOH (n = 6).

I.m. injections of CH_3COOH or oleum crotonis produce local necrosis or inflammation. Figure 3 shows that about 10 hours after injection of CH_3COOH an increase of kininogen content up to the 3 fold starts. The altitude of kininogen level remains nearly unaltered for more than 7 days. Animals which were treated with NaCl-solution show no reaction.

Figure 4 shows the effect of injected oleum crotonis. Changes of kininogen concentration in serum are shown in the course of 11 days after i.m. injection. The kininogen levels of a control group, which was treated with oleum arachidis remained unchanged; after injection of oleum crotonis the kininogen level is similar to that after laparatomy. The drop of kininogen in the course of the days after maximal kininogen increase is significant.

The size of kininogen elevation in serum is direct proportional to the extent of tissue damage. Figure 5 shows the size of kininogen level in correlation to the amount of injected CH_3COOH. There is a significant correlation between the amount of injected CH_3COOH and the altitude of the resulting kininogen level in serum. In contrast to lipoid-soluble CH_3COOH there is no significant dose-response correlation

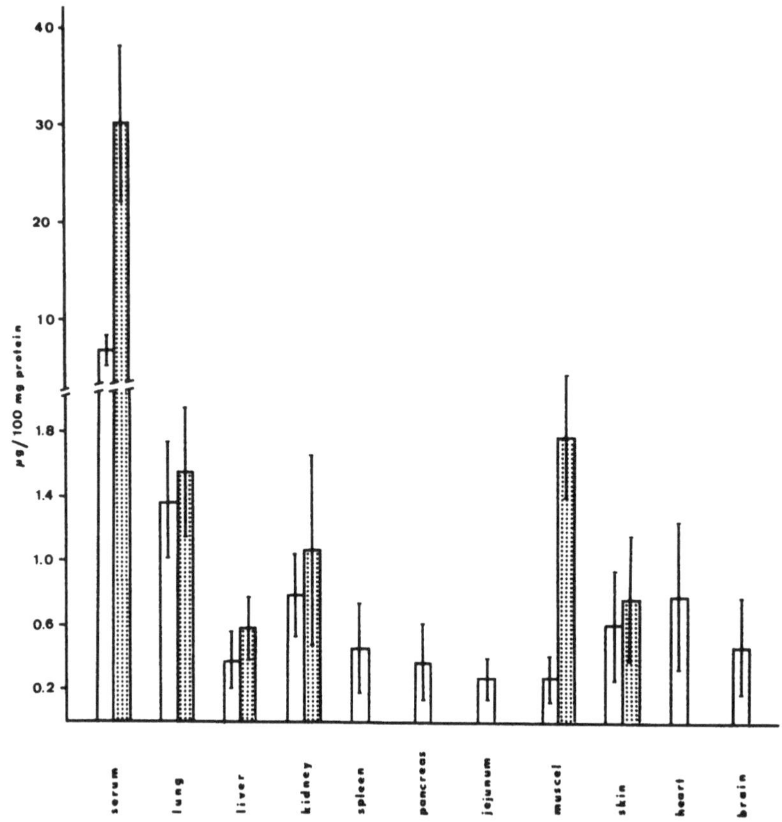

Fig.6: Kininogen in serum and some organs in rats, given in µg of trypsin-liberated bradykinin/ 100 mg protein. Spotted columns: 24 hours after i.m. injection of 0,2 ml CH_3COOH ($\bar{x} \pm s$, n = 4, blood content corrected).

after injection of oleum crotonis because this is only slowly absorbed.

The fact, that the rat is evidently able to double the amount of kininogen in serum within a short time induced us to search for the organ of enhanced synthesis. Figure 6 shows the amounts of kininogen we have found in some organs of untreated rats and in rats after injection of CH_3COOH. The kininogen content is shown in respect of the blood content of the organs which was estimated photospectrometrically. All explored organs contained kininogen in different amounts. In the untreated animals we found the highest kininogen concentrations in lung, kidney, heart and skin;

in correlation to serum these contents however are small. We have found no accumulation of kininogen in the liver which seems to be the organ of kininogen synthesis. After CH_3COOH injection the rise of kininogen took place only in serum and in damaged muscle tissue where we found a significant increase. All the other explored organs showed no significant change in kininogen content. Probably kininogen enrichment in damaged muscle tissue is due to an enhanced capillar permeability with following plasma invasion.

Simultanously with an increase of kininogen content in serum we were able to state an alteration in the compounds of serum proteins after injection of oleum crotonis. Figure 7 shows the levels of kininogen, albumine and α-globuline in the course of 11 days after injection of oleum crotonis and the correlat-

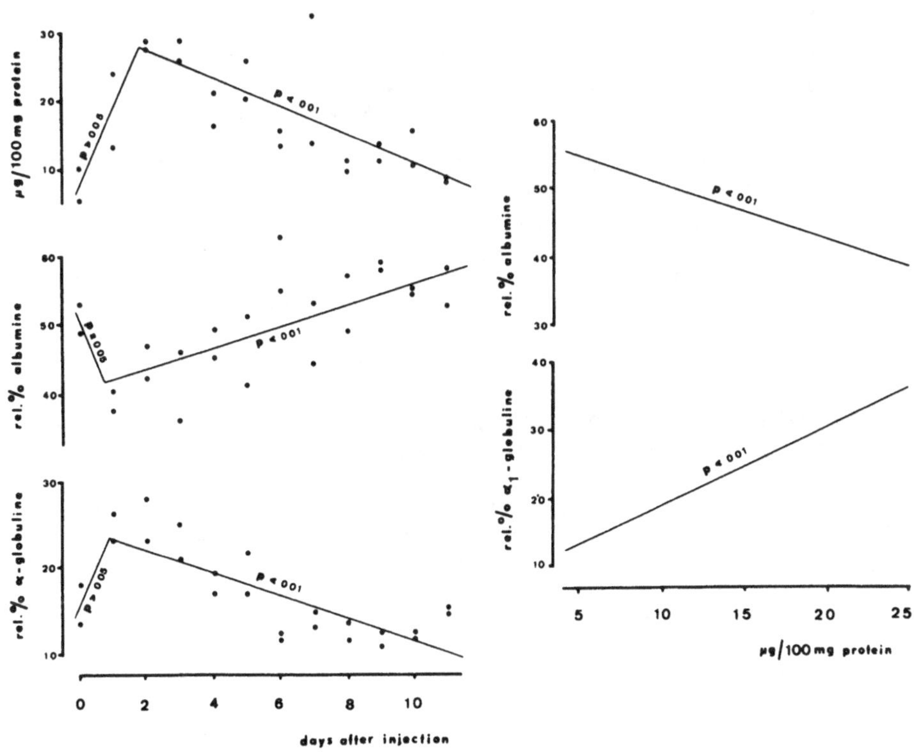

Fig.7: Kininogen (in µg of trypsin-liberated bradykinin/100 mg protein), albumin and α-globuline (in rel.%) in serum of rats after injection of 0,05 ml oleum crotonis.

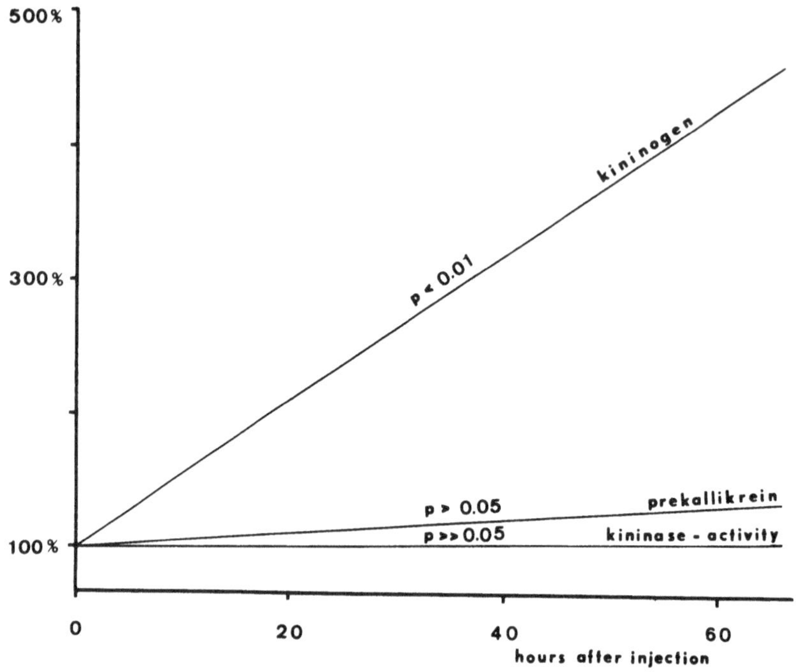

Fig.8: Kininogen, prekallikrein and kininase-activity in serum of rats after i.m. injection of 0,2 ml CH_3COOH (n = 6).

ions between kininogen and albumine and kininogen and α_1-globuline. These correlations are significant. While serum kininogen increased at about 200 - 300% α-globuline rises only to about 100%.

Besides kininogen estimation we explored whether prekallikrein or kininase-activity are altered after tissue injury. Figure 8 demonstrates the effect of CH_3COOH treatment on kininogen and prekallikrein content and kininase-activity. Besides the known increase of kininogen in the course of time after injection neither prekallikrein nor kininase-activity of the serum are changed significantly.

It is difficult to interprete the changes within the kallikrein-kinin-system we have found after tissue injury. Even the fact, that our method of kininogen estimation based on kinin liberation by the unphysiologic trypsin and used heat-acid-denatured serum aggrevates an explanation.

The significant correlation between the enhancement of kininogen and α_1-globuline refers to a real increase of kininogen synthesis after tissue injury. There is no accumulation of kininogen in the liver, it is probably released immediately after synthesis. The duplication of kininogen in serum within 24 hours seems to be a maximal rate of synthesis which can be compared with that rate which is found after depletion of serum kininogen by i.v. injected elac acid.

The increase of kininogen after extirpation of both kidneys is similar to the rise of angiotensinogen after nephrectomy; the latter fact is declared by reduced utilisation.

However the increase of serum kininogen after tissue injury and in inflammation is probably not due to a reduced kininogen utilisation. It seems to be possible to explain the augmented synthesis of kininogen on the basis of an increased utilisation; in the same way the augmented synthesis of fibrinogen after burns is declared. A participation of the kallikrein-kinin-system in inflammations is discussed by many authors. In spite of finding no significant change in prekallikrein concentration and kininase-activity after tissue damage an augmented kininogen utilisation is possible in the area of tissue damage and inflammation, perhaps by involving the leukocytes.

Further investigations will be necessary to prove or to exclude an active or passive role of kallikrein-kinin-system in inflammation. For example it would be important to know whether our findings will be confirmed when untreated plasma and serum kallikrein as kinin liberator are used or to which degree our findings can be influenced by an antiphlogistic treatment.

This work was supported by the Deutsche Forschungsgemeinschaft (SFB 51).

STUDIES ON THE MECHANISM OF KININ RELEASE INDUCED BY ANTILYMPHOCYTE GLOBULIN

John Bradley and Alfred Cuschieri

Departments of Medicine and Surgery

University of Liverpool, U.K.

INTRODUCTION

Antilymphocyte serum or globulin (ALG) has been used in man and animals as a means of immunosuppression in the transplantation of allogeneic tissue. Among the side effects encountered during its administration are fever, hypotensive episodes and anaphylactoid reactions. Kashiwagi et al (1968) found that the anaphylactoid reactions were characteristic: within 1-30 minutes of injection the patients became anxious and complained of low back pain. Nausea and involuntary defaecation accompanied cyanosis and hypotension. Recovery took place within 90 minutes. These reactions have been attributed to the presence of pre-formed reaginic or precipitating antibodies in the recipient against the donor of the ALG, or to the presence of aggregated gammaglobulin in the ALG.

The development of two anaphylactoid reactions to ALG was found in one patient to be associated with kininogen depletion and a rise in free plasma kinin levels. *In vitro* studies were made on the effect of this preparation of ALG upon the blood of the patient, P.D., and a normal male subject, C.

MATERIALS AND METHODS

Kininogen Determination

The kininogen level was determined by the method of Diniz

and Carvallho (1963).

Determination of Free Kinins

The free plasma kinin level was determined by the method of Talamo et al (1969) on a 9.0 ml of venous blood sample collected into a plastic syringe containing 9 mg of di-sodium EDTA.

Assay

An isolated rat uterus, previously stored at 4°C for a period of 18-24 hours, was suspended in a 5 ml bath of oxygenated de Jalon solution at 30-31°C. A four point assay on the samples was performed against standard synthetic bradykinin (sandoz B.R.S. 540). The de Jalon solution contained atropine sulphate (10^{-8}g/L), and brom-lysergic acid diethylamide (BOL, 10^{-8}g/L) and mepyramine maleate 10^{-8}g/L).

Passive Cutaneous Anaphylaxis

The method used was that described by Layton et al (1963). Samples of P.D's serum, collected at time of the marrow transplant and subsequently thereafter, were injected intradermally into the shaved abdomen of two Rhesus monkeys. A serum known to contain reaginic antibodies to horse protein was used as a positive control. Six hours later one monkey was injected i.v. with 4.5 ml 'Coomassie blue' (I.C.I. Limited, Cheshire) and with 0.5 ml (70 mg) horse ALG. At 48 hours the other monkey was injected with Coomassie blue and 0.5 ml ALG.

CLINICAL DETAILS

A 14-year old girl, P.D., developed bone marrow aplasia, and her condition deteriorated over several months. Because of frequent severe infections and continued marrow aplasia, an attempt was made at bone marrow transplantation using ALG as an immunosuppressive agent. Before the first intravenous dose of ALG was administered 5 mg of the preparation was injected intradermally. There was no immediate (Type I Gell and Coombs) or Arthus type (Type III Gell and Coombs) reaction. However, the administration of 2000 mg of ALG intravenously at a rate of 100 mg/minute produced an anaphylactoid reaction, one of the features of which was a prolonged period of diastolic hypotension. (See Figure 1).

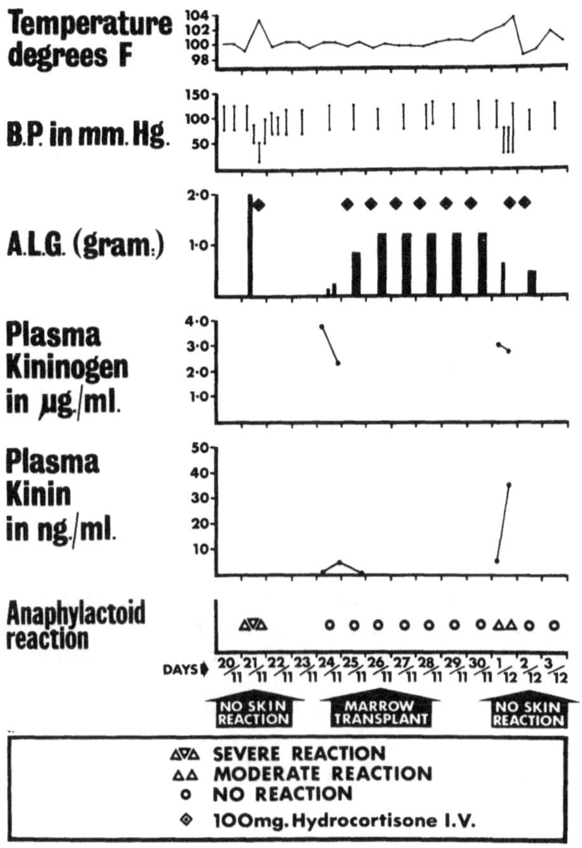

Figure 1 The relationship between Anaphylactoid reactions, ALG administration and kininogen and kinin levels in patient, P.D.

After a two-day rest period, ALG was administered again but at a slower rate (2-4 mg/minute) and in the amounts shown in Figure 1. It can be seen that the slow administration of ALG was associated with kininogen depletion but no appreciable rise in free kinin level of the plasma. However, on the fifth day after she had received a marrow transplant the patient had a second anaphylactoid reaction when the rate of i.v. infusion of ALG was inadvertently increased to 25 mg/minute. During this second anaphylactoid reaction the free plasma kinin level rose from 4 ng/ml to 32 ng/ml. On the following day the administration of

400 mg of ALG at a rate of 2 mg/minute produced no ill effects.

Though kinin liberation was demonstrated in vivo, the mechanism by which it occurred was not clear. There was no history of previous administration of horse serum or of atropic disease, and there was no immediate response to the intradermal administration of ALG. Reaginic antibodies to horse serum proteins were not detected by passive cutaneous anaphylaxis performed in the Rhesus monkey. No precipitating antibodies to horse serum proteins were detected by gel diffusion tests. Therefore, a series of in vitro studies were made using samples obtained from the patient, P.D., and from a normal healthy subject, C, known never to have received horse protein parenterally.

IN VITRO STUDIES

The kininogen content of venous whole blood was measured. To do this six ml of venous blood was withdrawn through a siliconised needle into a plastic syringe containing 10 units preservative-free heparin/ml blood (Evans Medical, Speke). This was subdivided into 3 aliquots of 2 ml each as follows:

(1) Control, to which nothing was added
(2) 20 µl ALG added
(3) 20 µl horse serum added

The tubes were incubated at 37°C for one hour and plasma extracted for kininogen.

From Table I it can be seen that whereas horse ALG produced kininogen depletion when added to both the blood from P.D. and from a normal healthy subject, the addition of horse serum did not. The fact that there was no reaction to normal hourse serum eliminated the possibility of human antibodies to horse protein as a cause of kinin liberation.

TABLE I

KININOGEN LEVELS (IN SYNTHETIC BRADYKININ EQUIVALENTS) OF WHOLE BLOOD AFTER ONE HOUR'S INCUBATION AT 37°C

	Kininogen in µg/ml		
Source	Control	ALG Added	Horse Serum Added
Patient	1.5	0.41	1.4
Normal Subject	2.7	1.45	2.65

(a) Determination of the Role of the Cellular Constituents

Two sets of experiments were made:

(1) The incubation of platelet-free plasma with ALG or horse serum:

(2) The incubation of mononuclear cells in autologous plasma with ALG or horse serum.

Kinin Liberation from Platelet-free Plasma: Blood was withdrawn through a siliconised needle into a plastic syringe containing 1 mg EDTA/ml blood. The sample was distributed into polycarbonate bottles surrounded by ice and all subsequent procedures were carried out at 4°C. The blood was centrifuged at 2,000 rpm for 20 minutes (1,000g). The plasma was then freed from platelets by centrifugation at 12,000 x g for 20 minutes. The supernatant plasma was divided into three 5 ml aliquots as follows:

(1) Control - nothing added

(2) 50 µl of ALG added

(3) 50 µl of horse serum added

These were incubated at 37°C for one hour and then extracted for kinins. The addition of horse ALG to platelet-free plasma of both the patient and the normal subject produced kinin release (as shown in Table II).

The kinin-like activity produced in samples incubated with ALG was completely inactivated by incubation with chymotrypsin (250 µg) for 1 hour at 37°C, thus confirming its polypeptide nature.

Though ALG liberated kinin from platelet-free plasma, it may or may not be the major or sole cause of kinin liberation. Antilymphocyte Globulin acts predominantly on lymphocytes and, therefore, the ALG-lymphocyte interaction may provoke the formation of kinins. To evaluate this a comparison was made of the amount of kinins liberated by ALG from plasma alone and from plasma-mononuclear cell suspensions.

TABLE II

FREE KININ LEVELS (IN SYNTHETIC BRADYKININ EQUIVALENTS) AFTER ONE HOUR'S INCUBATION AT 37°C

Source	Kinin Levels in µg/ml		
Human Platelet-free Plasma	Control	ALG Added	Horse Serum Added
Patient, P.D.	10.5	20.2	8.0
Normal, C.	2.8	16.0	3.4
Monomuclear Suspension			
Patient, P.D.	9.6	22.7	9.2
Goat Platelet-free Plasma	3.0	2.9	2.4

Kinin Liberation from Mononuclear Cell Suspensions: Fifty ml of blood was withdrawn through a sterile siliconised needle into a plastic syringe containing 1 mg EDTA/ml blood and distributed into plastic bottles in an icebath. The blood was centrifuged at 200 rpm for 5 minutes. The leucocyte-rich plasma was removed, centrifuged at 2,000 rpm for 10 minutes, and the supernatant plasma removed. The leucocytes were re-suspended in 1 ml of the patient's own plasma, and the suspension was carefully layered on to 3 ml of Ficoll-Triosil mixture in a polyethylene centrifuge tube. After centrifugation at 2,000 rpm for 20 minutes the lymphocytes could be seen as a distinct white band just below the plasma/Ficoll-Triosil interface. The lymphocyte layer was removed carefully and washed four times with TC199 (containing heparin, 10% foetal calf serum (Flow Laboratories) and EDTA (1 mg/ml) to inhibit kininases). After washing, the lymphocytes were suspended in 1 ml of autologous plasma and a white cell differential and viability count done on the suspension. The lymphocyte concentration was adjusted to 1×10^6 viable cells/ml. Three x 5-ml samples were put in plastic tubes as follows:

(1) Control - nothing added
(2) 50 µl ALG added
(3) 50 µl horse serum added.

After 1 hour's incubation at 37°C the total white cell and viability counts were determined. There was no significant change in either the total count or viability count. The samples were then centrifuged at 2,000 rpm

(1,000 g) for 20 minutes and the kinins extracted from the supernatant. The amount of kinins liberated from a mononuclear cell suspension in autologous plasma was the same as that liberated from the platelet-free plasma. (See Table II). Thus the action of ALG upon cells was not responsible for the kinin liberation in this case.

(b) Incubation of ALG and Goat Plasma

The administration of gammaglobulin preparations has been reported to provoke a shock-like state because of the presence of aggregated materials (Barandun et al, 1962). If the reactions were due to the presence of these, or any other non-specific agent, then the addition of ALG to non-human plasma should also provoke kinin liberation. There was no rise in free plasma kinin level when ALG was added to fresh non-glass contacted, platelet-free goat plasma. (See Table II).

(c) Gel Diffusion Studies on ALG

Apart from the antibodies to lymphocytes, other antibodies to erythrocytes, platelets, serum proteins, collagen, and basement membrane (Najarian and Simmons, 1971) may be present in ALG. Examination of the ALG revealed that antibodies were present to some serum proteins (α_2 macroglobulin, lipoproteins and transferrin) in sufficient concentration to make the precipitin line clearly visible.

Immuno-Electrophoresis

Human serum was placed in the central well. After electrophoresis ALG was placed in the upper trough and polyvalent rabbit anti-human serum protein antiserum in the lower trough. At least four precipitating antibodies to human serum proteins are present in the ALG.

DISCUSSION

The development of a shock-like state following the administration of ALG or ALS to humans is well recognised (Kashiwagi et al, 1968). It has been attributed either to the presence of human reaginic or precipitating antibodies, or to aggregated gammaglobulin in the ALG. Soon after the discovery of bradykinin in 1950, Beraldo found that shock-like states in dogs were associated with free kinins in the blood. Other shock-like states have been associated with activation of the kininogen-kinin system (Webster & Clark, 1959), and with a decrease of kininogen level of the blood (Diniz et al, 1961: Diniz & Carvallho, 1963). When, therefore, the patient, P.D., developed an anaphylactoid reaction after the rapid i.v. administration of ALG it appeared likely that there might be involvement of the kininogen-kinin system. This was found to be so and during a second reaction, a depletion of kininogen and an eightfold rise of plasma kinin level occurred. The kininogen depletion occurred when ALG was administered either at a fast (25-100 mg/minute) or a slow rate (2 mg/minute), but a significant rise in free plasma kinins only occurred when the rate of ALG administration was fast.

A rise of plasma kinin from 4 ng/ml to 32 ng/ml during the anaphylactoid reaction would appear to be sufficient to explain the signs and symptoms of the reaction, but the means by which kinin liberation occurred was not clear.

Reaginic antibodies (Mongar & Perera, 1962) may produce kinin liberation. In the patient, P.D., this was not the mechanism because:

(i) There was no positive skin reaction to intradermal ALG
(ii) The P.C.A. test in monkeys was negative
(iii) The occurrence of a second reaction immediately after the administration of 6,000 mg ALG in five days, and
(iv) The failure of horse serum to produce *in vitro* kinin liberation.

Antigen-antibody precipitates may provoke kinin liberation (Movat et al, 1968). However, the failure of horse serum to liberate kinins from whole venous blood of the patient and the normal subject eliminated this possibility.

Antilymphocyte globulin is directed against the mononuclear cells, and therefore some ALG-cell interaction may activate the kininogen-kinin system as already described in trypanosomiasis (Boreham and Goodwin, 1970). In this situation small amounts of kinins were liberated from the plasma substrate by trypanosome-antibody complexes, but not by the trypanosomes alone or by the rabbit-immune serum alone. Another possible pathway for kinin liberation is the cytotoxic effect upon cells causing liberation of intracellular enzymes which may activate the system (Greenbaum and Kim, 1967). The results obtained by incubating the mononuclear cells with ALG did not suggest that a cell-ALG interaction played a role in the *in vitro* liberation of kinins.

From the failure of ALG to generate kinins from fresh goat plasma it may be inferred that this is not a non-specific effect. Antibodies to human serum proteins were present in the ALG and demonstrable by gel diffusion tests. Movat et al (1968a, 1968b) have shown that the addition of antigen-antibody complexes to fresh guinea pig serum will lead to the formation of kinins. Therefore, horse antibody human protein complexes may well be responsible for the kinin release demonstrated *in vitro* and *in vivo*.

REFERENCES

1. Barandun, S., Kistler, P., Jeunet, F., and Isliker, H. (1962) Vox Sang. 7, 157.

2. Beraldo, W.T., Am. J. Physiol. (1950) 163, 283.

3. Boreham, P.F.L., and Goodwin, L.G. 'Bradykinin and Related Kinins' edited by F. Sicuteri, M. Rocha e Silva and N. Back, p. 539, New York - London, 1970.

4. Brocklehurst, W.E. and Lahiri, G.C., J. Physiol. (1962) 160, 15.

5. Coombs, R.R.A. and Gell, P.G.H. in Clinical Aspects of Immunology, 2nd Edition, Ch. 20, p. 575, Ed: Gell and Coombs, Blackwell, Oxford, 1968.

6. Diniz, L.R. and Carvallho, J.F., Ann. N.Y. Acad. Sci., (1963), 104, 77.

7. Greenbaum, L.M. and Kim, K.S. Brit. J. Pharmacol. (1967), 29, 238.

8. Kashiwagi, N., Brantigan, C.O., Brett Schneider, L., Groth, C.G., and Starzl, T.E. Annals of Int. Med. (1968), 68, 275.

9. Layton, L.L., Yamanaka, E., Greene, F.C. and Perlman, N.F., Int. Arch. Allergy (1963) 23, 87.

10. Mongar, J.L. and Perera, B.A.V. J. Physiol. (1962) 162, 70.

11. Movat, H.Z. and Di Lorenzo, N.L. Lab. Invest. (1968) 19, 187.

12. Movat, H.Z., Di Lorenzo, N.L. and Treloar, M.P. Lab. Invest. (1968) 19, 201.

13. Najarian, J.S. and Simmons, R.L. New Eng. J. Med. (1971) 285, 158.

14. Talamo, R.C., Haber, E., and Austen, K.F. J. Lab. Clin. Med., (1969) 74, 816.

15. Webster, M.E. and Clark, W.R. Am. J. Physiol. (1959) 197, 406.

ACTIVATION OF BRADYKININ SYSTEM IN ACUTE INFLAMMATION INDUCED BY ENDOGENOUS OR EXOGENOUS AMINES

Á.Gecse, E.Zsilinszky, J.Lonovics, L.Szekeres

Institute of Pathophysiology and Dermatologic Clinic,University of Szeged School of Medicine

Szeged,Hungary

The clinicopathological characteristics of acute inflammation are same in spite of the well known fact that several vasoactive mediator should play some role in the pathogenesis of inflammation /1,2,3,6,7,8/.The question is raised how is it possible that such a different injuries as 46.5°C thermic effect,freezing at -70°C,trypsin induce inflammation with similar features. This question should be explained well if a final common material should be found at the place of inflammation,locally.This substance should be able to induce itself the characteristic features of inflammation.

During our experiments the inflammation was not elicited by the above mentioned injuries with complex effect.Acute inflammation was induced in one hind paw of rats of R-Amsterdam strain weighing to 160-180 g,by subplantar injection of 100 µg histamine,1.0 µg serotonin or 100 µg dextran.Haematocrit determinations were carried out in each case and the results obtained in plasma were corrected with haematocrit.

Oedema formation was observed 30 min after the administration of vasoactive substances.The results are presented on Fig.1.The oedema formation was determined as mg weight increase of inflamed hind paw compared to the control one.The amount of injected vasoactive materials was choosen to induce similar inflammatory reaction.The weight increase was about 400-600 mg 30 and 60 min after the injection of histamine,5-HT and dextran, respectively.

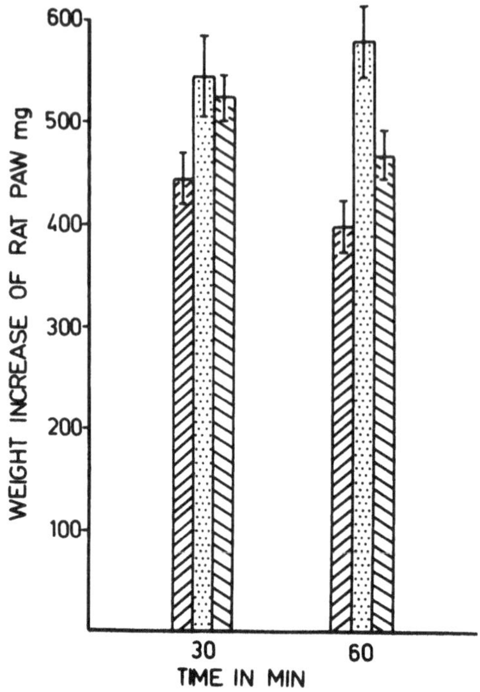

Fig.1. Oedema formation induced by histamine,serotonin and dextran. ▨ histamine, ⛶ 5-HT, ▧ dextran

The protein content was determined both in the plasma and in the interstitial fluid of inflamed rat paw which was previously perfused with cold saline to remove the blood from the vessels.The results are shown on Fig.2.While the plasma protein content decreased significantly at the same time the amount of proteins increased in rat paw 30 min after the injection of vaso-active substances.Further elevation was not observed in the 60th min.

Polyacrylamide gel electrophoresis was carried out to determine which protein components leaked to the extracellular extravascular compartment of paw during inflammation.The protein leakage involved not only the small molecular weight prealbumins,albumin but the alpha-2 macroglobulin and beta-1 lipoprotein

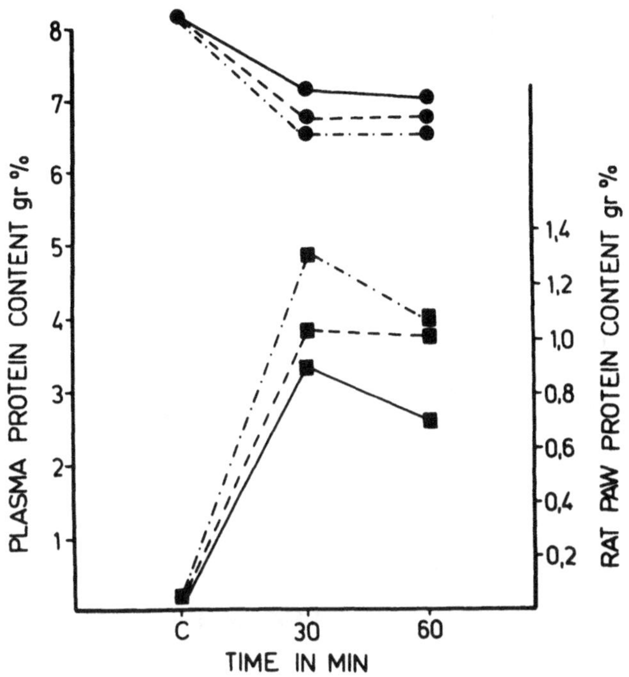

Fig.2.Changes of rat plasma and rat paw protein content after s.pl. injection of vasoactive substances. C is the control, ● plasma protein values, ■ rat paw protein values, —— histamine, --- 5-HT, -·-·- dextran.

with high molecular weight.The alpha-2 macroglobulin pathophysiological importance was discussed earlier /4/.The protein picture was similar in the interstitial fluid of inflamed rat paw either histamine,5-HT or dextran was applied as inflammatory agent.

The appearence of proteins with different molecular weight in the extracellular,extravascular compartment of rat paw conveyed the suggestion that the protein components of bradykinin system should also be there.

On Fig.3 the bradykininogen content of rat plasma and that of inflamed rat paw extract is presented.

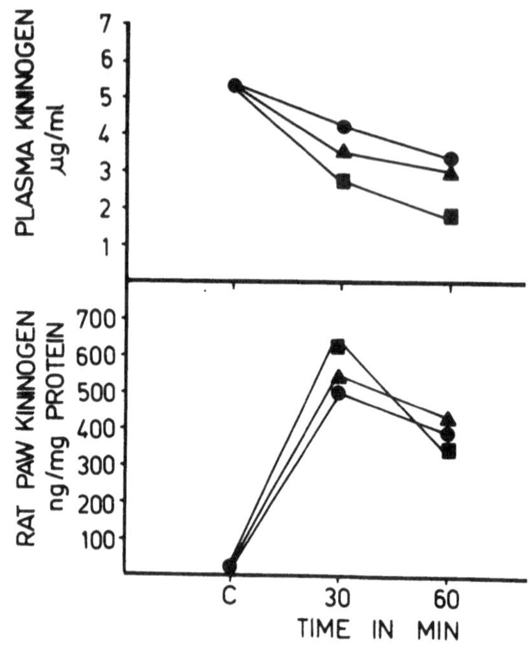

Fig.3.Kininogen content of rat plasma and that of rat paw after s.pl. injection of vasoactive substances. The bradykininogen is expressed as bradykinin formed by trypsin from plasma or rat paw extract.Trasylol was applied to antagonise the effect of trypsin.The upper figure shows the results obtained in plasma and the lower one presents the values determined in paw extract. ●——● histamine, ▲——▲ dextran, ■——■ serotonin.

The amount of plasma bradykininogen significantly decreased 30 or 60 min after the application of vasoactive substance compared to control.Insignificant amount of kininogen was found in the interstitial fluid of control paw.Huge amount of bradykininogen appeared in the extravascular extracellular compartment of paw 30 min after the injection of vasoactive materials.The decrease of kininogen level in plasma and the appearence of kinin substrate in inflamed rat paw gives only indirect evidence that the plasma-kinin system plays important role in the oedema formation induced by the amines.

For any substance to be considered to have particular mediator function certain criteria have to be met. Those which are frequently emphasized are the followings: 1./evidence for local formation or release of the substance under the conditions in which the response are observed and 2./the reproduction of the response qualytatively at least,by the hypothetical mediator substance.The first criterion arised the suggestion of the determination of bradykininogenase locally at the site of inflammation.

The results are presented on Fig.4.The activity of plasma bradykininogenase is illustrated on the upper figure,while the paw kinin-forming enzyme activity is presented on the lower one.

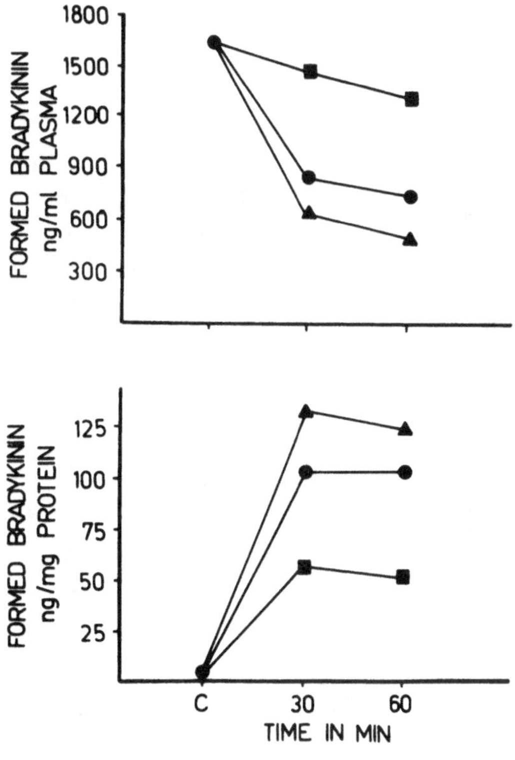

Fig.4.The activity of bradykininogenase in rat plasma and in rat paw.
The formed bradykinin is presented on the ordinate ng/ml glass activated plasma or ng/mg protein.Rat plasma kininogen was used as substrate for paw kininogenase.Cu-EDTA complex was applied as kininase antagonist.
●——●histamine,▲——▲serotonin,■——■dextran.

The activity of plasma kininogenase was found to be markedly reduced during inflammation either it was induced by histamine,serotonin or dextran.The control, not inflamed rat paw did not contain kininogenase while high amount of kinin-forming enzyme was found in the interstitial fluid of inflamed rat paw.

The activity of esterase was also determined by the method of Trautschold and Werle /9/ modified by Movat /personal communication/.The results are presented on Fig.5.30 min after the s.pl.injection of histamine,5-HT,dextran significant amount of enzyme was found at the place of inflammation.This enzyme should have come partly or totally from plasma because the enzyme activity was found to be significantly reduced there.

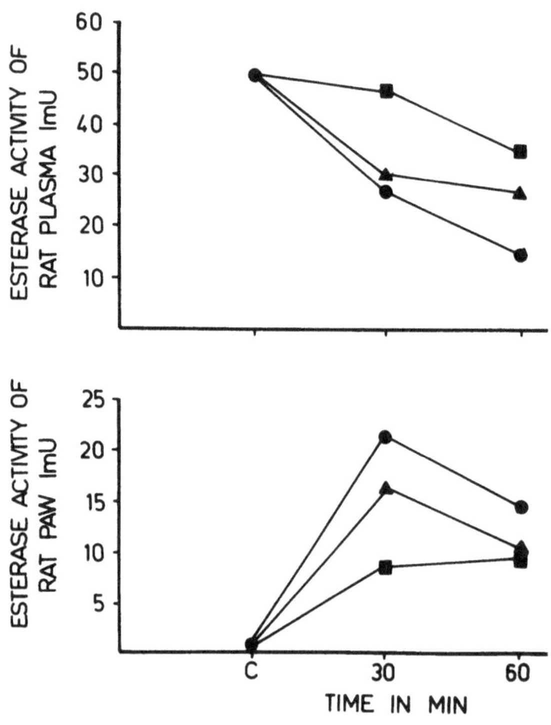

Fig.5.The activity of esterase enzyme of rat plasma and rat paw.
The upper figure showes the results in plasma and the lower one presents the values obtained in rat paw.BAEe was applied as substrate.●——●histamine,▲——▲5-HT, ■——■dextran.

The extravascular presence of bradykininogen and that of kinin-forming enzyme provides for the longer continuance of inflammation while this inflammatory reaction was found to be finished after a few hours. One answer is presented on Fig.6,where the specific activity of bradykininase is illustrated.Large amount of enzyme was found during inflammation.80-100 μg synthetic bradykinin was destroyed by the kininase during one min.It is probable that this enzyme plays important role in the suspension of inflammatory reaction.

The bradykinin system similar activation was observed during generalized dextran anaphylactoid reaction of rats.

The C-phenylglycine-n-heptyl ester /5/ by means of its antihistamine,antiserotonin effect inhibited the activation of kinin system during acute inflammation.

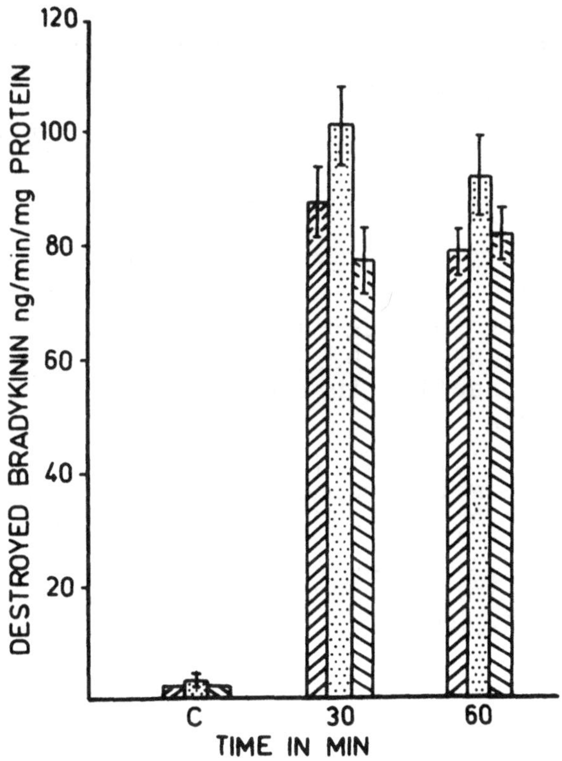

Fig.6.The activity of bradykininase in inflamed rat paw. C:control values,▨histamine,⚃serotonin,▧dextran.

Summarising our results it seems most likely that the histamine,serotonin and their liberator,dextran are only triggers of acute inflammation and the bradykinin system extravascular activation is an important factor in the oedema formation induced by the amines.

REFERENCES

1./ ARMSTRONG,D.,JEPSON,J.B.,KEELE,C.A.,STEWART,J.W.: Pain producing substance in human inflammatory exudates and plasma.J.Physiol./Lond./ 135:350-370 /1957/

2./ BROCKLEHURST,W.E.:The release of histamine and formation of a slow-reacting substance /SRS-A/ during anaphylactic shock.J.Physiol./Lond./ 151:416-435 /1960/

3./ CSABA,B.,SZILÁGYI,T.,DAMJANOVICH,S.,KÖVÉR,A.:Anaphylactic shock and peptone shock in the dog.I.The role of histamine in anaphylactic shock.Acta physiol.Acad. Sci.hung. 23:363-369 /1963/

4./ GECSE,Á.,ZSILINSZKY,E.,KARÁDY,S.:New vasoactive polypeptides isolated from ox alpha-2 macroglobulin. Adv.Exp.Biol.Med. 8:307-313 /1970/

5./ GECSE,Á.,ZSILINSZKY,E.,LONOVICS,J.,WEST,G.B.: C-phenylglycine-n-heptyl ester as an inhibitor of mediators of allergic reactions.Int.Arch.Allergy 41:174-179 /1971/

6./ GIROUD,J.P.,WILLOUGHBY,D.A.:The interrelations of complement and a prostaglandin-like substance in acute inflammation.J.Path. 101:241-250 /1970/

7./ PARROT,J.R.,WEST,G.B.:5-hydroxitryptamine and the anaphylactoid reaction in the rat.J.Physiol./Lond./ 136:27-41 /1957/

8./ ROCHA E SILVA,M.,ANTONIO,A.:Release of bradykinin and the mechanism of production of a "thermic edema /45°C/" in the rat's paw.Med.exp. 3:371-382 /1960/

9./ TRAUTSCHOLD,I.,WERLE,E.:Spectrophotometrische Bestimmung des Kallikreins and seiner Inhibitoren. Hoppe-Seylers Z.physiol.Chem. 325:48-58 /1961/

A STRUCTURE-ACTIVITY BASIS FOR VASOTROPIC PEPTIDE THERAPY IN SHOCK*

Burton M. Altura † and S.G. Hershey

Departments of Anesthesiology and Physiology, Albert Einstein College of Medicine, Bronx, New York 10461

Chemical analogues which enhance one or several biologic properties of important naturally occurring substances are regularly sought. Currently in this regard, there is intense interest in developing such analogues of the neurohypophyseal hormones (NHPH), oxytocin and vasopressin. Several which have recently been synthesized not only offer the possibility of increased insight into structure-activity relationships but also may provide for more precise therapeutic drug actions (17).

Vasopressin, an octapeptide normally present in the posterior pituitary gland, has two principal pharmacodynamic actions, one being to inhibit the renal excretion of osmotically free water and hence called the ADH or antidiuretic hormone effect, and the other being its pressor effect due to its constrictor action on the smooth muscle of the peripheral blood vessels. Of these, the antidiuretic action is currently thought to be more prominent in that it supposedly becomes evident with lower doses than the pressor effect (20, 30). But this latter, long established tenet now appears to be in need of some revision in light of recent studies (3, 5, 6, 8, 10). In addition, several vasopressin analogues have now been synthesized which exhibit increasingly selective vasoconstrictor actions (18, 19). In fact it is possible to raise the ratio of pressor to antidiuretic activity from approximately 1:1 for the naturally occurring human peptide (8-arginine vasopressin) to as great as 218:1 (2-phenylalanine-3-isoleucine-8-ornithine vasopressin) (19).

---

* This study was supported by research grants HE-12462 and HE-11391 from the National Institutes of Health, USPHS, Bethesda, Maryland.

† Recipient of Research Career Development Award 5-K3-GM-38, 603 from the N.I.H., USPHS.

Until the isolation and synthesis of vasopressin in 1953 by du Vigneaud (31,32) vasopressin was given little serious thought as a useful vasoactive drug because of its well-documented coronary constrictor action (20, 30). Interestingly, however, there is good evidence to indicate that at least two synthetic vasopressin analogues, PLV-2 (2-phenylalanine-8-lysine vasopressin) and POR-8 (8-ornithine vasopressin), may have significant antiarrhythmic properties in man (25-29). In addition, one of these analogues, namely PLV-2, was found to increase survival of rats and rabbits after several types of experimental shock (8, 13, 14, 16, 23, 24) as well as to modify reticuloendothelial system phagocytic function in normal and shocked animals (9, 11). Direct in-vivo microscopy, of the microcirculation, showed that PLV-2 maintained arteriolar and venular tone, sustained vasomotion, decreased shock-induced hypersensitivity to catecholamines and prevented microvascular injury (13, 14, 24). Thus, the latter observations suggested that the selective microvascular actions of PLV-2 might account for its dramatic effectiveness in increasing survival of animals in several types of experimental shock. These findings prompted the present structure-activity studies to determine: 1. Whether the pharmacologic effects and sites of action (arterioles, precapillary sphincters, metarterioles, venules, etc.) of these neurohypophyseal hormones (NHPH) in the microcirculation undergo change as their chemical structures are modified. And 2. Whether a favorable or unfavorable influence on survival can be accurately predicted from the actions of these peptides on the microcirculation.

METHODS

The studies reported below are limited to our findings in normal and shocked rats. Direct in-vivo microscopic observations at magnifications up to 1500 times were performed on mesenteries of lightly anesthetized rats (2). Different groups of rats were subjected to either hemorrhagic or intestinal ischemia shock using techniques similar to that described previously (14, 24). Briefly, in the case of hemorrhage, the animals were bled in a graded fashion over a 20-30 minute period, 3% by body weight. The blood was withheld from the animals for a 2 hour period, after which time 2 ml of normal Ringer's alone or containing one of four different vasopressin analogues in equi-pressor doses was then infused i.v. at a constant rate for 50 minutes. At the conclusion of infusion the shed blood was reinfused intra-arterially over a 20-30 minute period. Blood pressure was continuously monitored during these experiments and in the intestinal ischemia shock experiments. Intestinal ischemia shock was induced in rats by temporary ligation of the superior mesenteric artery for 60 minutes. One hour after removal of the arterial tie, 2 ml of Ringer's alone or containing the peptides was infused i.v. as in the hemorrhage experiments. All animals were observed for survival for 5 days. Mean survival time of the animals that eventually died was also monitored.

MICROCIRCULATORY ACTIONS OF NEUROHYPOPHYSEAL PEPTIDES

Table 1 summarizes the microcirculatory effects of the four different vasopressin analogues, namely, PLV-2, AVP (8-arginine vasopressin), POR-8

Table 1. Neurohypophyseal Peptide Constrictor Gradients of Reactivity in Rat Mesenteric Microcirculation*

Vessel Type	Vessel Size (μ)	Stimulus and order of vcs† AVP†† PLV-2	POR-8	POX
Precapillary Sphincter	6-12	2**	2	1
Metarteriole	15-20	3	1	2
Arteriole	20-33	4	4	4
Venule	48-90	1	3	3

* Data based upon compilation of results in present report and previous observations (see refs. 5, 6, 8, 10).

† vcs = vasoconstriction

** Numbers 1-4 refer to the relative order of sensitivity of the muscular microvessels towards a particular type of NHPH analogue

†† AVP = 8-arginine vasopressin; PLV-2 = 2-phenylalanine-8-lysine vasopressin; POR-8 = 8-ornithine vasopressin; POX = 2-phenylalanine-3-isoleucine-8-ornithine vasopressin.

and POX-85 (2-phenylalanine-3-isoleucine-8-ornithine vasopressin). Local application of either PLV-2 or AVP to the normal rat mesentery produces only venular constriction. Increasing doses of either of these latter two molecules yields a venular to arteriolar gradient of reactivity in the microcirculation which is the reverse of that exhibited by constrictor catecholamines (i.e., arteriolar to venular) (2, 5, 6, 33). Unlike the constrictor catecholamines, (e.g., epinephrine or norepinephrine) (6) or serotonin (unpublished findings), supra-maximal doses of PLV-2 do not produce occlusion of either the muscular

Table 2. Influence of PLV-2 on Survival of Rats after Hemorrhagic Shock*

Group	Milli-Units/min.	Survivors/ Total Rats	% Survivors	Survival Time (hrs.)
Control	2.0ml Ringers	6/16	37	40 ± 2.1
PLV-2†	0.245	8/17	47	47 ± 2.6**
Control	2.0ml Ringers	4/17	25	33 ± 3.8
PLV-2	0.49	16/32	50**	43 ± 0.8**
Control	2.0ml Ringers	8/18	44	22 ± 2.2
PLV-2	0.98	16/20	80**	37 ± 1.0**
Control	2.0ml Ringers	6/18	33	25 ± 2.4
PLV-2	1.96	13/20	65**	40 ± 2.5**

* Animals bled intra-arterially, 3% by body weight, over a 30 min. period. Blood was withheld from the animals for 2 hrs. PLV-2 in Ringers or Ringers alone was then infused i.v. over 51 min. followed by return of shed blood. All animals were observed for 5 days for survival.

† PLV-2 = 2-phenylalanine-8-lysine vasopressin (1 U = 18.2μg of active peptide).

** Significantly different from paired controls ($P < 0.05$) (Chi-square test).

Table 3. Influence of PLV-2 on Survival of Rats after Intestinal Ischemia Shock*

Group	Milli-Units/min.	Survivors/ Total Rats	% Survivors	Survival Time (hrs.)
Control	2.0ml Ringers	6/29	21	26.7 ± 2.3
PLV-2†	0.49	6/29	21	38.0 ± 4.0**
Control	2.0ml Ringers	4/25	16	27.7 ± 4.4
PLV-2	0.98	9/25	36††	44.4 ± 5.2**
Control	2.0ml Ringers	15/39	39	29.6 ± 2.9
PLV-2	1.31	19/29	66††	24.3 ± 2.3
Control	2.0ml Ringers	17/49	35	22.1 ± 1.8
PLV-2	1.63	22/39	56††	27.7 ± 3.1
Control	2.0ml Ringers	12/50	24	25.1 ± 2.0
PLV-2	2.04	12/25	48††	27.4 ± 2.7
Control	2.0ml Ringers	34/117	29	31.8 ± 2.7
PLV-2	3.06	33/69	48††	29.6 ± 3.5
Control	2.0ml Ringers	13/37	35	28.7 ± 2.6
PLV-2	6.12	8/27	30	24.9 ± 1.6

* Temporary ligation of superior mesenteric artery for 60 min. Survival determined at 5 days.

† PLV-2 = 2-phenylalanine-8-lysine vasopressin (1 U = 18.2μg of active polypeptide).

** Significantly different from paired controls ($P < 0.02$).

†† Significantly different from paired controls ($P < 0.05$).

Table 4. Influence of AVP on Survival of Rats after Hemorrhagic Shock*

Group	Milli-Units/min.	Survivors/ Total Rats	% Survivors	Survival Time (hrs.)
Control	2.0ml Ringers	4/18	22	31 ± 8.3
AVP†	0.245	3/20	15	39 ± 0.8
Control	2.0ml Ringers	4/14	29	36 ± 1.5
AVP	0.49	7/16	44	44 ± 6.2
Control	2.0ml Ringers	10/22	44	36 ± 2.3
AVP	0.98	11/23	48	28 ± 2.2**
Control	2.0ml Ringers	8/18	44	35 ± 1.2
AVP	1.96	2/18	11**	30 ± 1.1††

* Animals bled intra-arterially, 3.0% by body weight, over a 30 min. period. Blood was withheld from the animals for 2 hrs. AVP in Ringers or Ringers alone was then infused i.v. over 51 min. followed by return of shed blood. All animals were observed for 5 days for survival.

† AVP = 8-arginine vasopressin (1 U = 2.5μg of active polypeptide)

** Significantly different from paired controls ($P < 0.05$).

†† Significantly different from paired controls ($P < 0.01$).

Table 5. Influence of 8-Ornithine Vasopressin on Survival of Rats after Intestinal Ischemia Shock*

Group	Milli-Units/min.	Survivors/ Total Rats	% Survivors	Survival Time (hrs.)
Control	2.0ml Ringers	6/29	21	26.7 ± 2.3
POR†	0.49	13/30	43**	28.2 ± 3.3
Control	2.0ml Ringers	4/25	16	27.7 ± 4.4
POR	0.98	8/28	28	37.3 ± 4.5
Control	2.0ml Ringers	15/39	39	29.6 ± 2.9
POR	1.31	13/27	48	29.1 ± 3.0
Control	2.0ml Ringers	17/49	35	22.1 ± 1.8
POR	1.63	19/39	49	29.0 ± 3.5
Control	2.0ml Ringers	12/50	24	25.1 ± 2.0
POR	2.04	11/33	33	18.3 ± 1.9**
Control	2.0ml Ringers	34/117	29	31.8 ± 2.7
POR	3.06	21/56	38	25.0 ± 1.8**
Control	2.0ml Ringers	13/37	35	28.7 ± 2.6
POR	6.12	6/21	29	21.7 ± 2.1**

* Temporary ligation of superior mesenteric artery for 60 min. Survival determined at 5 days.

† POR = 8-ornithine vasopressin (1 U = 2.8µg of active polypeptide).

** Significantly different from paired controls ($P < 0.05$).

venules, arterioles or metarterioles (13). In other words, one could say that there is a ceiling on vasoconstriction with PLV-2. Supra-maximal doses of AVP, however, produce occlusion of arterioles and metarterioles as well as stasis and microvascular injury in the postcapillary venules. We have observed similar phenomena on many different types of isolated blood vessels from dogs and rats (3, 5, 7). That is, different NHPH analogues not only exhibit different relative affinities for the NHPH receptor, but exhibit significantly different maximal contractile responses. Such a molecule as AVP might, therefore, be expected to be of little value in the treatment of shock in rats. In point of fact this is so (see below).

Insertion of the amino acid ornithine in the 8th position of the vasopressin molecule, instead of the highly basic amino acid arginine, produces a gradient of reactivity in the rat mesenteric microcirculation which is entirely different from either AVP or PLV-2 (Table 1). Threshold doses of this molecule, Orn^8-vasopressin (POR-8), constrict metarterioles first and then precapillary sphincters. As the topical dose of this peptide is increased, the muscular venules constrict before the arterioles. High topical doses of Orn^8-vasopressin, however, occlude the arterioles similar to AVP, catecholamines and serotonin. Supra-maximal doses of Orn^8-vasopressin fail to occlude either the metarterioles or precapillaries (unpublished findings). But such occlusive effects on the arterioles, if seen in shocked animals, might be expected to prevent blood from entering either the capillaries or postcapillary venular vessels, thereby resulting in a molecule having little potential value in low-flow states (see below).

Table 6. Influence of 2-Phenylalanine-3-Isoleucine-8-Ornithine Vasopressin on Survival of Rats after Intestinal Ischemia Shock*

Group	Milli-Units/min.	Survivors/ Total Rats	% Survivors	Survival Time (hrs.)
Control	2.0ml Ringers	6/29	21	26.7 ± 2.3
POX†	0.49	7/29	24	33.8 ± 3.7
Control	2.0ml Ringers	4/25	16	27.7 ± 4.4
POX	0.98	5/24	21	36.8 ± 5.0
Control	2.0ml Ringers	15/39	39	29.6 ± 2.9
POX	1.31	15/32	47	24.2 ± 3.1
Control	2.0ml Ringers	17/49	35	22.1 ± 1.8
POX	1.63	10/39	26	27.4 ± 2.3
Control	2.0ml Ringers	12/50	24	25.1 ± 2.0
POX	2.04	5/50	10**	20.1 ± 1.5**
Control	2.0ml Ringers	34/117	29	31.8 ± 2.7
POX	3.06	8/58	14**	23.7 ± 2.6**
Control	2.0ml Ringers	13/37	35	28.7 ± 2.6
POX	6.12	3/26	12**	23.3 ± 1.1**

* Temporary ligation of superior mesenteric artery for 60 min. Survival determined at 5 days.

† POX = 2-phenylalanine-3-isoleucine-8-ornithine vasopressin (1 U = 8.3μg of active polypeptide).

** Significantly different from paired controls ($P < 0.05$).

Although local administration of 2-phenylalanine-3-isoleucine-8-ornithine vasopressin (POX-85) to the rat mesenteric microcirculation (Table 1) induces a different gradient of microvascular reactivity (i.e., precapillary — metarteriolar — venular to arteriolar) (5, 8), it is extremely potent and in low doses not only occludes precapillaries and metarterioles but arterioles and venules as well. Such a molecule should, at certain dose levels, exacerbate shock mortality since blood flow is not only prevented from entering the true capillaries (a necessity for tissue nutrition) but is prevented from returning to the heart by intense venular vasoconstriction.

INFLUENCE OF NHPH ANALOGUES IN EXPERIMENTAL SHOCK

Correlation of the above microvascular findings with established microcirculatory criteria for nutritional blood flow suggested that only PLV-2 could be expected to improve survival over wide dose ranges after shock. It was postulated that the other vasopressin analogues would either not influence survival or would probably predispose to increased mortality.

Tables 2 and 3 demonstrate that PLV-2 not only significantly increased survival rates by 100% after hemorrhagic and intestinal ischemia shock, over wide dose ranges, but also significantly increased the mean survival time of those animals that eventually went on to die. Each dose level of PLV-2 (as well as the other analogues below) raised blood pressure by successive 10 mmHg increments over Ringer-treated animals. Table 4 shows that doses of AVP, similar in potency to those of PLV-2 in terms of blood pressure, either exerted no effect, decreased mean survival time or actually exacerbated shock mortality after acute hemorrhage. Similar effects on survival rates were noted after intestinal ischemic shock (unpublished findings).

Although a very low dose of 8-ornithine vasopressin, second line in Table 5, did result in increased survival after intestinal ischemic shock, higher doses either resulted in no effect on survival or resulted in decreased survival time as demonstrated by the three last groups in the table. Similar findings were seen after hemorrhagic shock (unpublished findings).

Interestingly, 2-phenylalanine-3-isoleucine-8-ornithine vasopressin (as predicted from its microcirculatory actions) dose-dependently exacerbated mortality and decreased mean survival time after intestinal ischemia shock (Table 6). A similar result, namely exacerbation of mortality was observed after hemorrhagic shock (unpublished findings).

COMMENTS

The data presented here, as well as elsewhere (1, 3-16, 23, 24), strongly indicate that molecular changes in the chemical structure of vasopressin can yield vasotropic substances which will indeed exhibit discrete and selective microvascular effects. Our data not only suggest that the vasopressin group of polypeptides can produce selective microvascular actions, but clearly demonstrate that at least one such analogue, namely 2-phenylalanine-8-lysine vasopressin, can be effective over wide dose ranges in the treatment of highly lethal forms of experimental shock. Furthermore, these data indicate that analogues exerting primarily selective venular constrictor actions which are not occlusive, like PLV-2, are highly beneficial in the treatment of experimental shock, at least in rats and rabbits (8, 12-14, 16, 23, 24). Analouges (or drugs) which, however, either occlude arterioles, venules and/or precapillary sphincters and metarterioles may be detrimental in the treatment of shock. In addition, these experiments demonstrate that there is no correlation between mean blood pressure levels and survival since all of the neurohypophyseal peptides were administered in equi-pressor doses. That is to say, the pressor potency (at level of resistance arterioles) of these compounds is probably not as important as their microcirculatory actions relative to tissue perfusion (12, 13). Although clinical experience with PLV-2 in shock is as yet inadeauate for objective evaluation, the preliminary results look promising (21, 22). Finally, the present work when taken together with our previous observations on the microcirculation (5, 6, 8, 10, 13-15, 24) and on isolated blood vessels (1, 3-5, 7), strongly suggests that it may eventually be possible to accurately predict the value of synthetic analogues or homologues of vasoactive drugs in the treatment of low-flow states on the basis of their molecular structures and pharmacologic actions on small and large blood vessels.

CONCLUSIONS

Pharmacologic manipulation of the terminal vascular bed for correction of impaired tissue blood flow is a long recognized objective of shock and low-flow state therapy; failure of microcirculatory-dependent tissue perfusion is the accepted primary event leading to irreversibility. The cardiovascular dynamics which initiate and sustain low-flow relate essentially to an untenable balance in the pressure-resistance (P-R) relationships in critical portions of the peripheral circulation. It follows that use of vasotropic drugs which can rectify the unphysiologic P-R factors constitutes a valid therapeutic principle. The practical problem is not one of principle of therapy but availability of the appropriate vasoactive agents. The present study demonstrates that one such group of drugs, namely neurohypophyseal polypeptide analogues may, indeed, have the selective microvascular effects needed for pharmacologic manipulation of the microcirculation in low-flow states. The above studies suggest that it may eventually be possible to accurately predict the value of synthetic analogues of the neurohypophyseal hormones or of other classes of vasoactive drugs in the treatment of low-flow states solely on the basis of their chemical structures and profiles of microvascular actions.

REFERENCES

1. B.M. Altura, "Differential actions of polypeptides and other drugs on coronary inflow vessels," Am. Heart J., 72: 709-711, 1966.
2. B.M. Altura, "Evaluation of neurohumoral substances in local regulation of blood flow," Am. J. Physiol., 212: 1447-1454, 1967.
3. B.M. Altura, "Significance of amino acid residues in vasopressin on contraction in vascular muscle," Am. J. Physiol., 219: 222-229, 1970.
4. B.M. Altura, "Influence of magnesium and cysteine on vasopressin-induced contractions in various canine blood vessels," Experientia, 26: 1089-1090, 1970.
5. B.M. Altura, "Pharmacology of neurohypophyseal hormones and analogs on isolated vascular muscle and in the terminal vascular bed," in: J.A. Bevan, R.F. Furchgott, R.A. Maxwell and A.P. Somlyo, Eds., Physiology and Pharmacology of Vascular Neuroeffector Systems. Basel, Karger, pp. 274-290, 1971.
6. B.M. Altura, "Chemical and humoral regulation of blood flow through the precapillary sphincter," Microvascular Res., 3: October, 1971 (in press).
7. B.M. Altura, "Structure-activity relationships of neurohypophyseal polypeptides on different types of isolated mammalian blood vessels," in: These Proceedings.
8. B.M. Altura and S.G. Hershey, "Pharmacology of neurohypophyseal hormones and their synthetic analogues in the terminal vascular bed. Structure-activity relationships," Angiology, 18: 428-439, 1967.
9. B.M. Altura and S.G. Hershey, "Use of reticuloendothelial phagocytic function as an index in shock therapy," Bull. N.Y. Acad. Med., 43: 259-266, 1967.

10. B.M. Altura and S.G. Hershey, "Structure-activity relationships of neurohypophyseal polypeptides in the microcirculation: A molecular basis for shock therapy," in: B. Schildt and L. Thorén, Eds., Intermedes Proceedings 1968: Combined Injuries and Shock. Stockholm, Almqvist and Wiksell, pp. 247-259, 1968.
11. B.M. Altura and S.G. Hershey, "Influence of vasopressor drugs on reticuloendothelial phagocytic function in experimental shock," in: B. Schildt and L. Thorén, Eds., Intermedes Proceedings 1968: Combined Injuries and Shock. Stockholm, Almqvist and Wiksell, p. 185-193, 1968.
12. B.M. Altura and S.G. Hershey, "Structure-action relationships of vasopressin analogs in the microcirculation and their use in the treatment of experimental shock," Federation Proc., 29: 711, 1970.
13. B.M. Altura, S.G. Hershey and B.T. Altura, "Microcirculatory actions of polypeptides and their use in the treatment of experimental shock," Adv. Exptl. Med. Biol., 8: 239-247, 1970.
14. B.M. Altura, S.G. Hershey and V.D.B. Mazzia, "Microcirculatory approach to vasopressor therapy in intestinal ischemic (SMA) shock," Am. J. Surgery, 111: 186-192, 1966.
15. B.M. Altura, S.G. Hershey and B.W. Zweifach, "Effects of a synthetic analogue of vasopressin on vascular smooth muscle," Proc. Soc. Exptl. Biol. Med., 119: 258-261, 1965.
16. B.M. Altura, R. Hsu, V.D.B. Mazzia and S.G. Hershey, "Influence of vasopressors on survival after traumatic, intestinal ischemia and endotoxin shock in rats," Proc. Soc. Exptl. Biol. Med., 119: 389-393, 1965.
17. B. Berde, Ed., "Handbook of Experimental Pharmacology, vol. 23: Neurohypophysial Hormones and Similar Polypeptides." Berlin, Springer-Verlag, 1968.
18. B. Berde and R.A. Boissonnas, "Basic pharmacological properties of synthetic analogues and homologues of the neurohypophysial hormones," in: B. Berde, Ed., Handbook of Experimental Pharmacology, vol. 23: Neurohypophysial Hormones and Similar Polypeptides. Berlin, Springer-Verlag, pp. 802-870, 1968.
19. B. Berde, R.A. Boissonnas, R. Huguenin and E. Stürmer, "Vasopressin analogues with selective pressor activity," Experientia, 20: 42-44, 1964.
20. P. Brazeau, "Agents affecting the renal conservation of water," in: L.S. Goodman and A. Gilman, Eds., The Pharmacological Basis of Therapeutics. New York, Macmillan, pp. 874-885, 1970.
21. J.N. Cohn, F.E. Tristani and I.M. Khatri, "Systemic vasoconstrictor and renal vasodilator effects of PLV-2 (Octapressin) in man," Circulation, 38: 151-157, 1968.
22. S.G. Hershey and B.M. Altura, "Behandlung des Schocks durch Beeinflussung der peripheren Zirkulation mit vasoaktiven Wirkstoffen: eine mikrozirkulatorische Basis für die Therapie," Schweiz. Med. Wschr. 96: 1467-1471, 1516-1522, 1966 (2 parts).
23. S.G. Hershey, B.M. Altura and L.R. Orkin, "Therapy of intestinal ischemic (SMA) shock with vasoactive drugs," Anesthesiology, 29: 466-471, 1968.

24. S.G. Hershey, V.D.B. Mazzia, B.M. Altura and L. Gyure, "Effects of vasopressors on the microcirculation and on survival in hemorrhagic shock," Anesthesiology, 26: 179-189, 1965.
25. M.E. Hunter and R.A. Gordon, "Laboratory and clinical studies of 2-phenylalanine-8-lysine vasopressin (Octapressin)," Canad. Anaesth. Soc. J., 13: 40-45, 1966.
26. R.L. Katz, "Epinephrine and PLV-2: Cardiac rhythm and local vasoconstrictor effects," Anesthesiology, 26: 619-623, 1965.
27. R.L. Katz and G.J. Katz, "Surgical infiltration of pressor drugs and their interaction with volatile anesthetics," Brit. J. Anaesth., 38: 712-718, 1966.
28. O.V. Nielsen and N. Valentin, "Ornithine-8-vasopressin, a new vasoconstrictor used for haemostasis during operation for genital prolapse," Acta Obstet. Gynec. Scand., 49: 45-48, 1970.
29. K. Saameli, "The circulatory actions of the neurohypophysial hormones and similar polypeptides." Berlin, Springer-Verlag, pp. 545-612, 1968.
30. H.B. Ban Dyke, K. Adamsons and S.L. Engel, "Aspects of the biochemistry and physiology of the neurohypophyseal hormones. Recent Progr. Hormone Res., 11: 1-35, 1955.
31. V. du Vigneaud, H.C. Lawler and E.A. Popenoe, "Enzymatic cleavage of glycinamide from vasopressin and a proposed structure for this pressor-antidiuretic hormone of the posterior pituitary," J.Am. Chem. Soc., 75: 4880-4881, 1953.
32. V. du Vigneaud, C.H. Ressler, J.M. Swan, C.W. Roberts, P.G. Katsoyannis and S. Gordon, "The synthesis of an octapeptide amide with the hormonal activity of oxytocin." J. Am. Chem. Soc., 75: 4879-4880, 1953.
33. B.W. Zweifach, "Functional Behavior of the Microcirculation." Springfield, C.C. Thomas, 1961.

ACKNOWLEDGEMENTS

The authors are indebted to Miss J. Gargani, Mrs. C. Jablesnik, Miss Y. Waldemar and Mr. R.W. Burton for their very capable technical assistance throughout the course of these studies.

BRADYKININ METABOLISM IN SHOCK AND IN STATE OF SHOCK-RESISTANCE

S.Karády, J.Lonovics, E.Zsilinszky, Á.Gecse

Institute of Pathophysiology,University of Szeged School of Medicine,Szeged,Hungary

Because of the rather contradictory results in the literature with regard to the plasma bradykininogen content in shock of different etiology /1,3,5,7,8/ it seemed to us worthwhile to study the problem of brady-kinin metabolism in state of shock and in state of shock-resistance /stage of non specific resistance/.

The experiments were performed in 50 male albino rats of R-Amsterdam strain weighing 150-170 g.

Sublethal tourniquet-shock was induced by Stoner's /6/ method,the hind limbs of rats were excluded from the general circulation for two and a half hours.

The experiments were carried out in five groups of 10 animals.The severity of the shock was judged by determination the degree of drop in body temperature and rise in haematocrit value.The changes in the values of haematocrit and that of the total plasma protein concentration as well as that of the components of the kinin-system,that is the alterations of the blood level of kininogen,kininogenase and kininase were determined in the plasma.The blood samples were drawn from the aorta.

These parameters were determined in control rats /not subjected to tourniquet shock/,in state of shock /3 and 6 hours after the release of the tourniquet/ and in state of non specific shock-resistance/ 48 hours after the release of tourniquet/ and in another group

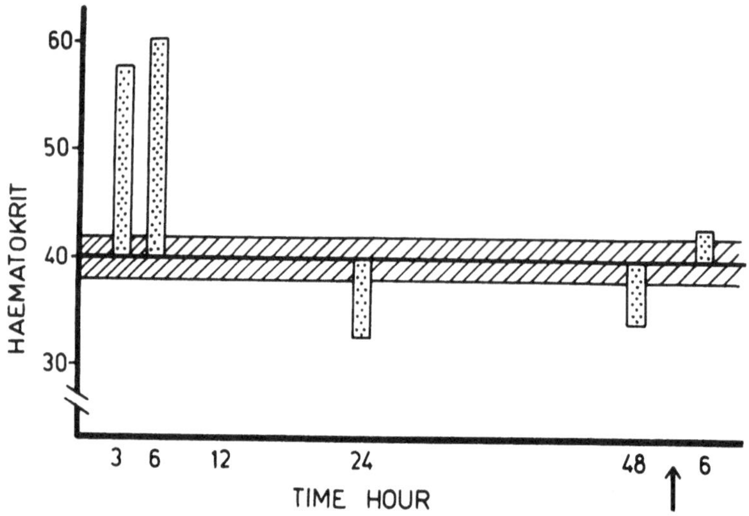

Fig.1.Changes of haematocrit in shock and in state of shock-resistance
The time elapsed after the release of tourniquet is presented on the abscissa.The arrow indicates the application of second tourniquet.The hatched area corresponds to the control value.

of rats rendered resistant 6 hours after the exposure to the second sublethal tourniquet stimulus.

The results are expressed in columns. The height of the columns indicates the mean value obtained in 10 rats belonging to the same group.For the statistical evaluation of the results the student "t" test was used. The error of standard deviation is also indicated on the figures.

Fig.1.shows the changes in the haematocrit values. One can see the significant haemoconcentration during shock and the significant haemodilution in the state of shock-resistance,and also in the latter state the slight haemoconcentration due to the repeated sublethal tourniquet 6 hours after its release.

The determination of the changes in haematocrit values seems to us important not only because this parameter expresses well the severity of the shock but also because the obtained results concerning other plasma parameters in such type of experiments,where haemoconcent-

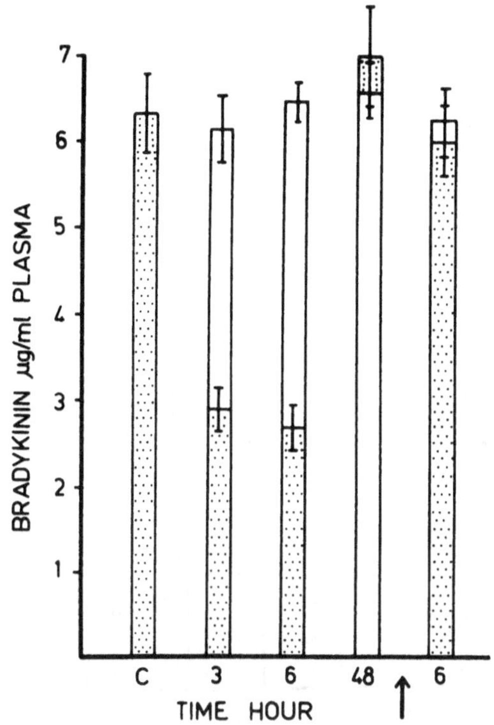

Fig.2.Plasma bradykininogen content of rats in state of shock and in state of shock-resistance.
C presents the control values. ⚃ values corrected with haematocrit, ☐ values not corrected with haematocrit.

ration and/or haemodilution can be expected,are misleading by not showing the true alterations without having them corrected according to the haemoconcentration or haemodilution occuring during the experimental procedure.

That can be well illustrated on the Fig.2.demonstrating the changes in the substrat's content of the kinin system,that is that of kininogen concentration, determined by the micromethod of Diniz /2/ and expressed as bradykinin ug/ml plasma without and after having made the necessary corrections of the obtained data according to the alterations of haematocrit values determined at the same time.

It appeares obvious,that the level of kininogen

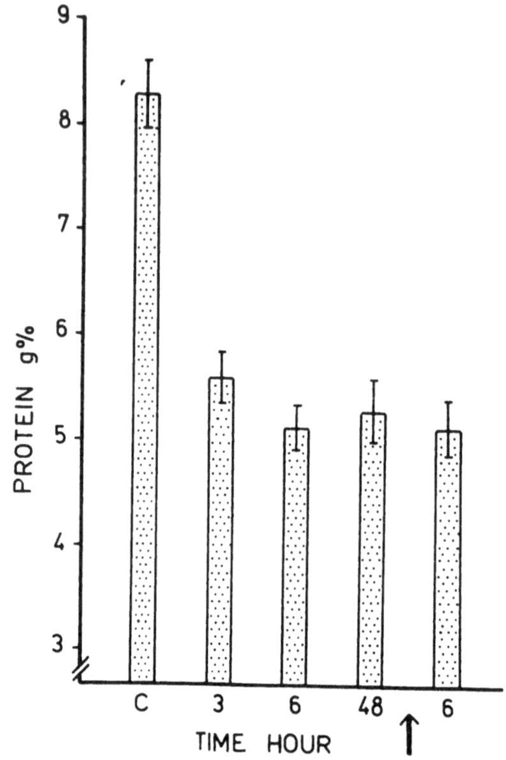

Fig.3.Changes of rat plasma total protein content in tourniquet shock.
The time elapsed after the release of tourniquet is presented on the abscissa.C indicates the control value.

significantly decreases in state of shock while in state of shock-resistance approaches the control value and is not significantly influenced by the application of the second sublethal tourniquet.

These experiments may also explain the discrepancies found in the shock literature as far as the plasma kininogen changes are concerned.

Urbanitz and Habermann /7/ state that the drop in plasma bradykininogen level in shock is not due to an increased kininogenase activity,it is rather a direct consequence of the loss of the total plasma protein taking place during shock.We have determined the alterations of the plasma total protein content in the course

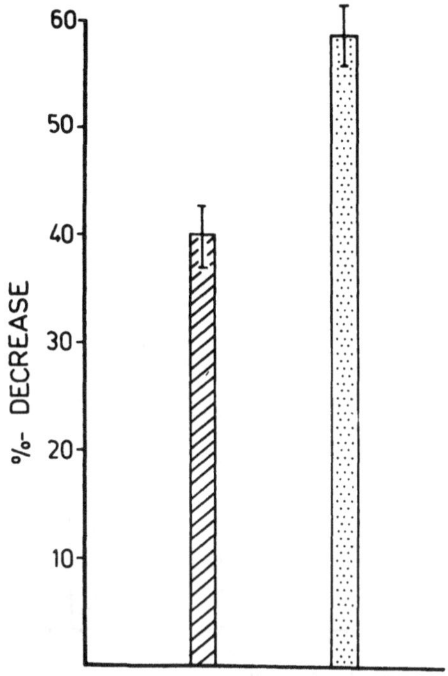

Fig.4.Percentage changes of plasma total protein and plasma bradykininogen in shock.
▨ total protein, ▩ plasma bradykininogen.

of tourniquet shock and in shock-resistance using Goa's /4/ micromethod.The results are shown on Fig.3.The columns represent the plasma total protein levels expressed in g%.

The Fig.4.demonstrates the results of the percentile comparison of the decrease in total plasma protein on the one part and that in plasma bradykininogen on the other.

It seems to be obvious from this figure that the percentage decrease in plasma bradykininogen is more significantly enhanced than that of the total plasma protein proving that the significant diminution of the plasma kininogen content cannot be explained simply by the loss of the total plasma protein,it is caused very

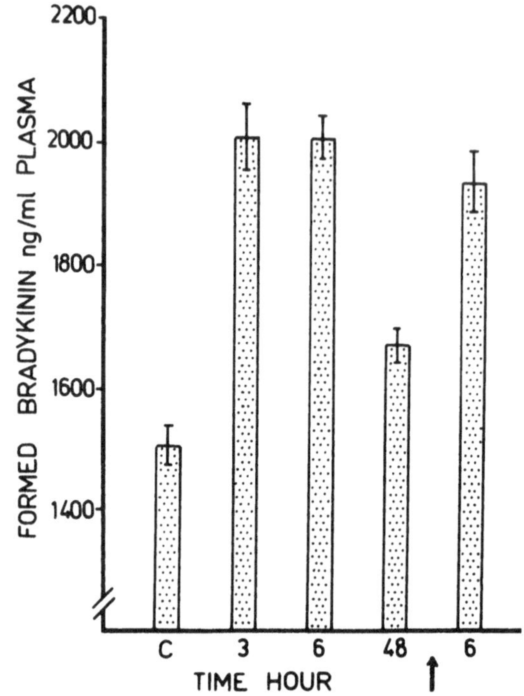

Fig.5.Alterations of plasma bradykininogenase activity in shock and in state of shock-resistance.
The time elapsed after the release of tourniquet is presented on the abscissa.C indicates the control value. The arrow illustrates the second application of tourniquet.

probably by the increased kininogenase activity shown on the Fig.4.

Fig.5.shows the changes in plasma kininogenase activity during shock and in state of shock-resistance. The values of the determination of kininogenase are expressed as formed bradykinin ng/ml plasma.

It can be seen that in shock-state the kininogenase activity was significantly enhanced compared to the control value,while the increase in the activity during the state of resistance was not significant,even if repeated sublethal tourniquet was applied.

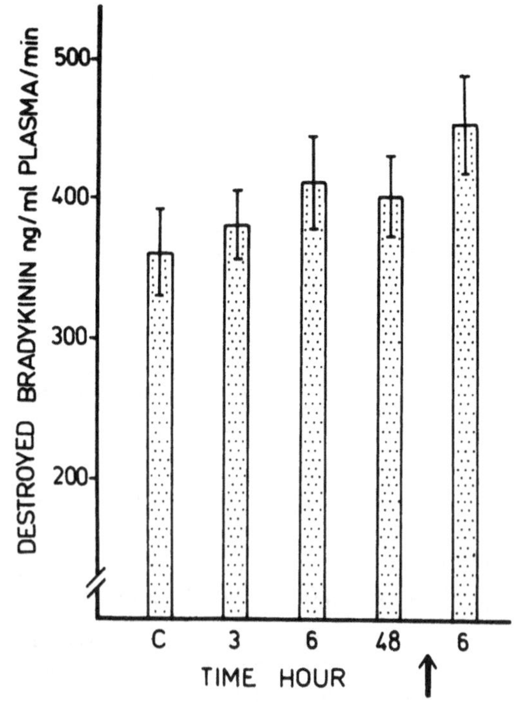

Fig.6.The activity of plasma kininase in shock and in state of shock-resistance.
C presents the control values.The arrow indicates the second application of tourniquet.

Fig.6.demonstrates the alteration of the plasma kininase activity during shock and in state of shock-resistance.The kininase activity was expressed as synthetic bradykinin ng destroyed by 1.0 ml plasma during 1 minute at 37°C and pH 8.0.

There was no significant change in the activity of plasma kininase in the course of tourniquet shock with the exception of the slight increase in kininase activity in state of resistance.

Summarizing the results of our investigations on bradykinin metabolism obtained in rats exposed to sub-lethal tourniquet shock using Stoner's method and in state of non specific shock resistance.It could be established that the plasma bradykininogen level decreases very significantly during shock /in higher degree than the decrease in total plasma protein content/. Plasma kininogenase activity was found to be significantly enhanced during shock, while the activity of plasma kininase did not exhibit any alteration. In the state of non specific shock-resistance the plasma changes in the components of the kinin system characteristic of shock have not been found.

REFERENCES

1./ ANKIER,S.I.,STARR,M.S.:The importance of plasma kinins in the anaphylactoid reaction in rats. Brit.J.Pharmacol. 31:331 /1967/

2./ DINIZ,C.R.,CARVALHO,I.F.:A micromethod for determination of bradykininogen under several conditions. Ann.N.Y.Acad.Sci. 104:77 /1963/

3./ GECSE,Á.,KARÁDY,S.,STARR,M.S.,WEST,G.B.:Release of bradykinin in shock.J.Physiol./Lond./ 178:8 /1965/

4./ GOA,J.:A microbiuret method for protein determination;determination of total protein in cerebrospinal fluid.Scand.J.clin.Lab.Invest. 5:218 /1953/

5./ STARR,N.S.,WEST,G.B.:Further evidence for the involvement of kinin in anaphylactic shock in the rat. Brit.J.Pharmacol. 37:178 /1969/

6./ STONER,H.B.:Studies on the mechanism of shock.The influence of environment on the changes in oxygen consumption,tissue temperature and blood flow produced by limb ischaemia.Brit.J.exp.Path. 39:251 /1958/

7./ URBANITZ,D.,HABERMANN,E.:In vivo investigations on the role of the kinin system in tissue injury and shock syndromes.Pharm.Res.Commun.1:201 /1969/

8./ WENT,M.,CSABA,B.:The role of bradykinin in anaphylactic shock in the rat.Acta physiol.Acad.Sci.hung. /in press/.

ACID DEPENDENT KININ-FORMING SYSTEM[1] IN MAMMALIAN MALIGNANT AND NORMAL TISSUE

Nathan Back and Robert Steger

Department of Biochemical Pharmacology
School of Pharmacy, State University of
New York at Buffalo, Buffalo, New York

Previous studies have demonstrated that a mammalian tumor (Murphy-Sturm lymphosarcoma) contains the following components of a vasoactive kinin-forming system: kininogen, a kinin-destroying enzyme, and a pre-kallikrein proenzyme capable of being activated by acetone (1). Levels of kininogen increased during tumor development as did kininase activity. Acetone-precipitated tumor extracts formed kinin when incubated with heated (80°C rat) or human plasma. This kinin-forming activity initially increased during tumor growth and then diminished to almost zero levels within 12-14 days following tumor transplant. The kallikrein activity was similar to that of rat plasma kallikrein with respect to pH profile, kinetics of kinin formation (using human plasma kininogen as substrate at pH 7.8), and action of plant, mammalian, and synthetic protease inhibitors.

The acetone-formed kallikrein, however, was unable to form kinin from kininogen present in the tumor homogenate extracts heated to 80°C. This posed the question as to the significance of a tumor kinin-forming system in which the kinin-forming enzyme is unable to digest its apparently appropriate substrate present in the same tissue site.

These present studies explore the conditions under which the tumor kininogen may yield kinin, and the protease forces involved in that reaction.

*1. Supported in part by USPHS grants No. HE-11492 and HE-03784, National Heart Institute.

Methods

Tumor Homogenate Extract

Implanation of Murphy-Sturm lymphosarcoma into Sprague-Dawley recipient rats and harvesting of the tumors 8 days after implant was carried out as previously described (1). An equal volume of physiologic saline was added to the tumor tissue and homogenized in a Votis blender. The homogenate then was centrifuged in a refrigerated centrifuge at 3000 rpm for 30 minutes to remove the fibrous material. The crude tumor extract was used as such after filtration through a gauze filter.

Glass activation of tumor extract. To each ml of tumor extract (60 mg/ml protein) was added 50 mg of Permutit (Folin Decalso, Permutit Co., N.Y., N.Y.) and 1 mg of 1-10 phenanthroline. The incubation mixture at neutral pH, was agitated for 30 minutes at 37°C and any kinin formed was assayed on the isolated rat uterus muscle perfused with modified Tyrodes solution containing a reduced concentration of calcium.

Acid-activation of tumor homogenate extracts. Effect of pH on kinin formation within the tumor homogenate extract was studied by adjusting 2 ml aliquots of tumor extract to a range of pH from 1.0 to 8.0 with sodium acetate-acetic acid buffer (1 molar) and 1 molar Tris buffer. A 2 mg quantity of 1-10 phenanthroline was added to the incubation mixture and the total volume adjusted to 2.3 ml with distilled water. The samples were incubated at 37°C for 30 minutes, readjusted to pH 7.0 and assayed for kinin activity.

Identification of kinin activity. Kinin activity further was identified by smooth muscle contracting activity on the isolated rat uterus, smooth muscle relaxing activity on the rat duodenum, and blood pressure lowering activity in the anesthetized dog. In addition, kinin was differentiated from other vasoactive mediators (notably angiotensin) by the action of crystalline trypsin and carboxypeptidase B. Those extracts that generated kinin-like activity (primarily at pH 4.0) were incubated (200 lambda quantities) for 5 minutes at 37°C with 50 micrograms of crystalline Trypsin. The pH then was adjusted to 7.0 with 0.1 ml of Tris buffer and 1 mg of soya bean Trypsin inhibitor was added to inhibit the trypsin activity. The smooth muscle activity on the rat uterus was studied. The effect of trypsin on angiotensin was studied in a similar manner using 200 lambda of 1×10^{-6} g angiotensin in the incubation mixture.

The effect of carboxypeptidase B on the kinin generated also was studied. Two ml of tumor homogenate extract (containing 2 mg of 1-10 phenanthroline) was incubated at pH 4.0 for 30 minutes at 37°C and readjusted to pH 7.0 with Tr is buffer. The mixture then was incubated for 5 minutes with 10 micrograms of carboxypeptidase B (Worthington Labs, Freehold, N.J.) and 200 lambda aliquots measured on the isolated rat uterus. These measurements were done both before and after incubation with carboxypeptidase B.

Time and Temperature Studies. The influence of time and temperature on kinin formation during incubation of the tumor homogenate extracts at pH 4.0 also was studied. Two mls of tumor extract were adjusted to pH 4.0, and then placed in an ice bath from which 20 lambda aliquots were removed at regular time intervals up to 120 minutes for bioassay on the perfused rat uterus. Two mls of tumor extract, after adjustment to pH 4.0, were incubated for 30 minutes at 0°, 25°, and 37°C and 5 lambda aliquots assayed for kinin activity on the rat uterus msucle. These studies also were done with purified tumor homogenate extracts.

Inhibitor study. Effect of various plant, mammalian, and synthetic protease inhibitors was studied on the kinin-forming capacity of the tumor homogenate extracts adjusted to pH 4.0. Two ml portions of tumor extracts containing 2 mg of 1-10 phenanthroline and varying concentrations of protease inhibitors were incubated together at pH 4.0 for 30 minutes at 37°C. The samples then were readjusted to pH 7.0 and assayed for kinin activity. Also included in the agents tested were EDTA, ammonium oxylate, sodium citrate, and heparin. The activity, as with all other studies described above, was compared with a synthetic bradykinin standard (Schwarz Bioresearch, Orangeburg, N.Y.).

Partial purification of acid-dependent kinin-forming activity. Purification of the kinin-forming activity of the tumor tissue was carried out with 235 grams of tumor. After addition of equal volume of saline, the tumor was homogenized and centrifuged at 3000 rpm for 30 minutes. The precipitate was discarded, and the supernatant was incubated for 3 hours at pH 4.0 with the addition of equal volume of 0.05 m potassium biphthalate. In this fashion, all of the tumor kininogen substrate was consumed. This preparation was centrifuged again for 30 minutes at 3000 rpm and the precipitate (containing the enzyme activity as well as most kinin bound under acid conditions) was reconstituted with phosphate buffer, 0.15 molar, pH 7.0. The precipitate thus reconstituted was dialyzed against 4 liters of phosphate buffer (pH 7.0) involving 3 changes thereby removing all the kinin activity. The preparation was centrifuged once again for 30 minutes at 3000 rpm and the supernatant saved. The precipitate was washed three times with phosphate buffer, pH 7.0, the three washings combined with the supernatant, and lyophilized. The resultant lyophilized preparation weighed 2.2 grams, was free of both kinin and kininogen substrate, and demonstrated kinin-forming enzyme activity, (scheme 1).

Activity on non-kininogen substrates. The purified enzyme preparation was studied on several non-kininogen substrates: hemoglobin, albumin, casein, alpha -, beta -, gamma-globulin. The hemoglobin preparation (Armour Labs, Chicago, Ill.) was diluted to

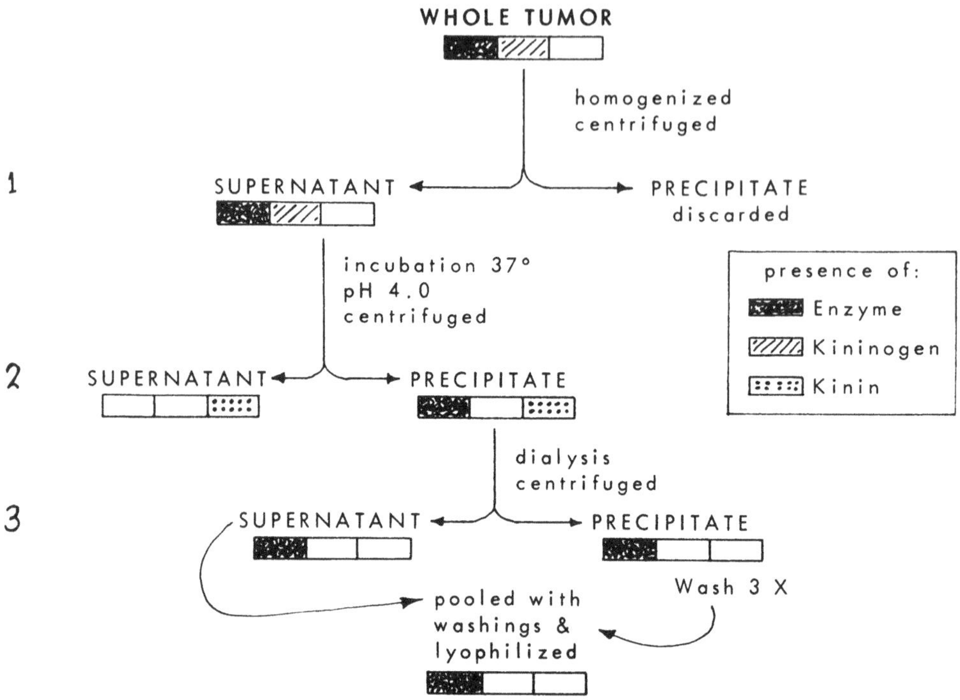

Scheme I Flow chart for partial purification of kinin-forming enzyme from tumor homogenate extract.

8% (w/v) and frozen until required for use (2). The incubation mixture contained 0.5 ml of the purified enzyme, 0.25 ml of the 8% hemoglobin made up to 1.0 ml with acetate-acid buffer, pH 4.0. The reaction was allowed to proceed for 120 minutes and then was stopped by the addition of 5 ml of 3% (w/v) trichloracetic acid. The contents were mixed thoroughly, incubated for an additional 10 minutes, and filtered through a Whatman #1 filter paper. Four mls of biuret reagent was added to 1 ml of the mixture and after incubation for 30 minutes at 37°C was read at 550 mu. The other substrates were also made up in an 8% (w/v) solution and assayed in a similar fashion.

Cell Fractionation Procedure. Two individual pools of tumor tissue, 25 grams and 5 grams, were fractionated to determine the cellular site for the acid-dependent kinin-forming activity[3,4]. To the tumor pool was added 10 times the volume of 0.25 m sucrose-0.0018m $CaCl_2$. This was homogenized at low speed in a Serval homogenizer. The homogenate was filtered through gauze and the filtrate placed over 2 volumes (500 ml) of 0.34 m sucrose-o.0018m$CaCl_2$. This was centrifuged at 600 x g for 10 minutes (yielding the nuclear fraction).

The pellet was collected and reconstituted to 25 ml with saline and centrifuged at 50,000 x g for two hours, and was saved as the microsomal fraction after reconstitution to 25 ml. The remaining supernatant contained the soluble cytoplasmic protein fraction.

All the fractions were properly reconstituted and studied for total protein, kinin-forming activity at pH 4.0, ability to digest hemoglobin, and kininogen content.

Results

Glass activation. Attempts to activate a kinin-forming system in the crude tumor homogenate extract (in the presence of 1-10-phenanthroline) with permutit were unsuccessful. These attempts were made at a neutal pH. No kinin formed either from within the homogenate mixture itself, or when assayed on cat or human plasma kininogen substrate. The homogenate extract is known to contain both a kininogen substrate that forms kinin by tryptic action, and a pre-kallikrein that is activated by acetone(1).

Acid Activation. Incubation of the tumor homogenate extract at various pH conditions for 30 minutes at 37°C resulted in significant formation of a smooth muscle stimulating activity (rat uterus) at pH 4.0, Fig. 1.

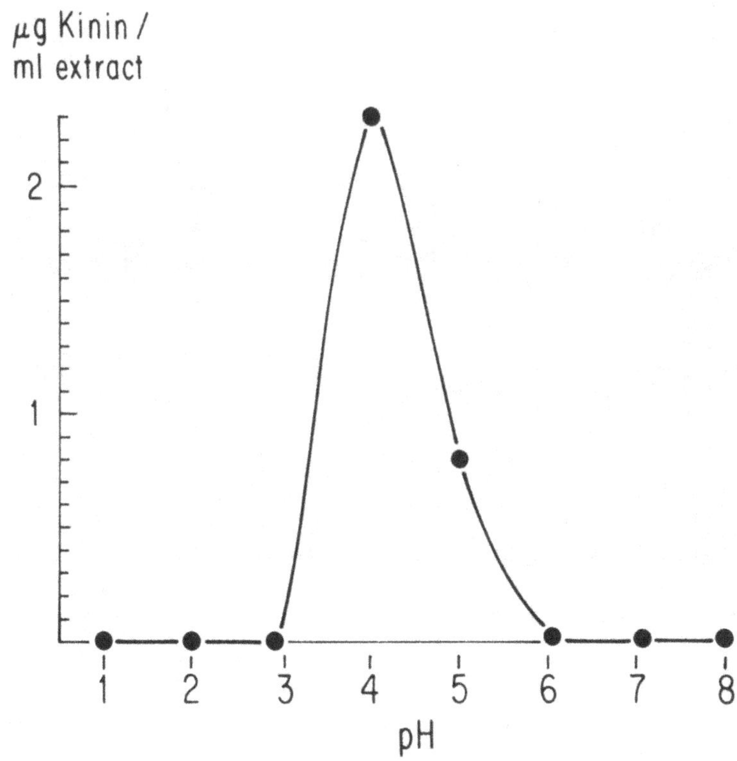

Figure 1: Effect of pH on kinin-forming activity of tumor homogenate extract.

This activity, in terms of kinin units per ml of extract, was equivalent to 23 micrograms of kinin. Slight kinin activity was generated at pH 5.0 but no activity at pH values less than 4.0 and greater than 5.0. This activity was generated from within the tumor extract itself, and did not require the addition of a substrate.

Identification of the kinin activity. The acid-activated smooth muscle stimulating activity was studied further to determine its exact nature. The smooth muscle stimulating activity generated contracted the rat uterus, Fig. 2, equivalent to 2×10^{-7} gm of synthetic bradykinin. The rat duodenum muscle was relaxed by both bradykinin and the activated tumor homogenate extract. The non-activated extract had no effect on either muscle. The activity was differentiated from angiotensin by the fact that angiotensin contracts the rat duodenum as well as the rat uterus muscle.

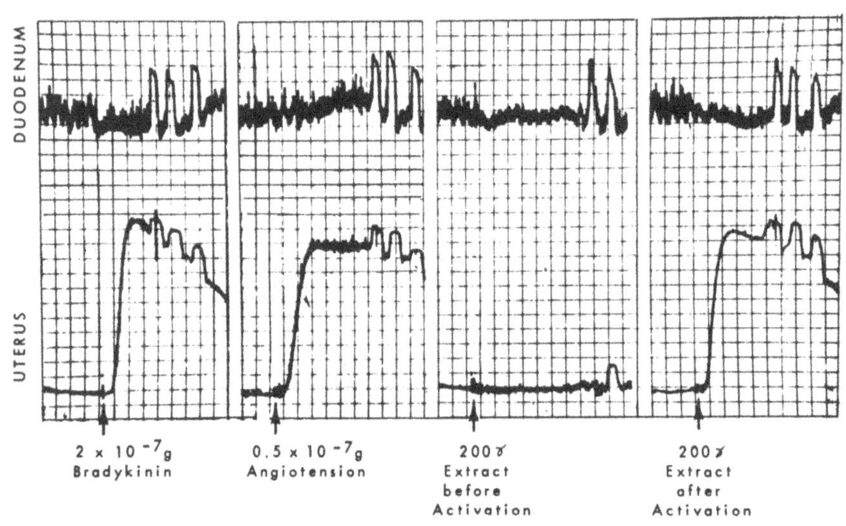

Figure 2: Effect of acid-activated tumor homogenate extract on rat uterus and rat duodenum.

The activated tumor extract also caused a fall in blood pressure in the anesthetized dog, Fig. 3, whereas the non-activated preparation had no such hypotensive effect.

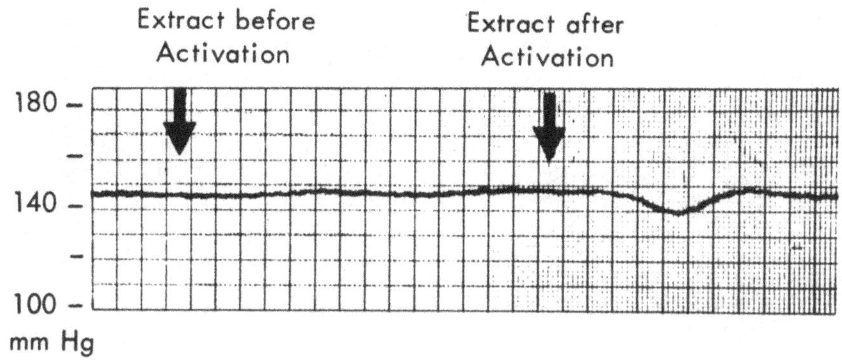

Fig. 3. Effect of acid-activated tumor homogenate extract on dog blood pressure.

The smooth muscle contracting activity on the rat uterus was not destroyed following incubation of the activated extract with crystalline trypsin, Fig. 4. Incubation of trypsin with angiotensin did destroy the muscle stimulating activity of that mediator.

The peptide character of the smooth muscle stimulating activity formed within the tumor homogenate extract was revealed when the activity was destroyed during incubation with carboxypeptidase B, Fig. 5.

These identification studies support the proposal that the activity formed on acidification of the tumor extract was a peptide kinin.
Inhibition Studies. None of the protease inhibitors was capable of inhibiting the formation of kinin during the activation process, Fig. 6. Inhibitors studied at concentrations shown to have inhibiting action on kinin formation (5) included the mammalian inhibitor TRASYLOL and KK (lyophilized mammalian protease inhibitor), plant inhibitors soya bean and lima bean trypsin inhibitors, avian inhibitor ovomucoid, and synthetic inhibitors epsilon amino caproic acid (EACA) and (AMCHA). In addition, the following agents did not have any inhibitory effect: EDTA, ammonium oxylate, heparin and sodium citrate. However, 3-phenylpyruvic acid at low molar con-

centrations did inhibit kinin formation, Fig. 7. This inhibition was non-competitive, Fig. 8.

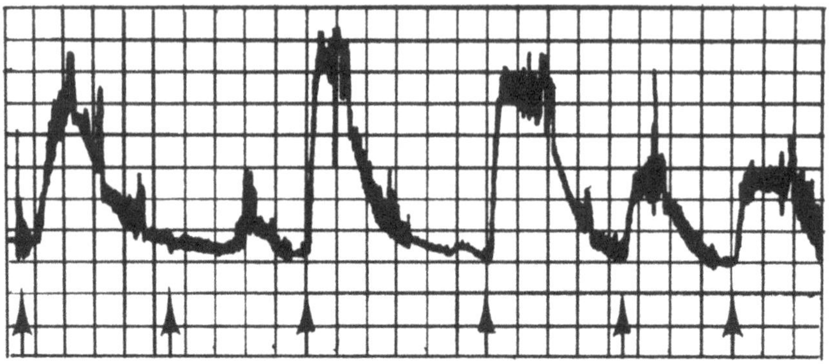

Figure 4: Effect of crystalline trypsin on rat uterine stimulating activity of angiotensin, bradykinin, and acid-activated tumor homogenate extract.

Studies with non-malignant tissue. Extracts of rodent non-malignant tissue homogenates were studied for their ability to form kinin under acid conditions similar to those applied to the tumor tissue. The tissue homogenates were prepared in the same manner as the tumor tissues, and acidified to pH 4.0. One hundred lambda aliquots were removed after 30 minutes of incubation, and assayed for kinin activity on the isolated rat uterus. Table 1 summarizes the results. As seen, kinin-forming activity under these conditions was present in the kidney, liver, lung, and spleen whereas muscle, skin, heart, white blood cells and red blood cells were devoid of activity. Incubation of the activated homogenate extracts of the tissues showing activity with carboxypeptidase B destroyed the activity. Various Murphy-Sturm lymphosarcoma preparations were equally as active, Fig. 9.

Partial purification of kinin-forming activity. Table 2 summarizes the results of the purification of the tissue homogenate extract. The whole tumor contained both kininogen and enzyme. Following

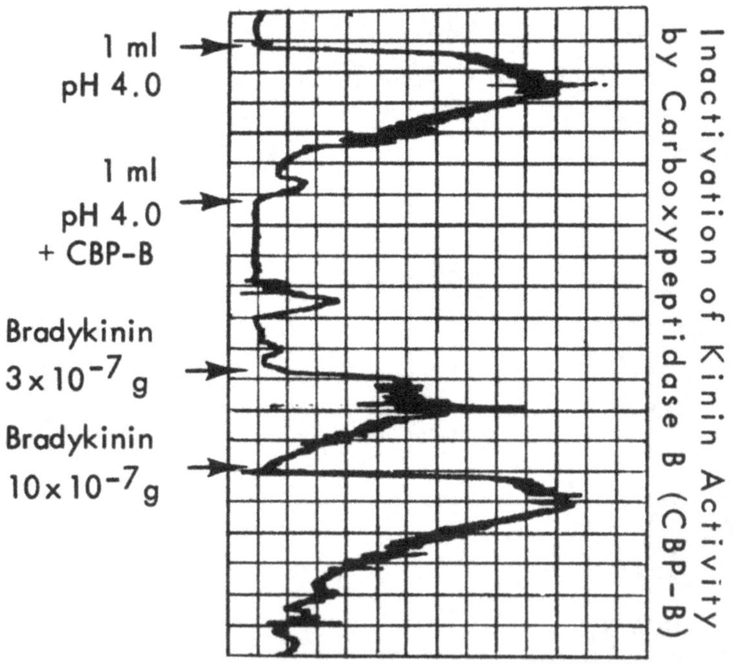

Figure 5: Effect of carboxypeptidase B (CBP-B) on rat uterine muscle stimulating activity of bradykinin and acid-activated tumor homogenate extract.

centrifugation supernatnat #1 also contained the substrate and enzyme. Incubation of the supernatant for 3 hours at pH 4.0, 37° C consumed all the substrate and after centrifugation, the supernatant #2 had some kinin activity while precipitate #2 had both enzyme and kinin activity. Dialysis of precipitate #2 and centrifugation yielded supernatant #2 with enzyme activity and precipitate #3 with enzyme activity. Washing of the precipitate #3 transferred the enzyme activity to the washings which were combined with supernatant #3 and lyophilized for use as the "purified" enzyme preparation.

Studies with the purified enzyme preparation. Apart from forming kinin from tumor tissue kininogen, the enzyme also was capable of forming kinin from 80°C heated rat plasma substrate, Fig. 10. Tested was 100 lambda of a 2.2 ml incubation mixture that contained 5.0 mg of enzyme, 0.2 ml rat serum, 0.8 ml acetic acid-sodium acetate buffer, and 1.0 ml of Tris buffer (pH 7.8) containing 1 mg (0.2 ml) soya bean trypsin inhibitor and 1 mg 1-10 phenanthroline. The kinin formed was equivalent to $3x10^{-8}$ gm bradykinin.

It was found that the enzyme digested both hemoglobin and alpha globulin, but had no effect on albumin, casein, beta and gamma globulin.

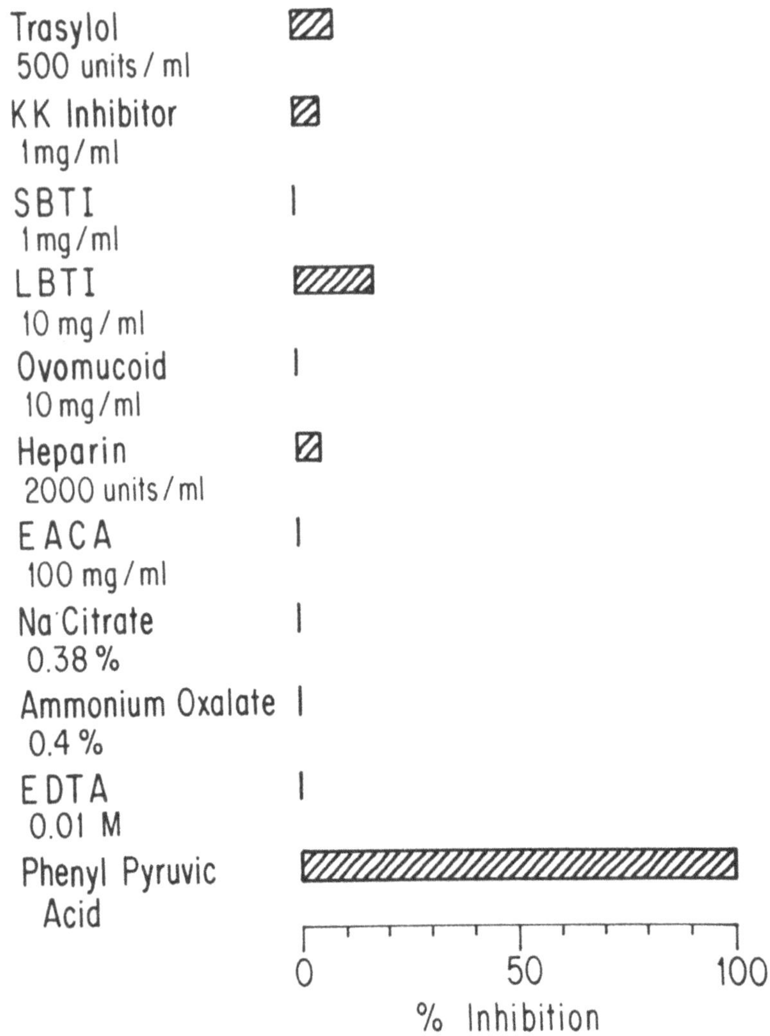

Figure 6: Effect of protease inhibitors, heparin, sodium citrate, ammonium oxalate, EDTA, and phenylpyruvic acid on kinin-forming activity of tumor homogenate extract. (SBTI=soya bean trypsin inhibitor; LBTI= lima bean trypsin inhibitor; KK=lyophilized mammalian kallikrein inhibitor; EACA=epsilon amino caproic acid; AMCAH (not shown tested at dose of 10 mg), =4-aminoethyl cyclohexane-1-carboxylic acid.)

Time and temperature studies with the enzyme are seen in Figs. 11 and 12. Optimum time for kinin formation at 37°C was 120 minutes while optimum temperature in the time range studied was 37°C.

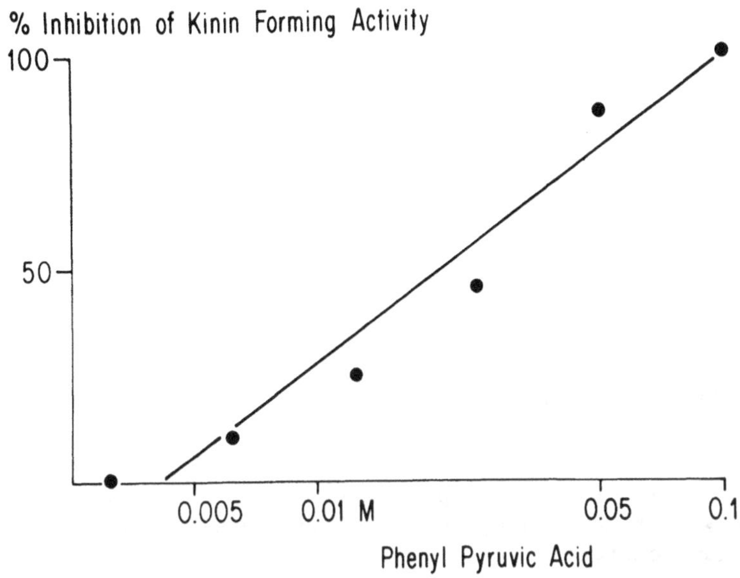

Figure 7: Effect of 3-phenyl pyruvic acid on the kinin-forming activity of tumor homogenate extract.

<u>Cell localization studies</u>. Ultracentrifugal studies of tumor cells localized the major part of the enzyme activity and kininogen content in the same cellular compartment, the soluble protoplasmic protein fraction Fig. 13. This fraction also digested hemoglobin and contained the highest protein content of all other fractions. Hogeboon's estimate of protein content in liver cells places the amount at 40% in the soluble protoplasmic protein fraction, 10% in the nuclear portion, 25% in the mitochondrial fraction and 20-25% in the microsomal fraction (4).

Discussion

Studies with transplanted Murphy-Sturm lymphosarcoma revealed the presence of components of a kinin-forming system. These components included kininogen, a kinin-destryoing enzyme (kininase), and a pre-kallikrein capable of being activated by acetone(1). Other tumor systems have been found to have fibrinolysin and kinin-forming capacities (6) and the implications of these findings in relationship to tumor growth have been considered (1,6).

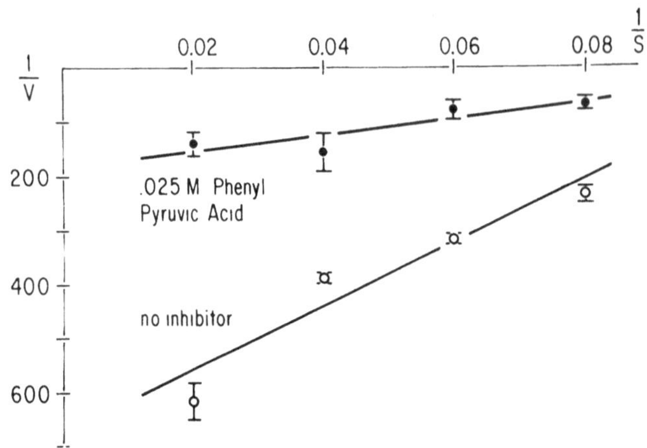

Figure 8: Lineweaver-Burk plot showing non-competitive inhibitory aspect of 3-phenylpyruvic acid on kinin-forming activity of tumor homogenate extracts.

TISSUE	pH 4.0	pH 7.8
KIDNEY	+	0
STRIATED MUSCLE (ABDOMINAL)	0	0
LIVER	+	0
SKIN	0	0
LUNG	+	0
SPLEEN	+	0
HEART	0	0
TUMOR	+	0

Table 1: Rat tissues showing kinin-forming activity following acid activation of homogenate extracts at pH 4.0. Absence of activity may indicate absence of proenzyme, absence of substrate, or both. +=activity, 0=no activity.

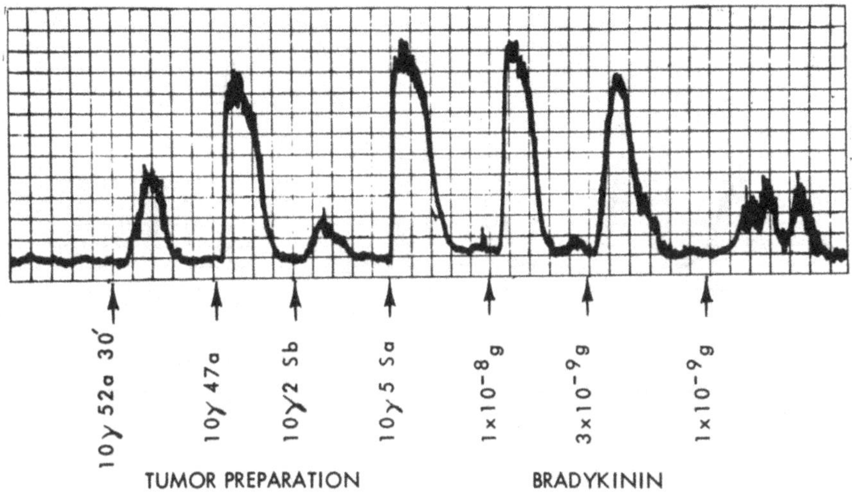

Figure 9: Effect of various tumor homogenate extract preparations followin g acid activation on smooth muscle activity of rat uterus.

The kinin-forming enzyme in the Murphy-Sturm lymphosarcoma, however, was not able to digest the kininogen substrate of that tumor. It would appear that the enzyme was designed to act upon extracellular plasma substrate, thus relating perhaps to the control of the microcirculation within the developing tumor. While glass activation did not activate an enzyme that could utilize the tumor kininogen as substrate, acidification activated a very potent enzyme that did digest the kininogen. This acid proteinase was most effective at pH 4.0, and appeared to have the characteristics of a cathepsin.

Catheptic activity has been found in tumor interstital fluid (7), human gastric carcinoma (8), renal carcinoma (9), and Jensen sarcoma (10). Greenbaum (11) has shown that spleen cathepsins at acid pH (3.8 -5.0) are capable of forming kinins. Furthermore, Greenbaum also has shown that malignant white cells (leukemia L-1210) as well as normal leukocytes and those present in inflammatory exudates have a kinin-forming system that does not resemble the customary kallikreins (12). This polymorphonuclear leucocyte kinin-forming system was not inhibited by any of the classical protease inhibitors as TRASYLOL, and plant trypsin inhibitors.

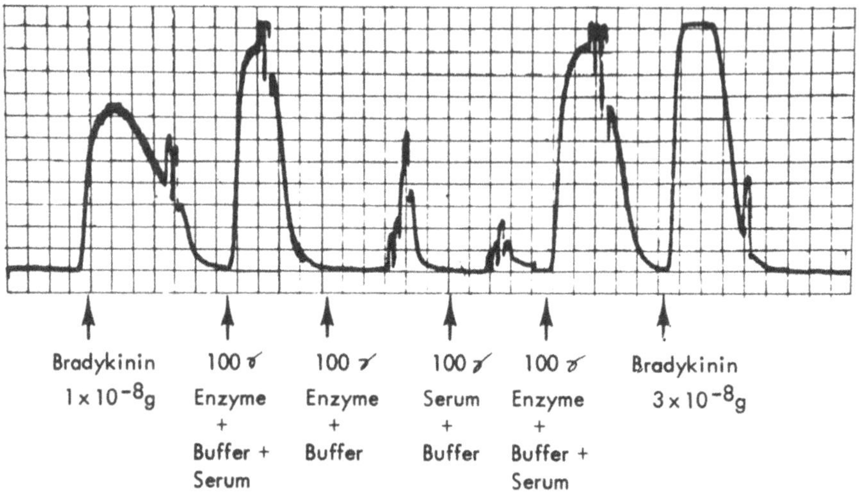

Figure 10: Ability of a kinin-forming enzyme from tumor homogenate extract to form kinin from rat plasma substrate. Contractions are rat uterus muscle.

The acid proteinase kinin-forming enzyme in the Murphy-Sturm lymphosarcoma most closely resembles a cathepsin with respect to optimal pH activity, substrate utilization, and action of protease inhibitors. The fact that this enzyme digests hemoglobin but not casein or gamma globulin strongly suggests that it is a cathepsin D. (13). Greenbaum localized a major portion of the kinin-forming activity of alveolar macrophages in the extra-lysosomal subcellular fraction.

The significance of this acid proteinase capable of utilizing a substrate that forms a vasoactive kinin is at present speculative. It is possible that in the course of tumor growth, the enzyme system provides a continued source of smaller peptides and amino acids for their growth needs, and only by chance are these peptides vasoactive. The tumor cell, however, may provide itself with a mechanism for altering its cell membrane permeability when intracellular conditions dictate, as might occur in the production of an unsatisfactory acid medium. This altered permeability, as effected by kinin, may allow for the equilibration of the unfavorable intracellular conditions with the interstitial environment, and thus provide for continued tumor growth. The possible role of this system in the control, alteration, and development of the tumor vasculature also cannot be excluded in view of the possibility that the kinin, although short-lived, may have an extracellular action as well. The acid proteinase also is capable of forming kinin from plasma kininogen substrate.

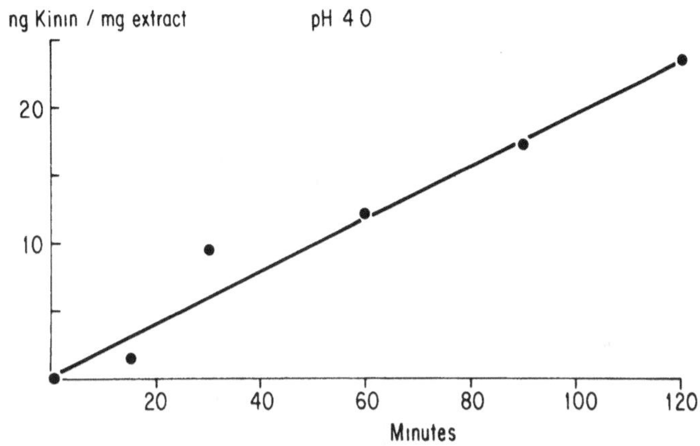

Figure 11: Kinetics of kinin formation during incubation of partially purified tumor homogenate enzyme and tumor kininogen (at pH 4.0).

Summary

1. A kinin-forming enzyme is present in the transplanted Murphy-Sturm lymphosarcoma that is active optimally at pH 4.0.
2. The enzyme is capable of digesting kininogen substrate present in the tumor.
3. The kinin formed contracts the smooth muscle of the rat uterus, relaxes the rat duodenum, lowers blood pressure in the dog, is not destroyed by crystalline trypsin, and is destroyed by carboxypeptidase B.
4. Partial purification of the enzyme was accomplished using precipitation and centrifugation techniques.
5. The partially purified enzyme also was capable of forming kinin from rat plasma kininogen substrate. Hemoglobin and alpha globulin also were digested by the enzyme but not casein, albumin, or beta and gamma globulin.
6. The enzyme was found present in rodent spleen, lung, liver,

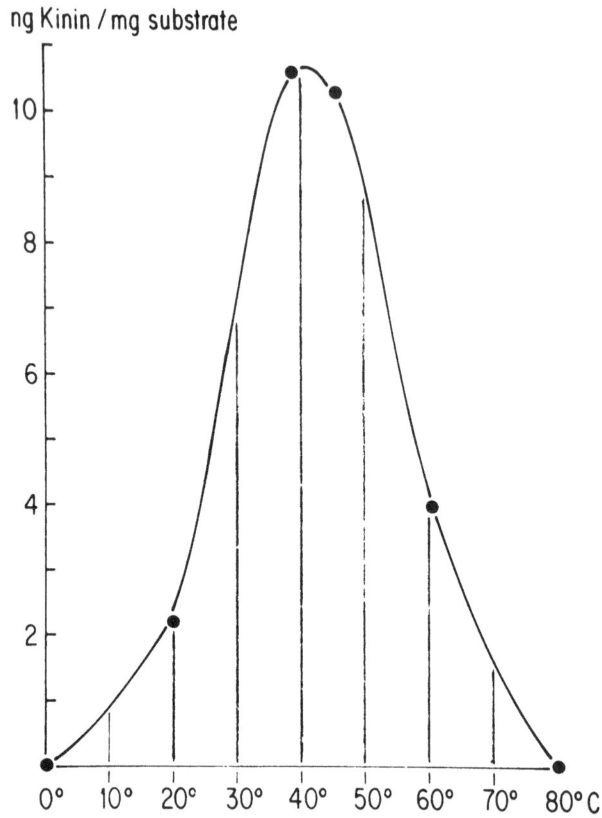

Figure 12: Effect of temperature on kinin-forming activity of partially purified enzyme from tumor homogenate extract.

and kidney but not heart, skin, muscle, white blood cells and red blood cells.

7. The enzyme was not inhibited by TRASYLOL, EACA, AMCHA, soya bean and lima bean trypsin inhibitor, heparin, EDTA, ammonium oxylate, and sodium citrate. However, 3-phenylpyruvic acid did inhibit the enzyme non-competitively.

8. The enzyme appears to be localized to the greatest extent in the soluble protoplasmic protein fraction of the cell along with the kininogen substrate.

9. All present evidence suggests that the enzyme most closely resembles a cathepsin D.

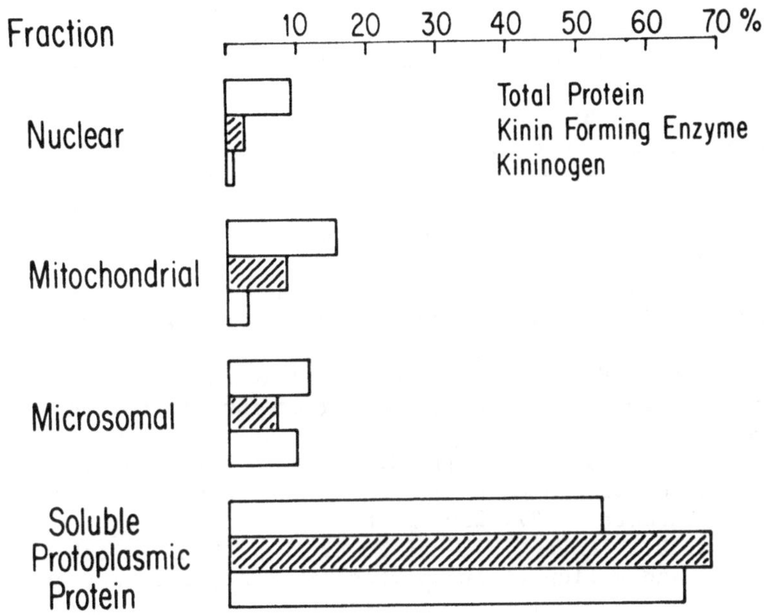

Figure 13: Distribution of total protein, kinin-forming enzyme and kininogen in tumor cell fractions.

References

1. Back, N., and Steger, R.: Characterization of a pre-kallikrein activity in developing transplanted mammalian tumors in Bradykinin and Related Peptides. Eds. Sicuteri, F., Rocha e Silva, M., and N. Back. Plenum Press, New York, 1970,pg. 225.

2. Barrett, A. J.: Lysosomal acid proteinase of rabbit liver. Biochem. J. 104: 601, 1967.
3. Shibko, S. and Tappel, A.L.: Rat liver microsomes. Biochem. J. 95: 73, 1965.
4. Hogeboon, G.H., Schneider, W.C., and Streibich, M.J.: Localization and integration of cellular function. Cancer Res., 13: 617, 1953.
5. Back, N., and Steger, R.: Effect of heparin on the kinin-forming activity of trypsin, plasmin, and various kallikreins. Proc. Soc. Exper. Biol. and Med. 133: 740, 1970.
6. LeBlanc, P., and Back, N.: The kallikrein-kinin system in the growth of the Ehrlich ascites tumor. Fed. Proc. 29: 807, 1970.
7. Sylven, B., and Bois-Svensson, I.: On the chemical pathology of interstitial fluid I.Proteolytic activities in transplanted mouse tumors. Cancer Research 25: 458, 1965.
8. Etherington, D.J., and Taylor, W.H.: Investigation of the cathepsins of human gastric carcinomata. Biochem. J. 115: 43P, 1969.
9. Schersten, T., Wahlquist, L., and Johansson, L.G.: Lysosomal enzyme activity in liver tissue from patients with renal carcinoma. Cancer, 23: 608, 1969.
10. Kampschmidt, R.F., and Wells, D.: Acid hydrolysis activity during the growth, necrosis, and regression of the Jensen Sarcoma. Cancer Research 28: 1938, 1968.
11. Greenbaum, L.M. and Yamafuji, K.: The in vitro inactivation and formation of plasma kinins by spleen cathepsins. Brit. J. Pharmac. Chemother. 27: 230, 1966.
12. Greenbaum, L.M. Chang, J., and Freer, R.: Kinin metabolism in normal and malignant luekocytes. in Bradykinin and Related Kinins. Cardiovascular, Biochemical, and Neural Actions. Ed. Sicuteri, F., Rocha e Silva, M., and Nathan Back. Plenum Press, 1970, pg. 39.
13. Lebez, D., Turk, V., and Kreger, I.: The possibility of differentiation of cathepsins on the basis of their ability to degrade various protein substrates. Enzymologia 34: 344, 1968.

ROLE OF VASOACTIVE PEPTIDES AND LOCAL HORMONES IN EXPERIMENTAL COLD INJURY IN MONKEYS

R. B. Arora, S. K. Gupta, K. Rajanna, H. H. Siddiqui, and M. Tariq

Department of Pharmacology, All-India Institute of Medical Sciences, Ansari Nagar, New Delhi-16 (INDIA)

Progress in the knowledge of frostbite has been slow, owing to the scarcity of cases in peace time and the difficulty of planned study during war. Frostbite is a condition of local cold injury resulting from exposure to cold well below the freezing temperature. Vasoconstriction is the first reaction to the cold exposure (Lewis, 1941). Initially arterioles and small arteries are involved and ultimately the whole vascular tree is affected (Sullivan and Towle, 1957). The spasm in arterioles and venules of the affected part and adjoining areas slow down the blood flow (Kulka et al, 1958). If the stasis persists thrombus formation occurs, leading to anoxia and death of the tissue (Zweifach, 1963). These changes are supposed to be brought about partly by the direct action of cold on the tissue and partly by the release of chemical mediators (Rocha E Silva, 1964). Therefore, the role of these mediators have been studied in monkeys after experimental cold injury. Since the inactivating enzymes play an important role in control of the level of these local hormones, the levels of bradykinase, angiotensinase and histaminase have also been estimated.

METHODS

PRODUCTION OF EXPERIMENTAL COLD INJURY

Experimental cold injury was produced in Rhesus monkeys of either sex, weighing 4-6 kg at -30°C for 1 hr in a specially designed cooling chamber described by Talwar et al (1966) with some modifications (Arora et al, 1971a). The monkeys were actively immunized against tetanus (ATS 1500U/5 kg) given intramuscularly before the experimentation. The shaved hind limb up to the upper one third of the leg was exposed to cold after which the animals were removed to the cages and the frozen leg was allowed to thaw at room temperature of 28 - 30°C.

ESTIMATION OF PEPTIDES AND LOCAL HORMONES

Local hormones were estimated in blood before, immediately after and on 3rd, 6th, and 10th day of the experimental cold injury.

Bradykinin: Bradykininogen was extracted from the plasma according to the method of Diniz and Carvalho (1963) and the liberated bradykinin was estimated on isolated guinea-pig ileum against synthetic bradykinin (BRS 640). For the estimation of bradykinin in the tissue, muscles surrounding the tendo-achilese (1 g) was homogenized with 10 ml of 0.2% acetic acid. Rest of the procedure was same as given for plasma.

Angiotensin: Blood angiotensin was extracted according to Osborn's method (1969) and was assayed on isolated rat colon according to Regoli and Vane (1964) against synthetic angiotensin (Hypertensin). For estimation of angiotensin in the tissue, 1 g of the muscle was homogenized with 40 ml ethanol and shaken vigorously. Remaining procedure was same as described for blood.

5-Hydroxytryptamine (5-HT): 5-HT in plasma and tissue (1 g) was extracted according to Amin et al (1954) and assayed on isolated stomach strip of rat (Vane, 1957; Lin and Yeoh, 1965) against standard 5-hydroxytryptamine (Serotonin) creatinine sulphate complex.

Histamine: Histamine in blood and tissue (1 g) was extracted and estimated on guinea-pig ileum according to the method of Code and McIntire (1960) against standard histamine acid phosphate.

ESTIMATION OF INACTIVATING ENZYMES

Bradykininase: Bradykininase activity was estimated by the method described by Abe et al (1969). Plasma mixed with synthetic bradykinin was incubated at 37°C. The samples drawn at different timings after incubation were assayed against standard bradykinin and the decay curve of bradykinin in each plasma sample was obtained.

Angiotensinase: Angiotensinase activity was estimated by the method of Landesman et al (1963). The decay of angiotensin added

TABLE I

EFFECT OF EXPERIMENTAL COLD INJURY ON PLASMA AND TISSUE LEVELS OF BRADYKININ, ANGIOTENSIN, HISTAMINE, AND 5-HYDROXYTRYPTAMINE IN MONKEYS

Groups	Bradykinin level		Angiotensin	Histamine level		5-HT level	
	Plasma (ug/ml)	Tissue (ug/g)	Level (Blood) ng/ml	Blood (ug/ml)	Tissue (ug/g)	Plasma (ng/ml)	Tissue (ug/g)
Normal	1.64 ± 0.025	1.60 ± 0.10	0.63 ± 0.01	2.67 ± 0.03	1.90 ± 0.07	4.33 ± 0.53	1.14 ± 0.25
Immediately after cold injury	*** 1.19 ± 0.022	*** 0.80 ± 0.10	*** 0.97 ± 0.01	*** 1.73 ± 0.02	2.11 ± 0.08	5.35 ± 0.54	** 0.71 ± 0.11
On 3rd day	*** 0.87 ± 0.02	*** 1.72 ± 0.02	*** 1.72 ± 0.02	*** 0.86 ± 0.02	** 2.23 ± 0.3	*** 20.60 ± 0.69	*** 0.43 ± 0.06
On 6th day	*** 0.44 ± 0.021	*** 0.48 ± 0.07	*** 0.91 ± 0.06	*** 0.58 ± 0.01	2.19 ± 0.27	*** 8.43 ± 0.80	*** 0.61 ± 0.08
On 10th day	1.42 ± 0.08	*** 0.95 ± 0.08	0.71 ± 0.08	2.42 ± 0.24	2.0 ± 0.26	4.05 ± 0.27	0.99 ± 0.07
Normal	1.64 ± 0.025	1.60 ± 0.10	0.63 ± 0.01	2.67 + 0.01	1.90 ± 0.07	4.33 + 0.53	1.14 + 0.25
	**	***	***	***			**

The values shown are means ± S. E.

*P 0.05
**P 0.01
***P 0.001

to plasma sample was estimated on isolated guinea-pig ileum against standard angiotensin in a manner similar to the above.

Histaminase: Histaminase activity was estimated by the method of Spencer (1963). The procedure adopted was same as described for bradykininase and angiotensinase. Assay of the surviving histamine was done on isolated guinea-pig ileum against standard histamine acid phosphate.

RESULTS

One hour duration at -30°C has been used to produce experimental cold injury in monkeys because it has been reported (Talwar et al, 1966) that it produces a standard reproducible type of injury. Oedema develops on thawing the limb and reaches a maximum by 24 hrs of cold injury. It is followed by blister formation and ulceration. The gangrene sets in within 3-6 days resulting in the falling off the gangrenous area within 10-12 days (Arora et al, 1971a).

EFFECT ON VASOACTIVE PEPTIDES AND LOCAL HORMONES

Local hormones and the activity of the inactivating enzymes were estimated before and after experimental cold injury. The normal level of liberated bradykinin from the plasma was 1.64 ug/ml. Immediately after the cold injury the level decreased to 1.19 ug/ml and it continued to decrease on 3rd and 6th day of the cold injury. The concentration on the 3rd day was 0.87 and 6th day 0.44 ug/ml (Table 1). The fall was significant immediately after, on 3rd and 6th day (P 0.001). The tissue level of liberated bradykinin decreased from the normal of 1.6 ug/g to 018 ug/g immediately after the injury and was significantly low throughout the experimental period (Table 1).

Angiotensin level in the blood increased to 0.97 ng/ml immediately after the cold injury and was maximum (1.72 ng/ml) on the 3rd day and remained high up to the 6th day of the injury. The increase in the angiotensin level was significant immediately after, on 3rd and 6th day (P 0.001). However in the tissue, we failed to demonstrate any angiotensin-like activity (Table 1).

Blood histamine level decreased in a manner similar to bradykinin. Immediately after the cold injury the level of histamine decreased to 1/73 ug/ml and continued to decrease up to the 6th day of the injury. The fall was significant immediately after, 3rd and 6th day of the cold injury (P 0.001). Histamine concentration in the tissue increased from the normal of 1.9 ug/g to 2.11 ug/g immediately after the cold injury and remained high up to the 6th day of the experimental period (Table 1).

TABLE 2

EFFECT OF EXPERIMENTAL COLD INJURY ON PLASMA BRADYKININASE, ANGIOTENSINASE, AND HISTAMINASE ACTIVITY IN MONKEYS

Groups	Bradykininase activity (%)		Angiotensinase Activity (%)		Histaminase activity (%)	
	30 sec	60 sec	20 min	40 min	20 min	40 min
Normal	24.01 ± 1.81	48.54 ± 1.71	43.81 ± 2.67	82.58 ± 3.12	28.91 ± 1.73	55.98 ± 1.82
Immediately after cold injury	46.52 ± 1.82***	74.62 ± 1.89***	44.55 ± 0.74***	41.01 ± 1.97***	58.43 ± 1.08***	88.20 ± 1.62***
On 3rd day	37.78 ± 1.22***	58.86 ± 0.96***	23.36 ± 0.91***	52.08 ± 1.72***	49.22 ± 0.96***	79.50 ± 1.59***
On 6th day	31.27 ± 1.59**	53.14 ± 1.82*	33.81 ± 2.12**	65.51 ± 1.72***	41.13 ± 0.71***	71.78 ± 1.27***
On 10th day	30.85 ± 1.59**	50.31 ± 1.05	40.83 ± 1.71	79.16 ± 0.87	36.62 ± 0.84***	67.37 ± 2.02***

The values shown are means ± S. E.

*P 0.05
**P 0.01
***P 0.001

The normal concentration of 5-HT in the plasma was 4.33 ng/ml which increased to 5.35 ng/ml immediately after the cold injury. However the maximum increase was observed on the 3rd day (20.60 ng/ml). The 5-HT remained elevated up to the 6th day of the injury. In contrast to the plasma, the concentration of 5-HT decreased in the tissue immediately after the cold injury and remained low up to the 6th day (Table 1).

EFFECT ON THE INACTIVATING ENZYMES

The results are presented in Table 2. The percentage destruction of the local hormones was taken as the measure of the enzymatic activity.

Before cold injury the percentage destruction of bradykinin after 60 seconds incubation was 48.54%. Immediately after the cold injury it increased to 74.62% and on the 3rd and 6th it decreased to 58.86 and 53.14% respectively but it was still more than control.

Whereas the percentage destruction of angiotensin after 40 mts incubation in normal monkeys was 82.52%. The activity decreased to 41.01% immediately after the cold injury. The activity remained significantly low on 3rd day (52.08%) and 6th day (65.51%).

The percentage destruction of histamine after 40 mts incubation of normal monkeys was 55.98%. The activity increased to 88.20% immediately after the cold injury and remained high on 3rd (79.50%), 6th (71.78%), and 10th (67.37%) day after the experimental cold injury.

DISCUSSION

Vasoactive peptides have been recognized to play an important role under various pathophysiological conditions. In the present study, attempts have been made to investigate the role of vasoactive peptides and local hormones. Studies on bradykinin has been carried out by extracting the substrate, bradykininogen from plasma and finally assaying the liberated bradykinin from the substrate by the action of trypsin (Diniz and Carvalho, 1963). The results are reported in terms of bradykinin liberated from the substrate. Immediately after the cold injury bradykinin level decreased and remained low up to the sixth day of the cold injury (Table 1). It is at present difficult to correlate the change in the bradykininogen level to the bradykinin concentration. Because the fall in the substrate could be either due to reduced synthesis of the substrate in the liver or other tissues or due to increased release of the active peptide (Periti and Gasparri, 1969; Urbanitz, et al, 1970). Tissue injury (Rocha e Silva and Antonio, 1960) and local ischaemia

(Bavazzano et al. 1969) has been reported to be kininogen depressant. Since there occurs a parallel fall in the kininogen level in the tissue of the monkey's tendo-achiles, in addition to the observation that exogenous administration of synthetic bradykinin (BRS 640) produces beneficial effect in the prevention of gangrene (Arora et al, 1969), it seems probable that a fall in the level of bradykinin occurs after experimental cold injury. Similarly histamine level decreased in the blood but increased in the tissue. The maximum fall in both the potent vasodilators occur on the 3rd and 6th day of the cold injury, the period during which the maximum gangrene formation occurs. During the same period significant increase in the concentration of angiotensin and 5-HT occurs in plasma. These observations therefore indicate that immediately after the cold injury the vasoconstriction is probably due to the local effect of cold and decreased concentration of vasodilators like bradykinin and histamine which is aggravated by the significant increase in the circulating level of angiotensin and 5-HT.

The increased kininase activity (Table 2) may be an important factor in decreasing the bradykinin level because the increase was significant on 3rd and 6th day of the injury, the period during which the kinin level was very low. Similarly the increased plasma histaminase activity might be playing a role in decreasing the level of histamine although the increased histaminase activity may or may not be related to the histamine release from the injured site. It has also been suggested that the fall in the histamine may be indirectly through the endogenous secretion of cortisone and thus inhibiting the synthesis of histamine (Schayer et al, 1954). Cortisol level in the plasma of monkey increased immediately after the cold injury possibly due to hyperactivity of the sympathetic system (Arora et al, 1971b) which in part may account for the fall in histamine level. However the cortisol level was normal on 3rd, 6th, and 10th day of cold injury whereas plasma histamine level remained low up to the 6th day of the injury. The angiotensinase activity decreased during the cold injury period (Table 2). Hence the increased level of angiotensin throughout the experimental period of cold injury may be due to the decreased angiotensinase activity during the cold injury. It seems also possible that kininase- II enzyme which inactivates bradykinin and liberate hypertensive peptide, angiotensin- II from angiotensin- I (Yang et al, 1971) may be having a role in decreasing the level of bradykinin and elevating that of angiotensin- II in experimental cold injury.

REFERENCES

Abe, K., Watanabe, K. N., Mouri, T., Seki, T., Miwa, I., Tokano, T., Yoshnaga, K., and Torika, T. (1969): Pharmacol. Res. Commun. 1, 127.

Amin, A. H., Crawford, T. B. B., and Gaddum, J. H. (1954): J. Physiol. 126, 596.

Arora, R. B., Jagota, S. C., Gupta, S. K., and Aggarwal, O. P. (1969): Pharmacol. Res. Commun., 1, 265.

Arora, R. B., Siddiqui, H. H., and Tariq, M. (1971a): Ann. Ind. Acad. Med. Sci., 7, 151.

Arora, R. B., Tariq, M., and Siddiqui, H. H. (1971b): Pharmacol. Res. Commun., (accepted).

Babazzano, A., Sidell, N., Michelacci, S., and Sicuteri, F. (1969): Pharmacol. Res. Commun., 1, 137.

Code, C. F. and McIntire, F. C. (1960): In Methods of Biochem. Analysis, ed. Glick, D., Interscience Publishers Inc., New York, 3, 49.

Diniz, C. K. and Carvalho, I. F. (1963): Ann. N. Y. Acad. Sci., 104, 77.

Kulka, J. P., Roos, T. B., Damin, G. J., and Blair, J. R. (1958): U. S. Army Res. Lab. Air Report No. 326.

Landesman, R., Biron, P., Castellanos, R., Larussa, R., and Wilson, K. H. (1963): Obstet. Gynec., 22, 316.

Lewis, T. (1941): Brit. Med. Jour., 2, 795.

Lin, R. C. Y. and Yeoh, T. S. (1965): J. Pharm. & Pharmacol., 17, 524.

Osborn, E. C. (1969): J. Pharm. & Pharmacol., 21, 862.

Periti, P. and Gasparri, F. (1969): Pharmacol. Res. Commun. 10, 182.

Regoli, D. and Vane, J. R. (1964): Brit. J. Pharmacol., 23, 351.

Rocha e Silva, M., (1964): Ann. N. Y. Acad. Sci., 116, 899.

Rocha e Silva, M. and Antonio, A. (1960): Med. Exptl., 3, 371.

Schayer, R. W., Smiley, R. L., and Davies, K. J. (1954): Proc. Soc. Exp. Biol. Med., 87, 590.

Spencer, P. S. J. (1963): J. Pharm. & Pharmacol., 15, 255.

Sullivan, B. J. and Towle, L. B. (1957): Amer. J. Physiol., 189, 498.

Talwar, J. R., Gulati, S. M., and Kapur, B. M. L. (1966): Ind. J. Med. Res., 54, 643.

Urbanitz, D., Sailer, R., and Habermann, E. (1970): Adv. Exptl. Med. Biol., 8, 343.

Vane, J. R. (1957): Brit. J. Pharmacol., 12, 344.

Yang, H. Y. T., Erdos, E. G., and Levin, Y. (1971): J. Pharmacol. Exp. Therap., 177, 291.

Zweifach, B. W. (1963): Feder. Proc., 22, 7351.

PREKALLIKREIN AND KALLIKREIN INHIBITOR IN PLASMA OF PATIENTS AFFECTED BY RECENT MYOCARDIAL INFARCTION *

F.Sicuteri, F.M.Antonini°, P.L.Del Bianco, G.Franchi and C.Curradi

Department of Clinical Pharmacology and of Gerontology°,

University of Florence - Florence, Italy.

Different causes may carry to the shock. The shock credibly has a pathological substrate, that is a damage of the microvessels, particularly capillaries and venules, vasa vasorum included; the sufference of the microvessels concerns mainly the endothelial and basal membrane. Pathogenetically the basic mechanism of the shock is the capillary damage by hypoxia. Hypoxia, in order to establish a capillary damage, must be calibrated in intensity and duration; differently hypoxia, if too important and acute, provokes death following an acute brain damage; if instead too slight, hypoxia does not damage sufficiently the microvessels.

According to the causes, the shock may be grouped in three classes: thus one may distinguish an autogenic, hematogenic and dynamogenic shock (Fig. 1). A calibrated and prolonged hemorrhage may be able to induce the hematogenic shock, a long lasting graduated release of hypotensive substances is the common cause of the autogenic shock. Finally a progressive failure of the heart realizes the dynamogenic shock. Shock from myocardial infarction is commonly considered as a typical dynamogenic shock.

Some years ago we proposed that cardiogenic shock from myocardial infarction must be considered as a mixed shock; in part as dynamogenic, in part as autogenic (Fig. 2). Infact during infarction proteases may be released from the damaged myocardium and so a long lasting generation of kinins takes place. We observed that the kininogen during myocardial infarction decreased (Sicuteri et al., 1967; 1970). These findings were confirmed by others (Wiegershausen et al., 1970, Massion et al., 1969, Pitt et al., 1969), both in man and in animals. A kallikrein inhibitor, Trasylol, seems to benefit patients with cardiogenic shock (Sotgiu, 1964; Palma and Brusadelli, 1969; Massion et al., 1969).

The present paper aims to enlarge our observations, by evaluating also prekallikrein (kallikreinogen) and kallikrein inhibitor.

METHODS

We considered some patients with myocardial infarction hospitalized at the

* With the support of the National Council of Researches, Rome, Italy.

ETIOLOGY OF SHOCK

1) Hematogenic shock { *(hemorrhage)*

2) Dynamogenic shock { *(cardiac insufficiency)*

3) Autogenic shock { *(burns*
necrosis
trauma
antigen-antibody reactions)

Fig. 1 - According to the original aetiology, all shocks may be classified in 3 groups (Sicuteri et al., 1970)

Coronary Unit of the Institute of Gerontology of the University of Florence. In all these patients the spontaneous esterase, prekallikrein and kallikrein inhibitor were evaluated, according to the kaolin contact technique (Sherry and Colman, 1968; Colman et al., 1969; 1969; 1969). This technique is based upon the sharping increase of the spontaneous arginine esterase activity of the plasma following the contact with kaolin. Kaolin, as it is known, when mixed with plasma, provokes an increase of the plasmatic arginine esterase. It is to notice that the striking increase of the arginine esterase activity induced by the kaolin contact is due mainly to an increase of the kallikrein activity, and not, or in very small extent, if so, to other esterase activity (Sherry and Colman, 1968).

In normal subjects the maximum of esterase activation (about 150 μ moles of TAME/ml of plasma/hour) is reached after 1 minute from the kaolin plasma contact. The difference between the maximum of activation (1 min.) and the spontaneous esterase activity corresponds to the prekallikrein level. After 1 minute from the kaolin contact the esterase activity will decrease: 5 min from the kaolin contact in the normal subjects the esterase activity appears reduced of 50%; this reduction is due to the activity of the kallikrein inhibitor; by using a particular calculation, the amount of kallikrein inhibitor may be evaluated. This technique of kaolin activation appears suitable in those clinical conditions in which a trouble of the kinin system may be postulated. The present investigation concerns 22 normal subjects and 16 patients who had recently had a myocardial infarction. Spontaneous

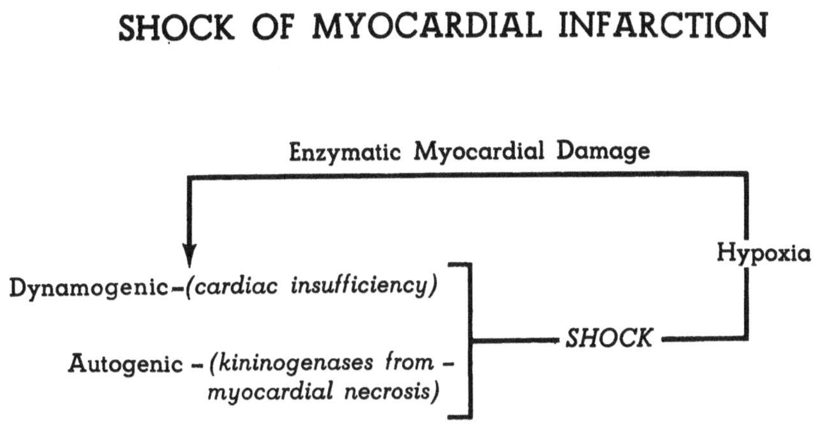

Fig. 2 - The cardiogenic shock seems to be the only shock having a mixed aetiology: dynamogenic (calibrated cardiac insufficiency) and autogenic (release of proteases from damaged myocardium).

esterase activity, kallikreinogen and kallikrein inhibitor were evaluated with a total of 70 determinations. In the majority of the patients with myocardial infarction prekallikrein and kallikrein inhibitor were measured nearly for two weeks, beginning from what was considered as the first day of the disease. Other enzymatic activities, as the transaminases etc. were carried out.

RESULTS

Present results may be summarized as follows:

1) In the normal group (22 determinations) the levels of the spontaneous esterase, kallikreinogen and kallikrein inhibitor correspond to those previousely reported (Fig. 3).

2) In the myocardial infarction group (48 determinations) the average of all samples indicates that: a) spontaneous esterase activity appears similar to that of the normal group; b) the 1 min. kaolin activated arginine esterase, poorly increases; c) arginine esterase, 5 min. after kaolin activation, is similar to that of normal subjects (Fig. 3 and 4).

3) The level of 1 min. kaolin activated esterase, lowers gradually, reaching the minimum at the eighth day of the disease (Fig. 5, 6 and 7).

COMMENT

The present results support the hypothesis of a some derangement of the kinin

KALLIKREINOGEN VALUES IN NORMAL SUBJECTS AND IN PATIENTS WITH MYOCARDIAL INFARCTION

	n. of cases	n. of determinations	Spontaneous esterase activity		Kaolin activated kallikreinogen 1 min.		5 min.	
			mean	E.S.	mean	E.S.	mean	E.S.
Normal	22	22	2,8	± 0,60	106,9	± 3,99	56,68	± 2,10
Infarction	16	48	4,35	± 0,65	27,4	± 3,14	49,1	± 3,37
Significance (T Student)			P>0,1		P<0,001		P>0,1	

Fig. 3 - General average of values of plasma kallikreinogen and kallikrein inhibitors in normal subjects and in patients with a recent myocardial infarction. Kallikreinogen only appears significantly depressed.

KALLIKREINOGEN ACTIVATION CURVES IN NORMAL SUBJECTS (MEAN OF 22 CASES) AND IN PATIENTS WITH MYOCARDIAL INFARCTION (MEAN OF 48 DETERMINATIONS)

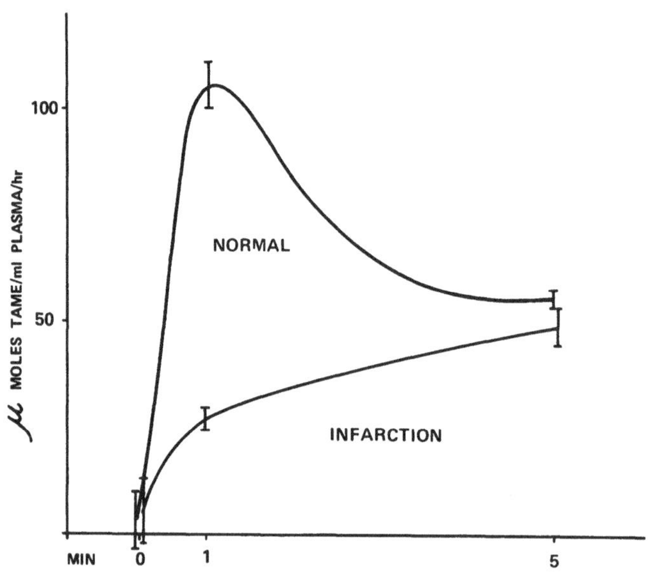

Fig. 4 - The dynamics of the kallikrein activation by kaolin appears deeply alterated in the patients with myocardial infarction. The plasma of the infarction group exhibits a strongly reduced ability to activate esterase at a minute from the kaolin contact, in comparison with normal subjects. This should indicate a minor disponibility of kallikreinogen, probably as a consequence of its activation to kallikrein. No changes on the esterase activity and on the kallikrein inhibitor.

ESTERASE ACTIVITY FOLLOWING INFARCTION

day of disease	number of determinations	spontaneous esterase activity	Kaolin activated Kallikreinogen	
			1 min.	5 min.
1°	2	9,0	54,5	58,0
2°	9	5,3	43,1	42,8
3°	8	2,7	36,6	55,7
4°	6	4,5	22,8	44,0
5°	4	4,0	26,7	49,5
6°	3	4,0	16,3	58,3
7°	2	2,5	20,5	48,0
8°	3	2,3	9,6	35,0
9°	3	3,3	17,0	43,6
10°	2	6,5	24,0	57,5
11°	3	2,0	10,0	53,3
16°	3	8,3	11,6	58,6
Mean	48	4,35	27,4	49,3
S.E.		± 0,65	± 3,14	± 3,37

Fig. 5 - The daily evaluation of the kaolin activable esterase shows that the decrease of kallikreinogen is gradual and reaches the minimum level on the eighth day of the disease.

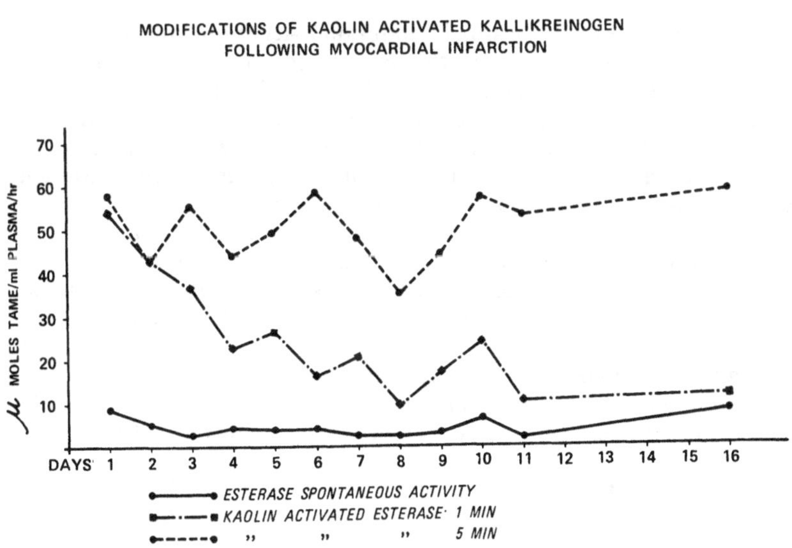

Fig. 6 - Graphic reproduction of the values of the table on fig. n. 5.

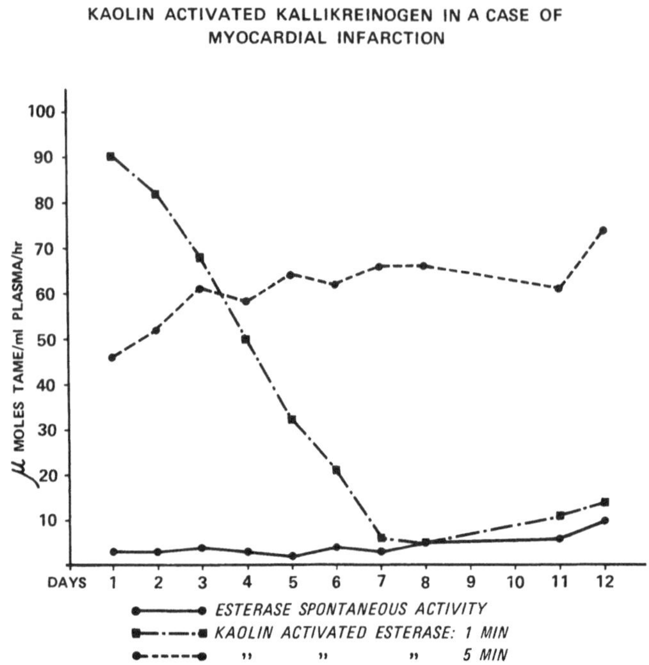

Fig. 7 - The dynamics of the esterase (kallikrein) activation by kaolin in a case of myocardial infarction: to notice the gradual and remarkable lowering of the kallikreinogen.

system following an acute myocardial infarction. According to this hypothesis previous investigations in man and in the animal indicated a lowering of plasma kininogen in spontaneous and experimental infarction. Enzymatic studies as that of esterase kaolin activable are expected to be the most suitable to focuse the changes in a complex system as that of kinins. Infact, by using the esterase kaolin activable technique, the derangement of the kinin system in myocardial infarction appears more evident. Of three parameters offered by this method, spontaneous esterase, kallikrein inhibitor and prekallikrein, the last alone appears strikingly modified: plasma sampled from patients with recent infarction loses day by day the ability of activating esterase (kallikrein) when mixed with kaolin. This loss is gradual and arrives to lower levels still, being near to zero, on the eighth day of the disease.

Presently the most acceptable interpretation of the reduced activability of esterase during a recent myocardial infarction may be summarized as follows:

1) Minor disponibility of plasma kallikreinogen, as consequence of a long lasting transformation in kallikrein. In other words this should indicate a strong activation of the kinin system.

2) Increased activity of the kallikrein inhibitor in plasma with the consequence of stronger inhibition of the activability in vitro of esterase by kaolin contact. In this case plasma kallikreinogen could not be necessarily reduced.

3) Reduced disponibility and/or activability of esterase by kaolin mixture in vitro;

in this case alone the amount of plasma kallikreinogen could be normal.

Since our patients have not been treated with heparin and/or with kallikrein inhibitors the possibility of an interference of some pharmacological action is quite poor.

SUMMARY

By using the kaolin activable esterase method, prekallikrein (kallikreinogen) and kallikrein inhibitors were evaluated in 16 patients (48 determinations) with recent myocardial infarction, and in 22 normal subjects (22 determinations). No changes were observed in the spontaneous esterase and kallikrein inhibitor: differently in patient group prekallikrein progressively decreased and reached the lowest level on the eighth day of the disease.

The significant gradual reduction of prekallikrein seems to support the hypothesis of an implication of the kinin system in myocardial infarction.

REFERENCES

Colman R.W., Mason J.W. and Sherry S.: "The kallikreinogen-kallikrein enzyme system of human plasma". Ann. Int. Med., 71, 763, 1969.

Colman R.W., Mattler L. and Sherry S.: "Studies on the pre-kallikrein (kallikreinogen) -kallikrein enzyme system of human plasma. I. Isolation and purification of plasma kallikreins". J. Clin. Invest. 48, 11, 1969.

Colman R.W., Mattler L. and Sherry S.: "Studies on the pre-kallikrein (kallikreinogen) -kallikrein enzyme system of human plasma. II. Evidence relating the kaolin-activated arginine esterase to plasma kallikrein". J. Clin. Invest. 48, 23, 1969.

Massion W.H., Blumel G. and Peschl L.: "The role of the plasma kinin system in cardiogenic shock and related conditions". Proceed. Int. Symp. on "Proteases and Antiproteases in Cardioangiology", Vietri sul Mare, Italy, June 1969. F.K. Schattauer Verlag, Stuggart, N.Y. 81, 1969.

Palma G. and Brusadelli S.: "Prospettive terapeutiche con inibitori enzimatici nell'infarto miocardico". Proceed. Int. Symp. on "Proteases and Antiproteases in Cardioangiology", Vietri sul Mare, Italy, June 1969. F.K. Schattauer Verlag, Stuttgart, N.Y. 1969.

Pitt B., Mason J., Conti C.R. and Colman R.W.: "Activation of the plasma kallikrein system during myocardial ischemia". Proceed. Int. Symp. on "Cardiovascular and neuro-actions of Bradykinin and Related Kinins". Fiesole (Florence), 21-25 July 1969. Advances in Exper. Med. Biol. 8, 403, 1970.

Sherry S. and Colman R.W.: "Observations on the plasma kallikreinogen (prekallikrein)-kallikrein enzyme system". Trans. Ass. Am. Physicians. 81, 40, 1968.

Sicuteri F., Franchi G., Del Bianco P.L. and Del Bene E.: "A contribution to the interpretation of shock and pain in myocardial infarction". Mal. Cardiovasc. 8, 3, 1967.

Sicuteri F., Del Bianco P.L. and Fanciullacci M.: "Kinins in the pathogenesis of cardiogenic shock and pain". Shock: Biochemical, Pharmacological an Clinical aspects, Plenum Press, 315, 1970.

Sotgiu G.: "Azione dell'inibitore delle proteasi (Trasylol) nell'infarto miocardico". Boll. Sci. Med. 136, 279, 1964.

Wiegershausen B., Henninghausen G., Klausch B., Korber H.G. and Hauzeur E.: "Plasma kininogen level in myocardial infarction". Proceed. Int. Symp. on "Cardiovascular and neuro-actions of Bradykinin and Related Kinins", Fiesole, (Florence), July 1969. Advances in Exper. Med. and Biol. 8, 403, 1970.

PLASMATIC KININOGEN IN ACUTE HEPATITIS AND IN LIVER CIRRHOSIS *

Del Bianco P.L., Fanciullacci M. and Franchi G.

Department of Clinical Pharmacology, University of Florence,

Florence, Italy

The first researches on the kininogen level in plasma indicated that the probable site of synthesis of this precursor was the liver (Diniz et al., 1961; Diniz and Carvalho, 1963). This observation was based upon the fact that, being the liver the main site of the protein synthesis, it was natural to beleive that also the polypeptide sequence of kininogen was realized in this organ. We still lack, however, of direct evidences of this situation in man; as regards to the animal, Bryan et al. (1971) were able to prove in the rat that the synthesis of kininogen occurs in the liver. However, in man we still lack of evident proofs. Diniz et al. (1961, 1963) noticed that in serious damage of the hepatic functions, namely cirrhosis, the plasmatic kininogen resulted to be depressed.

It is known, on the other hand, that the behaviour of the α_2 globulin fraction in liver cirrhosis is not constant, also if in most part of the cases we register a decrease of this proteic plasmatic fraction.

Successively we also observed that the level of kininogen in the blood from the inferior vena cava of subjects who had undergone cardiac catheterism for diagnostic purposes, and resulting without cardiopathy, was equal to that of blood sampled from the cubital vein (Pannuccio et al., 1966). We therefore considered useful to extend these first findings, observing the behaviour of kininogen in subjects suffering from liver cirrhosis. The research was extended to cases of acute virous hepatitis.

METHODS

Plasmatic kininogen was assayed by the technique of Diniz and Carvalho (1961). The study was carried out in patients of the Department of Medicine of the University of Florence hospitalized for liver cirrhosis and acute virous hepatitis.

The length of the observation was from 20 to 60 days during which the patients were carefully observed with suitable laboratory tests.

The kininogen value in all the patients with liver cirrhosis was assayed at the

* With the support of the National Council of Researches - Rome -

moment of hospitalization. In 9 cases the assay was carried out at the end of hospitalization after the conventional therapy (Fig. 1). For comparison the kininogen level of 36 normal subjects chosen amongst the students of the same department was assayed.

The results are summarized in figure 2. The kininogen medium values in the liver cirrhosis are significantly lower than in normal subjects. The medium value of kininogen in the 26 cases of liver cirrhosis is 1.49 mcg/ml against 2.6 mcg/ml of normal subjects.

There does not exist, however, a significant difference between the normal subjects kininogen and that of patients with acute virous hepatitis whose kininogen medium value results 2.1.

The total plasma proteins are significantly decreased in the patients with liver cirrhosis. A such difference does not involve the α_2 globulins that in some cases are increased.

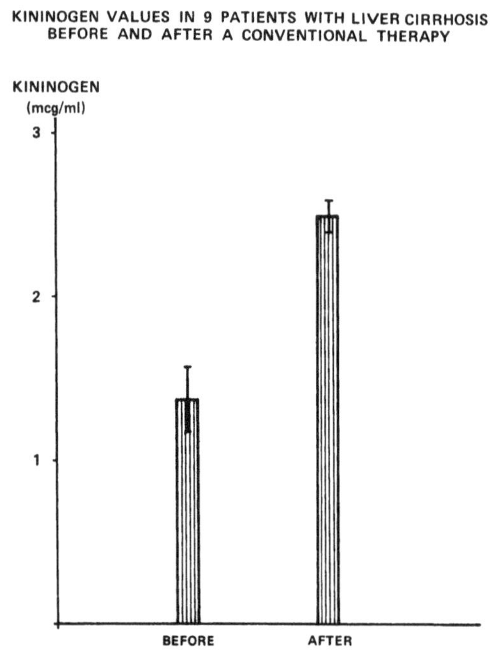

KININOGEN (mean of 9 cases)	1,37	2,40
S.E	± 0,2	± 0,1
SIGNIFICANCE (T. Student)	P< 0,005	

Fig. 1 - It is observed that after a conventional therapy, the kininogen level in patients affected by hepatic cirrhosis, returns towards normal levels, with statistically significant difference in regards to the values detected before the therapy

KININOGEN VALUES IN PATIENTS SUFFERING FROM CIRRHOSIS AND HEPATITIS

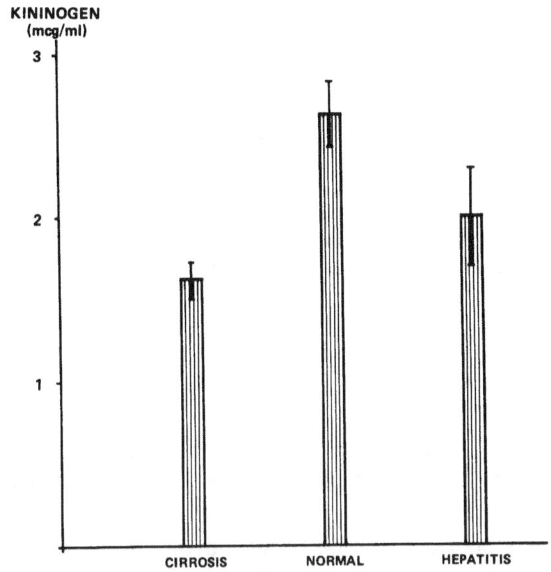

	CIRROSIS		NORMAL		HEPATITIS
NUMBER OF SUBJECTS	26		36		11
KININOGEN (medium values)	1,49		2,6		2,1
S.E.	± 0,1		± 0,2		± 0,3
SIGNIFICANCE (T. Student)		$P < 0{,}001$		$P > 0{,}05$	

Fig. 2

No significant variations in the total plasma proteins of the patients with acute virous hepatitis have been observed.

COMMENT

From the data referred different considerations may be made. Firstly, in conditions of serious hepatic damage, as in cirrhosis, there exists a significant decrease of the plasmatic kininogen, with no reduction of the α_2 globulins. This could be due to an increased release of kinins. On the other hand our previous non-published observations put in evidence the fact that in course of hepatic cirrhosis, the kinin level in the venous blood sampled from the cubital vein is normal. We should therefore also consider the possibility that the serious damage of the liver cells may compromise specifically the polypeptidic chain synthesis, considering that there are no significant changes in the α_2 globulins.

In acute virous hepatitis there are no significant decreases of the plasmatic kininogen. In this situation, on the other hand, the damage of the cells is limitated, and the left over parenchyma may be functionally sufficient to maintain within normal levels the kininogen. This result, though it is indirect, could favour the hypothesis of synthesis of kininogen in the liver.

SUMMARY

The plasmatic kininogen level has been studied in patients suffering from hepatic cirrhosis and virous hepatitis. It results significantly decreased in hepatic cirrhosis, while it is normal in virous hepatitis. This result may be interpreted as datum favouring the fact that the kininogen synthesis occurs in the liver.

REFERENCES

Bryan F.T., Ryan J.W. and Nimeyer R.: "Bradykininogen synthesis by liver". This Symposium

Diniz C.R., Carvalho I.F., Ryan J. and Rocha e Silva M.: "A micromethod for the determination of bradykininogen in blood plasma". Nature (Lond.) 192, 1194, 1961.

Diniz C.R. and Carvalho I.F.: "A micromethod of estimation of bradykininogen under several physiological and pathological conditions". Ann. N.Y. Acad. Sci. 104, 77, 1963.

Panuccio P., Cresci F. and Viroli L.: "Hypotensive peptides". Proc.Int. Symposium. E.G.Erdös, N.Back and F.Sicuteri Ed. Springer Verlag, N.Y. 1966, p. 534 (Discussion).

Kallikrein on Arterial Hypertension:1. Kallikrein-Like Activity in the Urine of Figure-in-Eight Ligature in one Kidney

H. Croxatto, Mary Luz San Martìn and J. Roblero
Laboratorio Fisiologìa Universidad Catòlica de Chile
Santiago, Chile

Most data at present point to the kidney as the most important final arbiter in determining the long term basal level of arterial pressure, but the renin angiotensin theory cannot explain all the facts (Guyton et al., 1969). Kallikrein obtained from different sources induces vascular reactions which are opposite to that elicited by renin. Kinins promote in the kidneys an increase of blood flow and a striking natriuresis (Webster and Gilmore, 1964). On the other hand kallikrein is present in the renal tissue and shows pharmacological properties which are similar to that found in the urine (Croxatto and Noé, 1970). These results lead us to investigate whether urinary kallikrein undergoes changes in the rat which develops renal hypertension when the circulation of the kidney is impaired. This study brought out clear evidence that a typical decrease in kallikrein activity of the urine is present in any rat which develops hypertension caused by unilateral nephrectomy and a figure-in-8 ligature applied to the contralateral organ.

METHODS

Four successive series of experiments were undertaken in adult male rats. (140-160 g) In the three first series 114 Wistar rats and in fourth series 30 Sprague-Dowley rats were used, distributed in different groups: a) 42 normal; b) 35 uninephrectomized; c) 17 figure-in-8 ligature in one kidney and d) 50 uninephrectomized and with figure-in-8 ligature in the remnant kidney. In the latter group the operation to produce hypertension was carried out according to the Grollman procedure (1941). In the first stage of the operation, a figure-in-8 ligature was made in one kidney, a week later the other kidney was removed. Blood pressure was measured in the rat tail at weekly intervals beginning 10 days after the kidney removal. The microphone technique described by Friedman and Freed (1949) was used. During the blood pressure determinations the animals were kept quiet under light ether anesthesia. Once or twice a week, the urine was collected by placing the rats in individual metabolic cages for 1 to 8 hours in the morning. No food was given but they had free access to drinking water during the collection period. In many experiments only the freshly voided urine was used for testing, but in some cases the urine was collected for several hours to carry out purification procedures. In the latter case, the urine was received under toluene to prevent microbe growth. Each sample of urine was measured, centrifuged and kept frozen when not immediately assayed.

Kallikrein testing: Individual samples of fresh voided urine, or pooled urine corresponding to the different groups, or urine fractions in different purification stages, were tested using two bioassays: a) contraction effect upon an isolated rat uterus and b) vasodepressor effect on the rat blood pressure of an anesthetized rat.

Isolated rat uterus: A rat uterine horn was submerged in a bath of 20 ml capacity containing Tyrode solution at 36°C, continuously oxygenated by air bubbling. The kallikrein activity was determined using two methods: a direct one and an undirect one.

Direct method: According to Beraldo et al. (1966), and Croxatto and San Martìn (1969), rat urinary kallikrein induces direct oxytocic effect on the rat uterus, effect which is dose dependant.

The oxytocic activity of different amounts of urine was compared to that induced by a standard solution of bradykinin. The activity of 1 ml of urine was expressed in ng of bradykinin which produced equivalent oxytocic effect. For instance, if 0.01ml of urine was required to produce an effect comparable to 5 ng of bradykinin it was assigned to 1 ml of urine a kallikrein activity of 500 ng of bradykinin.

Indirect method: The kallikrein activity can be measured, as well, by the ability of the urine to form kinins when it is conveniently diluted in Tyrode solution and incubated at 37°C in the presence of kininogen for 4 minutes. Doses of urine much lower than the threshold dose to produce an oxytocic activity by the direct method were required. The mixture of the urine plus substrate at the end of the 4 minutes of incubation was added to the bath. The magnitude of uterine contraction depends on the amount of liberated kinins. Since a perfect correlation was found between both the direct and indirect method to test the kallikrein activity, it was decided to use the data derived from the direct method to express the kallikrein activity. The kininogen used in the experiments was obtained from two different sources: one was obtained from human plasma fractionation° and the other one was dog serum incubated at 37°C for 90 hours, under sterile conditions. This serum has no direct oxytocic effect in the doses used and releases kinin very quickly in the presence of rat urine.

Blood pressure test : The kallikrein activity was also determined injecting the dialized urine or its purified fractions in the femoral vein of rats anesthetized with sodium pentobarbital. The blood pressure was recorded either with membrane manometer, calibrated in mm of Hg or with a Grass polygraph using a Stathan transducer connected to the carotid artery.

Purification procedures : In order to asses whether the oxytocic activity was due to the presence of a kallikrein enzyme, the following purification procedure was adopted: The centrifuged urine introduced in cellulose tubing was dialized at room temperature for 48 h, against tap water and 24 h against destilled water in the cold room. Usually toluene was added (2%) to prevent microbes growth when urine was exposed to room temperature. The small precipitate formed during dialysis was discarded by centrifugation. Kallikrein was precipitated from the supernatant by the addition of 4 volumes of cold acetone. The mixture was centrifuged in the cold room and the precipitate was dissolved in a small volume of 0.05 M ammonium acetate. The small amount of insoluble material was discarded through centrifugation. Further purification was acheived by gel filtration. A column (50 x 2.5 cm) of Sephadex G-100 embedded in ammonium acetate 0.05 M, pH 7.8 was employed. The column was charged with 10 ml of a solution corresponding to 150 ml of urine. The effluent solutions collected in separate volume of 8 ml were tested using either the isolated rat uterus or the blood pressure test. The effluent solutions which showed at a dose of 0.1 ml oxytocic and depressor effects were

° We are very much indebted to Miss M. Webster for her generous supply of human kininogen

mixed altogether and liophilyzed.

Inhibitory solutions: The kallikrein inhibitor, Trasylol Bayer (5.000 U p ml) and carboxypeptidase B (Biochemical Laboratories) were used to inhibit the oxytocic activity displayed by urine and its different purified fractions. 40 to 100 mcg of carboxypeptidase were introduced to the uterus bath prior the urine addition. Soya bean trypsin inhibitor at the doses of 2-5 mg added to the bath was also used.

RESULTS

The experiments confirm that the uterotonic activity of rat urine upon uterus is due to a non dialyzable substance, protein in character. Its capacity to set free kinins in the presence of kininogen runs parallel to its direct oxytocic effect. Trasylol (500-1000 U) inhibits the release of kinins when the enzyme and the substrate are mixed and in addition, it blocks the uterus contraction when it is added to the bath prior to urine or its active fractions (Fig. 1).

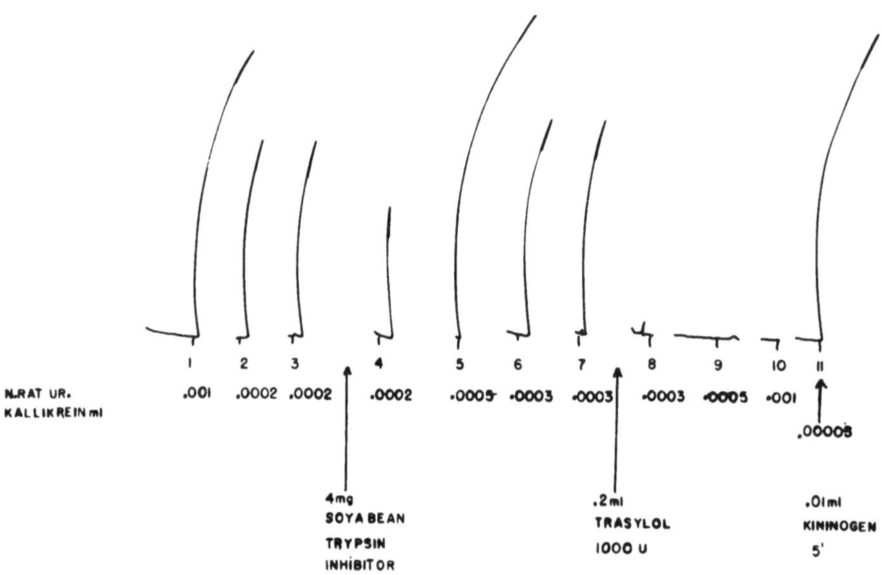

Fig. 1. Contractions of an isolated rat uterus (20 ml bath). Effect of 4 mg of soya bean trypsin inhibitor and Trasylol on the purified kallikrein (through Sephadex filtration). These inhibitors were introduced in the bath before the addition of kallikrein. In 11, a mixture of a subthreshold dose of kallikrein and kininogen incubated for 5 m was introduced in the bath containing Trasylol.

Carboxypeptidase (40 80 mcg) introduced into the uterus bath, inhibits the oxytocic effect of urine as well as bradykinin (Fig. 2).

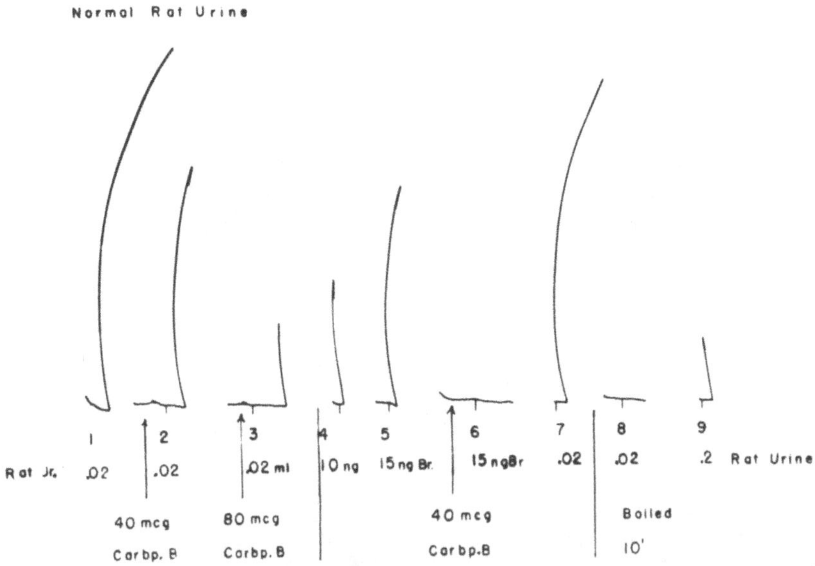

Fig. 2 Contractions of an isolated rat uterus produced by rat urine (in ml.) In 2,40 mcg carboxypeptidase B (c.p.B.) was added before 0.02 ml of urine were introduced. In 3,80 mcg of c.p.B. before 0,02 ml of urine was introduced. In 5,15 ng bradykinin. In 6,40 mcg c.p.B. was added before 15 ng bradykinin. After washing 0,02 ml of urine was introduced. 8, 0,02 ml of boiled urine and in 9, 0,2 ml of boiled urine for 10 minutes were added.

Soya bean trypsin inhibitor (2-5 mg) has only very weak inhibitory effect in the conditions that Trasylol hinders completely the reaction (Fig. 1). The action of Trasylol upon uterus is reversible and disappears after the first uterine washing.

The intravenous injection in the rat of the fresh voided, or its purified fractions, bring out vasopressor effects (Fig. 3) which are parallel to their uterotonic activities (Fig. 4). Vasopressor effects are also inhibited by the injection of Trasylol to the recipient rat (Fig. 5). These experiments not only provide strong support to the concept that a kallikrein enzyme is the responsible substance, but also they can be used as an accurate procedure to quantify the k.a.

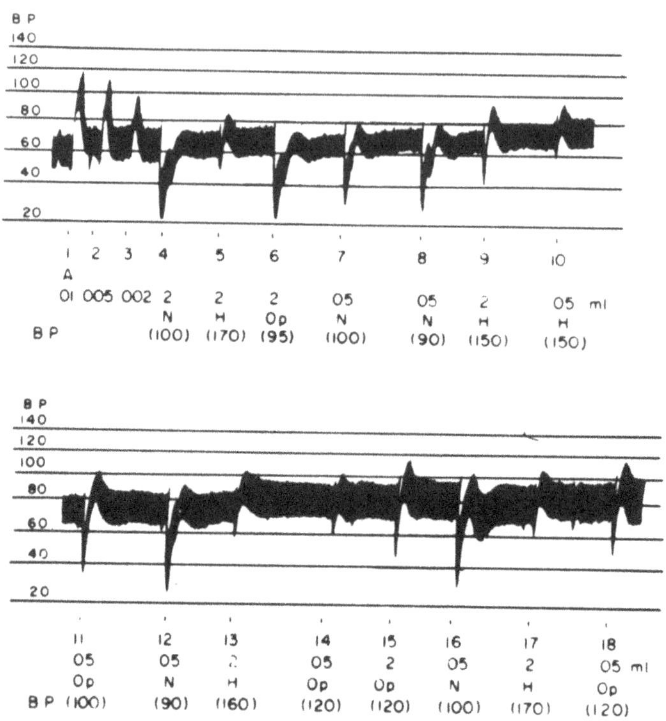

Fig. 3 Blood pressure changes produced by the injection of dialyzed urine of (N) normal rats; (OP) rats with figure-in-8 ligature in one kidney and the opposite untouched and (H) rats uniphrectomized and figure-in-8 ligature in the remnant. Figures in parenthesis indicate blood pressure (B.P.) in the rat whose urine was injected.

A, correspond to angiotensin II in mcg. The second row of figures indicate urine volume injected.

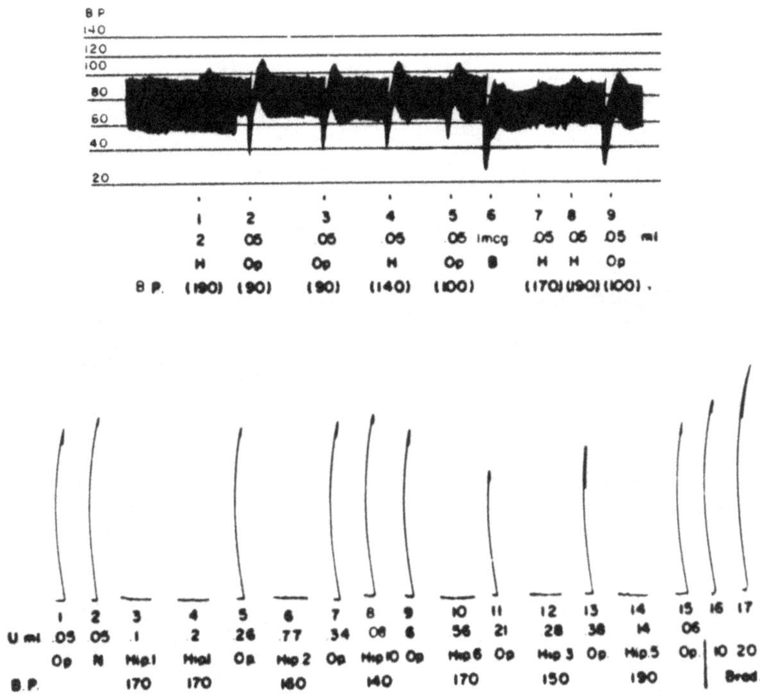

Fig. 4 Blood pressure changes and oxytocic effects produced by different samples of fresh voided urine of rats (OP) with figure-in-8 ligature in one kidney and the contralateral untouched; and rats (H) with uninephrectomy and a figure-in-8 ligature in the remnant. The second row of figures indicate the volume of urine injected. B indicates injection of bradykinin. B.P. correspond to the blood pressure of the rats whose urine was tested. The volume of urine added to the uterus bath was in each case, 1/10 of the amount excreted in 5 hours collection.

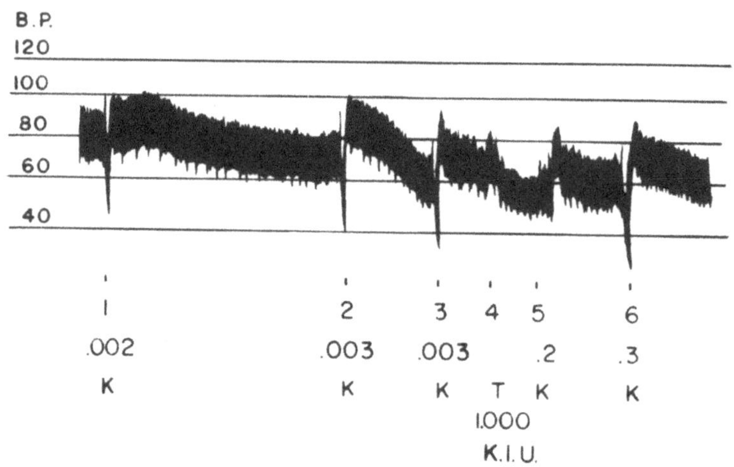

Fig. 5 Blood pressure of an anesthetized rat. Through the femoral vein, 0.002, 0.003 and 0.003 ml of purified kallikrein was injected. In 4, 1000 U. of Trasylol was injected. Afterwards 0.2 and 0.3 ml of kallikrein were injected.

In the urine of normal rats the k.a. per ml shows moderate variations from day to day. Apparently it is inversely correlated with the amount of excreted water. The highest water diuresis coincides with the lowest k.a. per ml. The average value for the k.a. found in 3 normal groups of Wistar rats was about 725 ng and for the Sprague-Dowley rats 425 ng of bradykinin per ml of urine (Table 1).

KALLIKREIN LIKE ACTIVITY IN 1ml OF URINE
EXPRESSED IN ng BRADYKININ OXYTOCIC EQUIVALENT EFFECT

NUMBER OF EXPERIMENT	DAYS AFTER OPERATION	STRAIN	GROUP OF RATS				
			Normal	Uninephrectomized	Fig of 8 in One Kidney	Fig of 8 Uninephrectomized Hypertensive	
							Blood Pressure
I	48	Wistar	745 ± 113 (4)	300 ± 48 (4)	280 ± 83 (4)	40 ± 10 (5)	158
II	90	"	950 ± 103 (4)		480 ± 112 (11)	27 ± 14 (6)	165
III	101	"	520 ± 82 (8)	358 ± 70 (6)	370 ± 92 (8)	30 ± 12 (6)	155
IV	70	Sprague Dowley	450 ± 72 (10)	310 ± 88 (6)		< 10 (8)	192 ± 18
	75	"		285 ± 81 (4)		< 10 (9)	188 ± 15.5
	84	"		200 ± 53 (9)		< 10 (10)	188 ± 15.5
	108	"	420 ± 83 (10)	270 ± 78 (9)		< 10 (8)	190 ± 15

Table 1 Mean values and s.d. of k.a. per 1 ml of urine of different groups of rats. Bioassays were performed using the urine collected in the indicated day after the kidney removal.

The reduction of kidney mass in the uninephrectomized significantly diminishes the amount of kallikrein excreted, particularly within the first month following the kidney estirpation (Fig. 6). In the group c) corresponding to the animals bearer of figure-in-8 ligature in one kidney with the other untouched, there was also an abrupt fall of k.a. immediately after the ligature but it went up to the levels observed in the uninephrectomized (Fig. 6 and table 1).

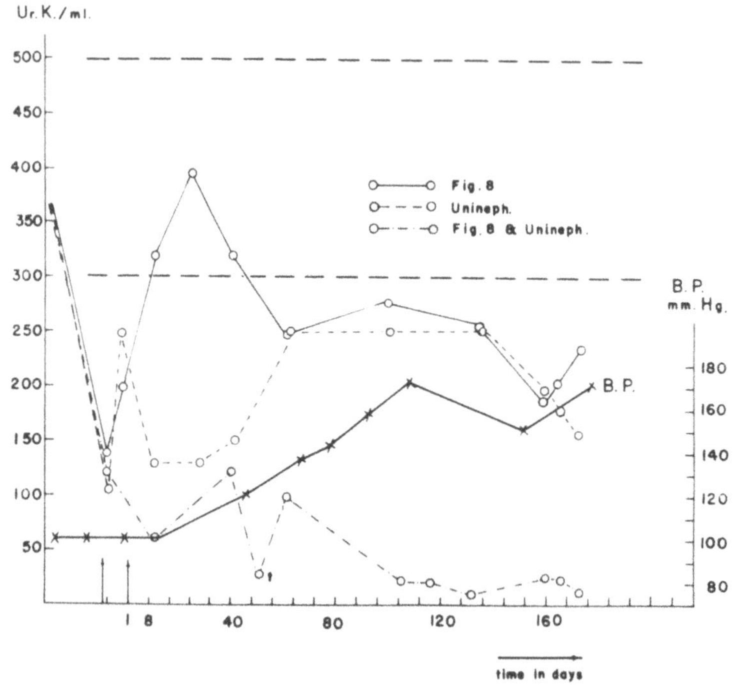

Fig. 6 Urinary k.a. expressed in ng of bradykinin per ml. in three groups of rats: 4 uninephrectomized, 4 with figure-in-8 ligature in one kidney and the contralateral one untouched; and 6 figure-in-8 ligature and uninephrectomized hypertensive rats. Abscissa indicate the days after the kidney removal. The curve B.P. indicates the blood pressure in mm Hg. of the last group of rats.

All the rats of the groups a, b, and c exhibited normal blood pressure through the experiment. In the group d) of rats with figure-in-8 ligature in the lone kidney there was a conspicuous drop of the k.a. in the first week, but later on this activity continues if the animal develops hypertension. As it is shown in table 1 and in figures 6 and 7, the lowest values were exhibited by the rats having the highest blood pressure levels. Rats which in two consecutive readings had 160 mm Hg blood pressure or more, the k.a. per ml was 10 times lower or even less than that found in any of the control rats.

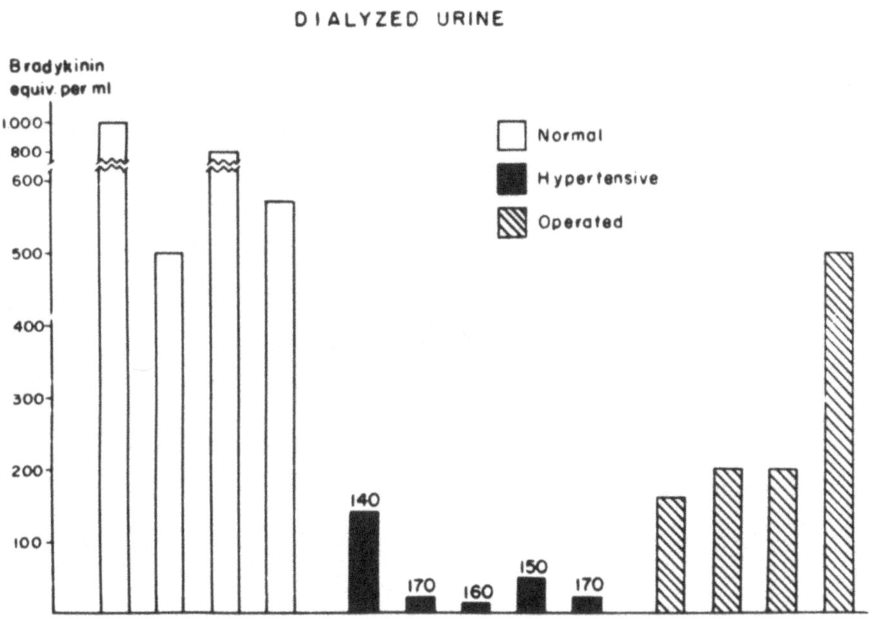

Fig. 7 The columns represent the k.a. per ml of dialyzed urine, in 4 normal, 5 hypertensive and 4 rats with figure-in-8 ligature in one kidney and the other one untouched. Figures above black columns correspond to the blood pressure in the respective rat.

The differences on urinary k.a. between the normotensive and the hypertensive rats can be seen in some typical experiments shown in figures 8 and 9. In the latter one a comparative study provides data showing that the striking differences between two urine pools of normotensive and hypertensive rats affect not only the k.a. activity per ml but also the kallikrein excretory rate correlated to body weight and time. Besides these parameters, the threshold dose expressed in ml of urine to produce a noticeable contraction of the uterus shows that 14 fold more urine in the hypertensive than of the control is required.

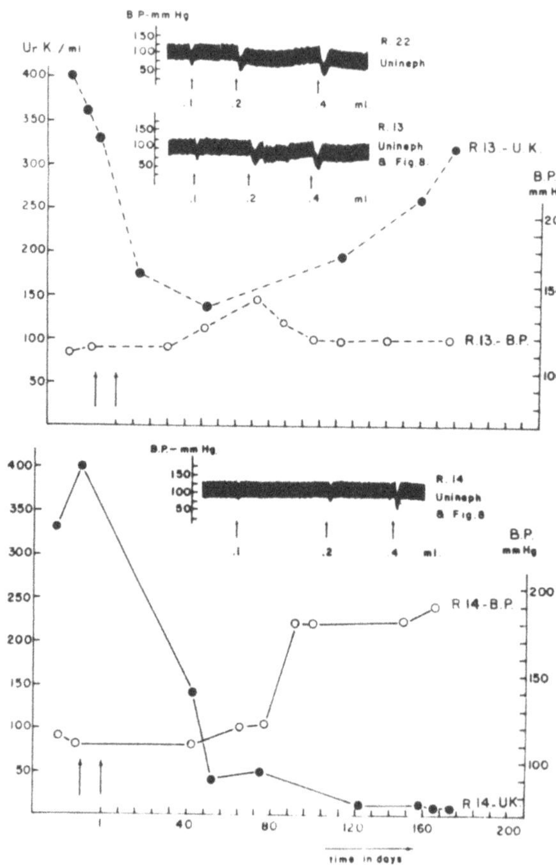

Fig. 8 Urinary k.a. expressed in ng of bradykinin per ml and blood pressure (B.P.) of rat 13 and rat 14. Abscissa indicate days after kidney removal. Both rats had a figure-in-8 ligature in one kidney and the contralateral was extirpated. Inserts in the upper part indicate the changes in the blood pressure of a recipient rat injected with different amounts of the urine of Rat 13, Rat 14 and Rat 22, uninephrectomized.

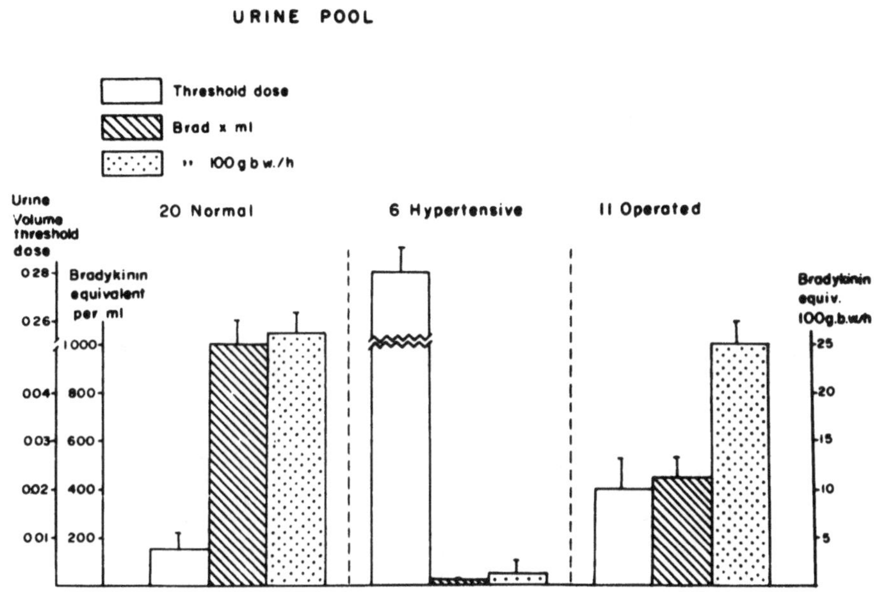

Fig. 9 Mean values of k.a. in 3 urine pools of 3 groups of rats: 20 normal, 6 hypertensive (mean blood pressure 165 mm Hg) and 11 rats with figure-in-8 ligature in one kidney and the contralateral untouched.

Threshold dose is the volume of urine to produce a noticeable uterine contraction.

In the same rats (d) with figure-in-8 ligature in the lone kidney, which do not develop hypertension, the k.a. is similar or somewhat lower than that recorded in the uninephrectomized rat (Fig. 8). In rats of the group d) with a moderate or fluctuating hypertension, the k.a. shows intermediate values. It changes from one to the other week, but the average activity is lower than in any of the normotensive controls. In 7 rats of the group d) which regained spontaneousley a normal or a near to normal blood pressure, the highly diminished k.a. observed during the hypertensive period, reached values which were closer to that found in the normotensive uninephrectomized rats.

The difference on k.a. obtained between the normotensive and hypertensive using the isolated rat uterus was confirmed in the blood pressure assay either with dialyzed urine or its purified fractions (Fig. 9 and 10).

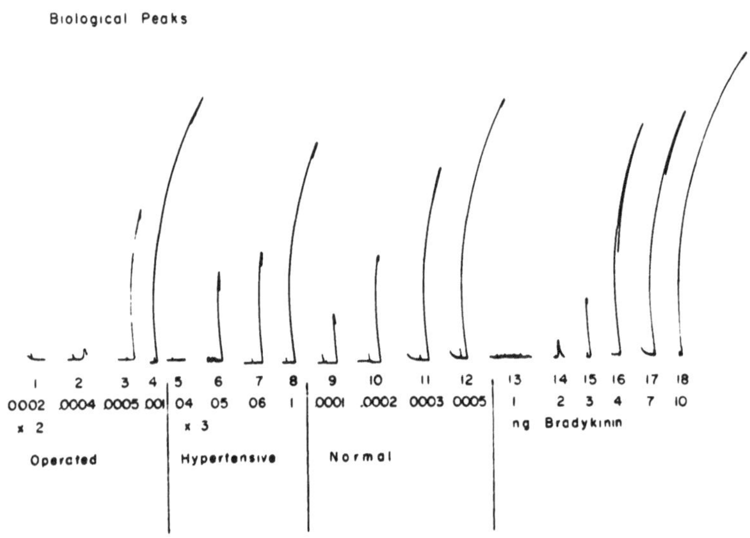

Fig. 10 Contraction of an isolated rat uterus produced by different amounts of purified kallikrein obtained from 3 groups of rats: (operated) with figure-in-8 ligature in one kidney and the other untouched; (hypertensive) with figüre-in-8 in the lone kidney, and normal rats. The k.a. of the operated rat urine must be multiplied by 2, and the k.a. of the hypertensive rats by 3, because of the final concentration of the respective solutions. Bradykinin in ng.

Purification by gel filtration on Sephadex G-100 established that a substance collected in a single peak gathered all the oxytocic and vasodepressor activities. The volume (50 ml) corresponding to the peak contains the substance which has direct effect upon the rat uterus muscle and sets free kinins in the presence of kininogen. (Fig. 11, 12 and 13). The purified fractions obtained from pooled urine through gel filtration, corroborates the differences in k.a. found among the groups studied when fresh voided or dialyzed urine was employed. (Fig. 14). Furthermore the comparative studies with the purified fractions incubated with kininogen supplied the evidence that the k.a. low values found in the hypertensive rat urine is due to a reduction in the concentration of the enzyme and not to the presence of a kallikrein inhibitor.

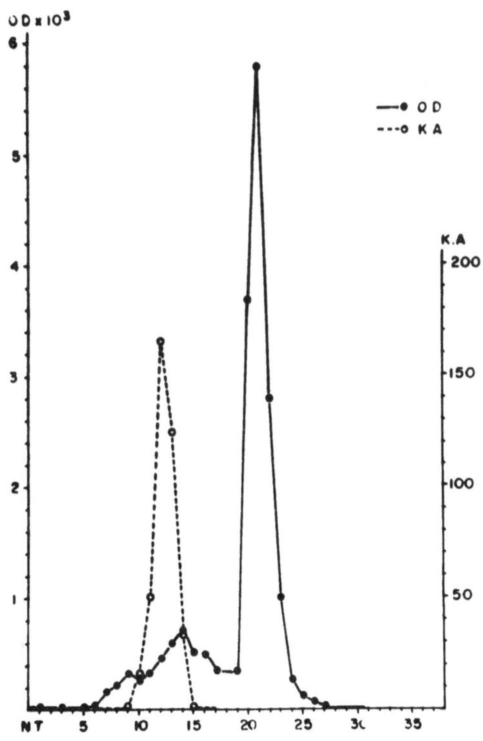

Fig. 11 Optical density (O.D. x 10^3) and kallikrein activity in the effluent solution from a column of Sephadex G-100. k.a. activity is expressed in ng of bradykinin. The column was charged with an extract of normal rat urine (80 ml).

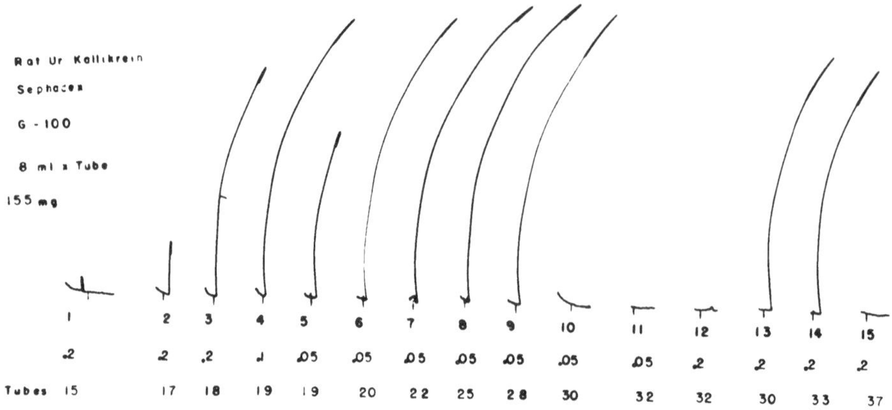

Fig. 12 Contractions of an isolated rat uterus produced by the direct action of the effluent solution from the Sephadex column. The second row of figures gives the amount of each solution in ml introduced in the bath. The lowest figures correspond to the tubes numbers. Notice that using 0.05 ml, the peak of the activity is between Tube N 17 and 28.

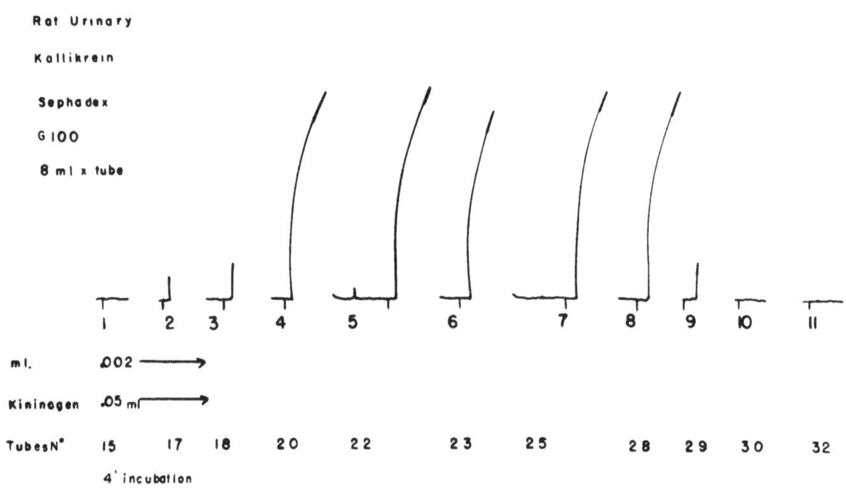

Fig. 13 Contractions of an isolated rat uterus, produced by the addition to the bath of mixture of kininogen and 0.002 ml of the effluent solutions coming out from a Sephadex column. Each mixture was incubated for 4 minutes. The lowest row of figures correspond to the tubes number of the fraction collector. The effects can be compared to that obtained by the direct action on the uterus by the same solutions (Fig. 12).

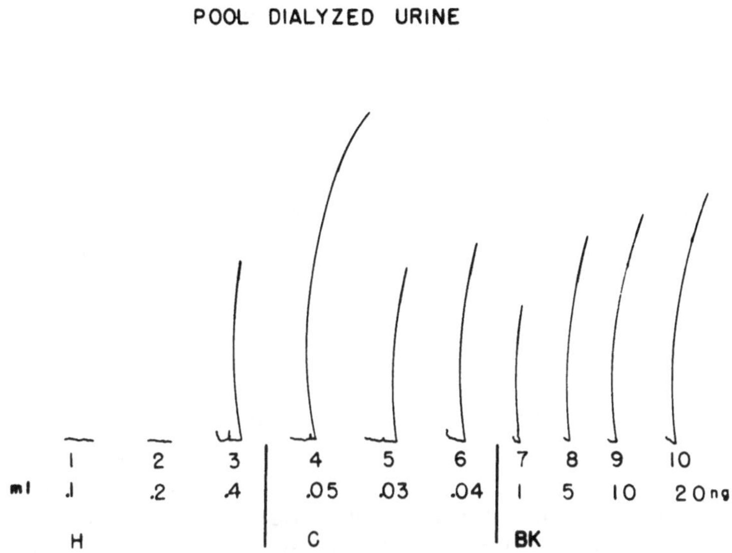

Fig. 14 Oxytocic effect of the dialyzed urine pool of a group of 10 normal (C) and 6 hypertensive rats (H) (160 mm Hg mean blood pressure). Bradykinin in ng.

DISCUSSION

The significant decrease in k.a. in the urine of the hypertensive rats raises the question as to whether the enzyme is involved in some way in the mechanism of renal hypertension. The experiments here described do not provide the evidence, but the data support the concept that in general there is an inverse correlation between the k.a. in the urine and blood pressure level. It is a constant fact that a very low k.a. is found in the urine of a hypertensive rat with a figure-in-8 ligature rat, particularly when high pressure level is maintained. Neither an explanation can be given about the origin of the urinary kallikrein, however parallel studies have shown that a kallikrein similar to the urinary enzyme is present in the renal tissue and can be extracted by the several methods proposed for the purification of kidney renin (Croxatto and Noé, 1970). It is possible to assume that kallikrein in the urine reflects the kidney ability to produce this enzyme. On the other hand it is reasonable to suppose that the circulatory changes induced in the organ by the Grollman procedure, can impair either its production or its activation. Determination of kallikrein in the blood both in the renal artery and vein might disclose whether the kidney is able to deliver kallikrein to systemic circulation as it does with renin. It cannot be discarded that an impairment of the metabolic function of the nephron, not related to the mechanism of hypertension is responsible of the paramount decrease in the k.a. in the urine. Recently it has been demonstrated that the kidney functional derangement observed on hypophysectomized rats coincided with a constant lowering in the k.a. (Croxatto and San Martìn, 1970).

SUMMARY

Kallikrein activity in the rat urine, tested either by its oxytocic activity upon isolated rat uterus or its depressor effect on the rat blood pressure, diminishes as a result of a reduction of the kidney functional mass. The kallikrein activity per ml of urine expressed in ng of bradykinin which produce the equivalent oxytocic effect, shows in the average of 725 ng, in the normal Wistar rats and 435 ng in Sprague-Dowley. In the uninephrectomized the averages found were 325 and 266 ng. In both strains respectively; in the figure-in-8 ligature in one kidney, leaving untouched the contralateral one it was 373 ng in the rats figure-in-8 ligature in one kidney followed by removal of the contralateral one, a procedure to produce hypertension, a more striking lowering in k.a. was found. When the animal develops a constant high blood pressure, this activity reaches the lowest values (less than 10 ng). In moderate hypertensive rats or in those which show pressure fluctuations, intermediate values are found as well k.a. fluctuates, but the average values are lower than in the normotensive. Rats with a figure-in-8 ligature in the lone kidney which do not develop hypertension or after a period of hypertension normalize their blood pressure, had a k.a. in the urine which is similar or somewhat lower than that found in the uninephrectomized normotensive rats.

REFERENCES

Beraldo W.T., Araujo L.R.L. and Maris-Guia M.: Amer. J. of Physiol. 211, 975, 1966.

Croxatto H., San Martìn M.L., Cerda P. and Croxatto H.: IX Congresso da Assoçao Latino Americana de Ciencias Fisiológicas Belo Horizonte, Brasil (1969).

Croxatto H. and Noé G.: unpublished observations.
Friedman M. and Freed S.C.: Proc. Soc. Biol. Exp. Biol. Med. 70, 670, 1949.
Grollman A.A.: Proc. Soc. Exp. Biol. Med. 27, 102, 1941.
Guyton C.A., Coleman G.T., Fourcada C.J. and Navar L.G.: Bull. of N.Y. Acad. of Medicine, 45, 811, 1969.
Webster M.E. and Gilmore J.P.: Am. J. Physiol. 206, 714, 1964.

THE EFFECTS OF BRADYKININ ON SINGLE RANVIER NODE OF NERVE FIBRES IN THE FROG*

Campese V.M., Mitolo-Chieppa D.
(2nd Chair -Dept. of Pharmacology- University of Bari, Italy)

Bradykinin has been found to be highly effective in producing pain when injected intraarterially or intraperitoneally in dogs and men (1-4), or when applied on the "cantharidin blister area" (5). The injection of bradykinin into the splenic artery evokes a discharge in the splenic-splanchnic nerve (6).

Lim (1967) suggests that "pain" receptors but not axons are excited by bradykinin. These receptors would be the free-branching nonmyelinated terminals of paravascular nerves which are found in the connective tissue space surrounding capillaries in the tested organs.

Aspirin, when applied on nerve fibres or axons, does not eliminate the impulse traffic, the periferal analgesic action being exerted on pain receptors or nerve endings (6).

It is not certain whether bradykinin stimulates pain receptors only. In fact, the slowly adapting pressure receptors with myelinated afferent fibres were also found to be excited by close-arterial injection of bradykinin into the skin of cats, as shown by recording the activity of cutaneous afferent fibres in the saphenous nerve (7). Moreover, Haefely et alii (8), Trendelenburg (9), found very recently that bradykinin causes a marked facilitation of synaptic transmission in the superior cervical ganglion of the cats. These authors

*This paper appears out of sequence. It would ideally have been presented in the Physiology section. Also, because it was received too late, its contents have not been indexed.

concluded that facilitation of ganglionic transmission could depend on a depolarizing action on postsynaptic neuron.

In the present research, we have investigated the actions and interactions of bradykinin and aspirin on the axons.

METHOD

The "circulated node" technique (10) was used throughout. Single nerve fibres were isolated from either the motor branch of N. tibialis to the gastrocnemius muscle or from the ramus cutaneous cruris posterior of Rana esculenta (11).

One single node of freshly dissected myelinated nerve fibre was placed in an opening of a polythene tube and adjusted by moving the fibre until it layed in the axial stream of Ringer's or test solution flowing in the tube. A special stop-cock located near the superfused node allowed the rinsing fluid to be changed in less than one second. The inter-nodes on both sides had short air-gaps (of about 0.5 mm), in which the external fluid was reduced to a small film of high resistance by drying. The neighbouring nodes were placed in side pools containing 1% novocain-Ringer solution to keep them inactive. The pulses were applied between the central node and one of its neighbours by means of Ag-AgCl electrodes connected through saturated KCl-Agar bridges. The membrane potential changes were measured from the other gap by means of a differential amplifier of high imput impedance in a negative feedback arrangement. The output potential of the amplifier was recorded on a dual-beam oscilloscope (Solartron). The experiments were performed at room temperature. Ringer sosution had the following composition (mM/l): NaCl 110.5; $NaHCO_3$ 2.4; KCl 2.5; $CaCl_2$ 1.8 .

The effects of bradykinin, kindly supplied by Sandoz (Basel), on repetitive activity, amplitude and duration of the action potential (AP) threshold and refractory period (RP) were examined. The duration of the AP was measured by the lenght of a horizontal line drawn from the firing level to the descending phase of the AP. The duration of RP was calculated by the time interval between the stimulus evoking the control potential and the stimulus evoking a potential 20% lower than control. The intensity of the stimulus for measuring RP was twice the threshold stimulus. The threshold was determined by means of short stimuli (200 µV).

RESULTS

Results are shown in the table.

Table - The effects of bradykinin (6 µg/ml) on single Ranvier node of isolated nerve fibres in the Frog.

REPETITIVE ACTIVITY		PA AMPLITUDE	PA DURATION	DURATION OF RELATIVELY REFRACTORY PERIOD	THRESHOLD	MEMBRANE POTENTIAL
Motor fibres	Sensory fibres					
		Mean ± S.E.	Mean ± S.E.	Mean ± S.E.		
evoked	enhanced	+ 9% (±1,06)	+ 50% (±6,88)	+ 6% (±3,05)	light enhancement	light depolarization

The repetitive activity has been investigated on 5 sensory and 3 motor fibres; the amplitude and duration of PA on 8 fibres; the relatively refractory period on 4 fibres; the threshold on 6 fibres.

The sensory fibres, when stimulated by electrical pulses of long duration (10-25 msec.) showed repetitive

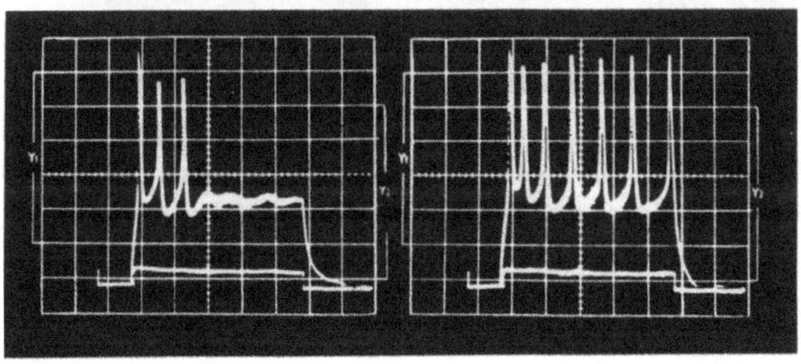

Fig. 1 - Effect of bradykinin (6 µg/ml) on repetitive activity in a sensory fibre. Abscissa : 5 msec/square. Ordinate: 20 mV/square.

activity, which did not occur in motor fibres.

Synthetic bradykinin (6 µg/ml) applied on single Ranvier node rised the frequency of repetitive firing of sensory fibres (Fig. 1), and evoked repetitive activity in motor fibres (Fig. 2).

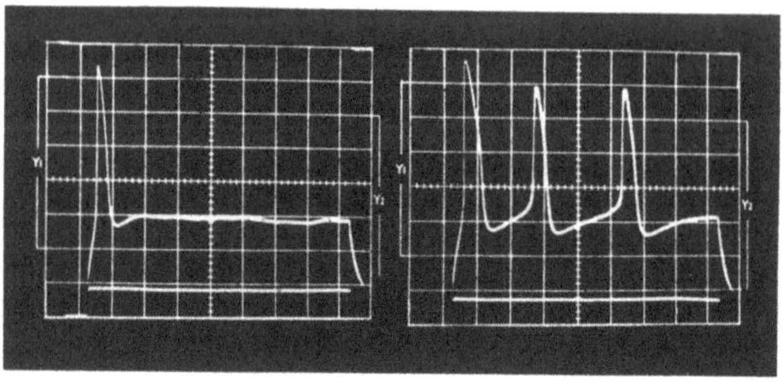

Fig. 2 - Effect of bradykinin (6 µg/ml) on repetitive activity in a motor fibre. Abscissa: 2 msec/square. Ordinate: 20 mV/square.

Moreover, bradykinin produced the following effects: rise of AP amplitude, increase in duration of AP due to prolongation of the repolarizing phase (Fig. 3),

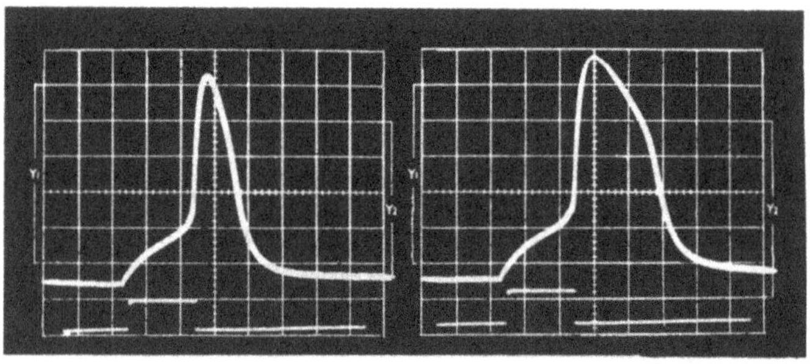

Fig. 3 - Effect of bradykinin (6 µg/ml) on AP amplitude and duration in a motor fibre. Abscissa: 0,5 msec/square. Ordinate: 20 mV/square

light enhancement of threshold as well as refractory period, and light depolarization of resting potential. Aspirin (5 x 10^{-3}) failed to inhibit the effects of bradykinin on isolated nerve myelinated fibres.

DISCUSSION

Sensory fibres, when stimulated with continuous current of long duration, react with repetitive activity i.e. an impulse traffic lasting as long as the stimulation period.
This type od reaction did not occur in motor fibres, in which a single action potential was recorded at the beginning of stimulation.

Repetitive or non-repetitive activity of a nervous fibre is related to the constant of accomodation. In fact, motor fibres, which have a little constant of accomodation, react with only one potential in response to continuous current. Sensory fibres, having a high constant of accomodation, react with repetitive potentials.

Our results suggest tha bradykinin enhances the constant of accomodation of sensory fibres as well as motor fibres. In fact, it makes repetitive the normally non-repetitive activity of the motor fibres, and enhances the frequency of the repetitive potentials of the sensory fibres.

Further investigations to understand the mechanism of action of bradykinin are being carried out. Preliminary data suggest that bradykinin acts also by facilitating the transmission of nervous impulses along the nerve fibre. Our findings about the effects of aspirin on nerve fibres are in agreement with Lim's (1964; 1967). In his studies, aspirin, when applied on the nerve fibres, did not eliminate the propagation of the AP from the pain receptors stimulated by injecting bradykinin into the splenic artery. In our investigations aspirin failed to inhibit the effects of bradykinin on isolated nerve fibres.

This observation, together with that of Lim, suggest that the action od aspirin is exerted on pain receptors, not on axons.

In conclusion, from the results of our research it seems that bradykinin acts on the axons, and that aspirin fails to inhibit the effects of bradykinin on the axons.

BIBLIOGRAPHY

1) Potter G.D., Guzman F., Lim R.K.S. -
Nature 193, 983, 1962

2) Guzman F., Braun C., Lim R.K.S., Potter G.D.
Rodgers D.W. - Arch.int.Pharmac. 149, 571, 1964

3) Lim R.K.S., Guzman F., Rodgers D.W., Goto K., Braun C., Dickerson G.D., Engle R.J.
Arch.intPharmac. 152, 25, 1964

4) Braun C., Guzman F., Horton E.W., Lim R.K.S., Potter J. D.,- J. Physiol. 155, 13P, 1961

5) Lim R.K.S.
Anestesiology 28, 106, 1967

6) Fjällbrant N., Iggo A. -
J. Physiol.156, 578, 1961

7) Haefely W., Hürlimann A., Thoenen H. -
Hypotensive Peptides - Proc.int.Symp. Florence -
Springer Verlag - N.Y. - 1966

8) Trendelenburg U. -
J. Pharmacol. 154, 418, 1966

9) Staempfli R. -
J. Physiol. (Paris) 48, 710, 1956

10) Staempfli R. -
Erg. Physiol. 47, 70, 1952

RELEASE OF SEROTONIN IN MAN INDUCED BY SOME POLYPEPTIDES*

Pentti Peltola

Kivelä Hospital, Helsinki

The common appearance of serotonin and bradykinin in some clinical syndromes, as in carcinoid, dumping syndrome and migraine, has raised the question whether the increase in one could be produced by an increase in the other or whether they are independent. It is known from the study of Hilton et al. (1956) that serotonin may increase the plasma content of catecholamines, and this in turn may liberate bradykinin-forming enzyme from the salivary gland. On the other hand, the possible release of serotonin through bradykinin seems not to be well established. We tried to study this problem by giving intravenous infusions of bradykinin and of another peptide, angiotensin, to man and by estimating the excretion of the metabolites of serotonin in urine.

METHODS

In earlier studies we have found, both in man and rat (Airaksinen et al. 1960, Leppänen et al. 1966), that after a rapid intravenous injection of serotonin, the excretion of its main metabolite, 5-hydroxyindoleacetic acid, increases in conjugated form, not as a free acid, as usual after an intramuscular injection or after a slow intravenous infusion. This increase in conjugated 5HIAA, mainly as sulfates and glycuronides, is a short-lived one, lasting from2 to 4 hours, and during this time the free 5HIAA excretion may remain normal or may even be decreased. Thus in the usual estimation of only the free 5HIAA, even a remarkably acute release of serotonin may be overlooked, even one in amounts that can give clinical symptoms.

In this study the 5HIAA excretion was measured in two-hour period by treating the urine samples with arylsulfatase and

*This paper is presented out of sequence. It would ideally have been presented in the Biochemistry section.

beta-glucuronidase as described earlier (Leppänen et al. 1966) and estimating the 5HIAA before and after treatment by the method of Udenfriend, and by controlling the 5HIAA by paper chromatography.

Bradykinin was infused during 20 minutes at rates of 0.2 and 1.0 ug/kg/min into 15 patients, and angiotensin into 8 normotensives and 15 hypertensive patients at rates increasing from 3 ng/kg/min to 20 ng/kg/min until a diastolic blood pressure increase of 20 mm Hg was reached.

RESULTS

As a result it was found that bradykinin in the higher dose caused a tenfold increase from 0-0.05 mg/2 hr to 0.3 mg/2 hr in the 2-hour excretion of conjugated 5HIAA, the free acid remaining practically unchanged.

With angiotensine a similar increase from 0.17 ± 0.10 to 0.75 ± 0.26 mg/2 hr was found, in both groups only with hypertensive doses. In a case of primary hyperaldosteronism, known to be very sensitive to angiotensin, a dose of 5 ng/kg/min increased the blood pressure and at the same time the excretion of conjugated 5HIAA sevenfold, whereas in a patient with renovascular hypertension even a dose of 20 ng/kg/min of angiotensin was ineffective in both respects.

DISCUSSION

On the basis of these studies it seems apparent that both bradykinin and angiotensin, the latter in hypertensive doses, cause an increased excretion of conjugated 5HIAA, and thus may release serotonin into the blood stream. On the basis of this study alone it cannot be decided from where the serotonin is derived. However, our earlier studies seem to indicate that it perhaps could be released from the main store of serotonin in the body, from the intestinal wall through the contraction of the intestine, which both of these peptides are known to cause. We have shown earlier (Kärki et al. 1960) that when in rabbit invivo the intraluminal intestinal pressure is increased, either through injection of air or liquid into an intestinal loop ligated at both ends or through causing it to contract by intravenous injection of neostigmine, the serotonin content in the corresponding intestinal vein is increased. In man a similar increase in the excretion of conjugated 5HIAA has been found in dumping syndrome (Peltola, 1968), possibly through the rapid emptying of the stomach into the small intestine, and also after neostigmine injection of patients with spastic colon (Peltola, in press).

As other possible sources may be mentioned the liberation of serotonin from platelets or secondarily after release of catecholamines. Studies on this problem are in progress.

The weak effect of angiotensin in patients with renovascular hypertension could be explained by the well-known decreased sensitivity of these patients to angiotensin in general.

These findings could explain the rather variable results in studies on the role of serotonin in dumping and migraine syndromes, when usually only the free 5HIAA has been estimated. They also could explain some of the rather unpredictable vascular reactions caused by some polypeptides.

REFERENCES

Airaksinen, M. M., Paasonen, M. K., and Peltola, P., New Aspects in the 5-hydroxyindoleacetic acid excretion after administration of 5-hydroxytryptamine, Ann. Med. Exper. Fenn. 38:237 (1960).

Hilton, S. M., and Lewis, G. P., The relationship between glandular activity, bradykinin formation and functional vasodilatation in the submandibular salivary gland, J. Physiol. 134:471 (1956).

Kärki, N. T., Paasonen, M. K., and Peltola, P., Factors influencing the release of 5-hydroxytryptamine from the rabbit intestine into the venous blood, Ann. Med. Exper. Fenn. 38:231 (1960).

Leppänen, V., Paasonen, M. K., and Peltola, P., Excretion of 5-hydroxyindoleacetic acid conjugates in neurocirculatory asthenia and after injection of 5-hydroxytryptamine, Ann. Med. Exper. Fenn. 44:424 (1966).

Peltola, P., Excretion of free and conjugated 5-hydroxyindoleacetic acid during provoked dumping syndrome, Ann. Med. Exper. Fenn. 46:505 (1968).

AUTHOR INDEX

(Underscored numbers indicate complete papers in this volume.)

SUBJECT INDEX

GPSR Compliance
The European Union's (EU) General Product Safety Regulation (GPSR) is a set of rules that requires consumer products to be safe and our obligations to ensure this.

If you have any concerns about our products, you can contact us on

ProductSafety@springernature.com

In case Publisher is established outside the EU, the EU authorized representative is:

Springer Nature Customer Service Center GmbH
Europaplatz 3
69115 Heidelberg, Germany

www.ingramcontent.com/pod-product-compliance
Lightning Source LLC
LaVergne TN
LVHW082022200926
848481LV00010B/2904

* 9 7 8 1 4 6 8 4 7 4 4 1 1 *